ESSENTIALS OF PARASITOLOGY

ESSENTIALS OF PARASITOLOGY

Marvin C. Meyer *and* **O. Wilford Olsen**

University of Maine
Orono, Maine

Colorado State University
Fort Collins, Colorado

Second Edition

WM. C. BROWN COMPANY PUBLISHERS
Dubuque, Iowa

Copyright © 1971, 1975 by Wm. C. Brown Company Publishers

Library of Congress Catalog Card Number: 74-83078

ISBN 0—697—04682—6

Third Printing, 1977

Printed in the United States of America

Contents

Section I. Protozoa

Sarcodina, 3; Mastigophora, 7; Opalinata, 15; Gregarinia, 15; Coccidia, 17; Ciliatea, 31; *References*, 33.

Section II. Helminths

Section III. Arthropoda

Section IV. Technique Procedures

Preface

This text is designed to meet the laboratory needs of advanced undergraduate and beginning graduate students taking a course in general parasitology. In some laboratories, parasitology is still taught largely from stained smears and whole mounts of parasites from man, with little or no contact with living material. The common complaint from students taking such a course is that there is an overwhelming amount of seemingly irrelevant and confusing classification, anatomical detail, and terminology; the student tends to become discouraged, to lose sight of the fact that the parasites he is examining were once living, and to find it difficult to relate them to the biological phenomenon called parasitism. A classificatory-anatomical approach to parasitism is not only sadly stereotyped, but it fails to engender real insight into these fascinating animals, how they live, their adaptive properties, and host-parasite relationships.

Nothing will serve to stimulate the student and vitalize the course more than actual contact with living specimens. Observing a live trypanosome, cercaria, or tapeworm in its natural environment is far more vivid to the student than stained mounts. Fixed and stained material is, however, necessary in the identification of certain parasites (notably plasmodia), to illustrate forms unavailable in the living state, and to supplement the study of viable material. It seems unwise, however, to use only fixed and stained material to illustrate certain stages which the student will rarely, if ever, see again in such a condition.

To vitalize the course, therefore, detailed instructions for working the cycles of a number of species are included with each of the larger groups. Also included are instructions for the autopsy of hosts, and the recovery, fixation, staining, clearing, and mounting of parasites. If the student masters the basic procedures, he can apply more sophisticated techniques later as the occasion demands.

This revised edition endeavors to maintain the same objectives as its predecessor. Corrections have been made throughout the text. Each section has been carefully revised to include the relevant information that has appeared since *Essentials of Parasitology* was first published four years ago. The lists of references have been updated. Several new illustrations have been added, and some figures from the preceding edition have been improved. New material includes sections on the protozoan *Toxoplasma gondii*, which occurs in up to half of the human adults in the United States; the nematode *Toxocara canis*, a readily available parasite of dogs (which are so resourceful in the different ways of acquiring patent infections); and, among the arthropods, the crustaceans *Lernaea cyprinaceae* and *Argulus catostomi*, both on fish, and the pentastomid *Porocephala crotali* from snakes.

Only enough classification is included in the text to give the student an understanding of the general relationships of the parasites considered. Complete classification is included in tabular form at the end of each of the major metazoan groups. An effort has been made to use scientific names which are generally accepted as correct. (Binomens that are not now accepted as correct, though commonly used, are given in parentheses.)

For a mastery of even the fundamental techniques, the student needs something specific to guide him through the maze of procedures scattered through the various volumes and journals dealing with microscopical techniques. A selection of standard, laboratory-tested methods for treating material, from recovery at autopsy to examination with a compound microscope, is given in a simple step-by-step form that is easy to use and designed to increase the student's ability to understand the general principles involved.

Each exercise consists of preliminary statements relating the parasite to the scheme of classification, and a basic description of the anatomy and life cycle of the more important and commonly available species. In addition, detailed instructions are included for working the life cycles of a number of species. In presenting the material for study, no attempt is made to proceed along phylogenetic lines. Instead, the first few exercises in each group are devoted to species which show the organ systems more clearly, e.g., *Clonorchis* in the Digenea, *Taenia* in the Cestoda, etc. A key to the common families of each major group of metazoa concludes each section.

A large number of exercises are included, to provide a variety of species for as many types of programs as possible. In most cases the species included, if all were used, would require more time than normally allotted to the group involved. This is intentional—to provide for the study of species which may be in local supply and to allow the instructor a certain latitude of choice. Except in minor details, the account for the species treated will be found equally applicable for closely related species. It should be remembered that these are suggestions only, and that each instructor should decide which species to use for intensive study, which to use for demonstration, and which to omit. The instructor is the best judge of how each will fit into his overall objectives and plans. Here again, the choice of life cycle experiments should be dictated by availability of material and by interest. These experiments may be worked in part or in their entirety. Naturally, species of medical significance are of great interest to students. But since it is difficult or impossible to provide most of these in quantities sufficient for class use, common related species, from fish and from other animals, are included.

Numerous line drawings are included to aid students in visualizing morphological characteristics of the different species. They are, unless otherwise indicated, freehand sketches from many sources, with such changes as necessary to clarify taxonomic features.

The formulae of the various reagents used and instructions for their preparation are included in the Appendix. The lists of references, some of which are cited in the text, suggest some of the more important books and relevant research papers. Many of these contain extensive references which students may consult for additional information on particular topics.

Grateful acknowledgment is due David C. O'Meara of the University of Maine, for taking the photomicrographs of Figures B and F on plate IV, and Figures A, H, and J on plate V. For permission to reproduce, gratitude is expressed to Maine Department of Inland Fisheries and Game, Augusta, for plates I, II, III; to Ward's Natural Science Establishment, Inc., Rochester, New York, for plates IV and V, exclusive of the 5 figures *supra* provided by O'Meara, and Figures 18, 101, and 102; to Thompson Biological Laboratory, Tulsa, Oklahoma, for Figure 28; to CCM: General Biological Supply House, Inc., Chicago, Illinois, for Figures 15, 58, 59, 60, 100, 147F, and 148; to Carolina Biological Supply Co., Burlington, North Carolina, for front cover and Figures 103, 146, and 147E; to J.M. Edney and A.O. Foster, with permission of Hunter, Frye, and Swartzwelder, A Manual of Tropical Medicine, 3rd ed., 1960, W.B. Saunders Company, Philadelphia, Pennsylvania, for Figure 147A-D-F; and to Massey M. Nakamura, with permission of *Focus* magazine, Bausch and Lomb, Rochester, New York, for Figure 150.

In conclusion, we wish to express our appreciation of the kindness of instructors and students who have helped in ferreting out errors and in suggesting changes in the text.

General References

Baer, J.G. 1951. Ecology of Animal Parasites. Univ. Ill. Press, Urbana, 224 p.

Belding, D.L. 1965. Textbook of Clinical Parasitology. 3rd ed. Appleton-Century-Crofts, New York, 1374 p.

Chandler, A.C., and C.P. Read. 1961. Introduction to Parasitology. 10th ed. John Wiley & Sons, Inc., New York, 822 p.

Cheng, T.C. 1973. General Parasitology. Academic Press, New York, 965 p.

Dogiel, V.A. 1962. General Parasitology. 3rd ed. Leningrad University Press (Transl. Z. Kabata. 1964. Oliver and Boyd, Edinburgh, 516 p.).

Faust, E.C., P.C. Beaver, and R.C. Jung. 1968. Animal Agents and Vectors of Human Disease. 3rd ed. Lea & Febiger, Philadelphia, 461 p.

Faust, E.C., P.F. Russell, and R.C. Jung. 1970. Craig and Faust's Clinical Parasitology. 8th ed. Lea & Febiger, Philadelphia, 890 p.

Garnham, P.C.C. 1971. Progress in Parasitology, Oxford Univ. Press, New York, 224 p.

Hyman, Libbie. 1951a. The Invertebrates. Platyhelminthes and Rhynchocoela, vol. 2, McGraw-Hill Book Co., New York, 550 p.

_____. 1951b. The Invertebrates. Acanthocephala, Aschelminthes, and Entoprocta, vol. 3, McGraw-Hill Book Co., New York, 572 p.

Noble, E.R., and G.A. Noble. 1971. Parasitology: The Biology of Animal Parasites. 3rd ed. Lea & Febiger, Philadelphia, 617 p.

Olsen, O.W. 1974. Animal Parasites: Their Biology and Ecology. 3rd ed. University Park Press, Baltimore, 525 p.

Rogers, W.P. 1962. The Nature of Parasitism. Academic Press, New York, 287 p.

Soulsby, E.J.L. 1968. Helminths, Arthropods and Protozoa of Domesticated Animals (6th ed. of Monnig's Veterinary Helminthology and Entomology). Williams & Wilkins, Baltimore, 824 p.

SECTION I

PROTOZOA

As a group of animals, the phylum Protozoa includes many species that are parasitic. These species infect all kinds of animals, including other protozoans. The higher animals—including man—are infected by many kinds of protozoan parasites. In some cases, they are among the most serious of parasites, being the cause of much morbidity and death.

A study of some common representatives will provide a background for understanding the different groups of parasitic protozoans.

SUBPHYLUM SARCOMASTIGOPHORA

Superclass Sarcodina   Class Rhizopodea

The rhizopod Endamoebidae contains the genera *Entamoeba, Iodamoeba, Endolimax, Dientamoeba*, which occur in the alimentary canal of man as well as in many other species of animals. Most of the species in these genera are harmless commensals but some are serious parasites.

The species are differentiated primarily on the basis of the nuclear morphology.

The life cycle generally consists of 4 stages: (1) ameboid trophozoites (feeding stage); (2) precysts; (3) cysts; and (4) metacystic trophozoites. (*Entamoeba gingivalis* and *Dientamoeba fragilis* are exceptions in which only a trophozoite stage occurs.)

Living trophozoites

Entamoeba gingivalis. This species occurs naturally between the base of the teeth and gums in many persons, whence ample material for study can be obtained. With the aid of a small mirror and the flat end of a clean wooden toothpick dipped in 95% ethyl alcohol and dried before use, material can be removed from around the base of the teeth for examination. If all students in the laboratory cooperate, several will be found to be infected and able to provide material for all to see.

Slide and cover glass should be warmed and the material from around the teeth liquefied with a drop of clear saliva. Examine the preparation with high power of the microscope with the diaphragm closed down to give contrast. When available, a phase contrast microscope is preferred. Embryonic tissue from the gums may be present and must be recognized.

The living trophozoites measure up to 35 μ* in diameter, averaging 10 to 20 μ. Active individuals show a band of clear ectoplasm surrounding the grayish granular endoplasm. The pseudopodia are constantly changing in shape and size as the organism moves around on the slide. Food vacuoles containing bacteria, leukocytes, and epithelial cells in various stages of digestion are present.

Entamoeba terrapinae. Living trophozoites usually are available in cultures from biological supply houses. Observe the live specimens, noting the clear ectoplasm and darker, more granular endoplasm whose food vacuoles contain bacteria and particles of media.

Other readily available species that may be collected for study are *E. ranarum* from frogs and *E. muris* from laboratory mice. Scrapings from the large intestine of either host should provide an abundance of trophozoites. Mount the scrapings in physiological solution and examine.

Stained Amebae

Stained specimens are necessary in order to observe the fine anatomical details of the nucleus on which differentiation of the species is based.

Entamoeba histolytica. This is the pathogenic species that invades the tissues of the human large intestine. It consists of large and small races with the large one being considered the more dangerous. Another small *histolytica*-like form with slight anatomical differences is known as *E. hartmani*. Other related amebae which may deserve consideration are *E. dispar* and *E. polecki*. For our purpose, efforts will be directed to recognition of *E. histolytica*, the pathogenic species.

---

*Micron, the designation of long standing, is the same as micrometer (symbol μm), the designation approved by the General Conference of Weights and Measures (and adopted by the U.S. Bureau of Standards) to replace micron.

Trophozoites: Properly fixed specimens stained in iron hematoxylin are roundish in shape with the reticulate cytoplasm bluish gray in color. It is differentiated into clearer ectoplasm and denser endoplasm. They usually measure 18 to 25 μ in diameter. Red blood cells are commonly ingested and usually undergoing digestion in the food vacuoles. The nucleus is spherical, measuring 4 to 7 μ in diameter, and is highly vesicular. There is a small centrally located endosome or karyosome. Small, evenly distributed chromatin granules appear on the inner surface of the thin nuclear membrane (Figure 1).

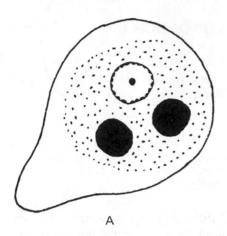 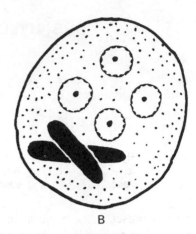

Figure 1. *Entamoeba histolytica,* the tissue-invading ameba of the human intestine, showing characteristic nucleus and chromatoidal bodies. (A) Trophozoite containing 2 red blood cells; (B) Normal tetranucleate cyst with thick chromatoidal bodies having smooth, rounded ends.

The presence of red blood cells, the delicate nuclear membrane lined with small chromatin particles, and minute, centrally located endosome are important diagnostic characteristics of the trophozoites of this species.

Cysts: The cystic phase consists of the precystic and encysted forms. The precystic entamebae expel the food material and assume a spherical shape. Since they resemble encysted forms, they are generally disregarded in diagnosing infections.

The cystic forms are spherical bodies 7 to 8 μ in diameter. The cyst wall is unstained and hyaline in appearance, the finely reticulated cytoplasm bluish grey, and the nuclear membrane and granules black. The nucleus is similar in appearance to that of the trophozoite. In addition to the nucleus, the cytoplasm of young cysts contains one to several large black chromatoidal bodies with rounded, smooth ends. Round spaces appearing in the cytoplasm are glycogen vacuoles. As maturation of the cyst progresses, the nucleus divides once to form 2 small nuclei, each of which divides to produce a total of 4 in the fully developed cyst. Occasionally, additional divisions of the nuclei result in aberrant forms with more nuclei. In ripe cysts, the chromatoidal bodies are lacking, having been absorbed as nutritive material.

The nuclear membrane lined with delicate chromatin granules, the minute centrally located endosome, 4 small nuclei, and thick chromatoidal bodies with smooth, rounded ends are important diagnostic characteristics of cysts of this species. Size in itself is of little value since the cysts of other species are similar.

Formalinized specimens stained in iodine are used in routine diagnoses. The cytoplasm is lemon or greenish yellow in color and the nuclear material and chromatoidal bodies refractile. Glycogen vacuoles, when present, are yellowish brown. The cyst wall does not stain.

Metacystic amebae: These are small forms produced by division of the tetranucleate cystic stage. They are similar to mature trophozoites, though smaller, and emerge from the cyst in the large intestine.

Entamoeba coli

This is a common nonpathogenic species occurring in the human colon.

Trophozoites: In most individuals, the body outline is somewhat rounded and may show division between ectoplasm and endoplasm. The food vacuoles contain bacteria and fecal debris but not red blood cells. The nuclear membrane is thicker than that of *E. histolytica*. The peripheral chromatin granules are coarser, often in block form. The endosome is much larger, about 1 μ in diameter, and usually located eccentrically. The nucleus is 5 to 8 μ in diameter (Figure 2).

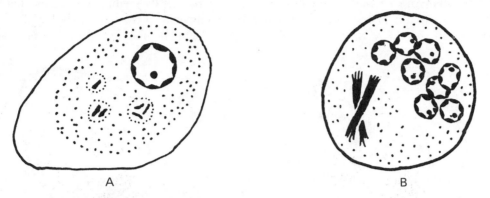

A   B

Figure 2. *Entamoeba coli* from the human intestine, showing characteristic nucleus and chromatoidal bodies. (A) Trophozoite; (B) Normal octonucleate cyst showing splinterlike chromatoidal bodies.

The constant presence of bacteria in the food vacuoles, the thick nuclear membrane, coarseness of the chromatin on the nuclear membrane, and eccentric location of the endosome are the diagnostic characteristics of the trophozoites of this species.

Cysts: The cytoplasm is finely reticulated, and, in young cysts with only 1 or 2 nuclei, a large well-defined glycogen vacuole may be present. The chromatoidal bodies are less abundant than in *E. histolytica*, and when present they are filamentous and irregularly shaped fragments with sharp, splintered ends. Usually there are 8 nuclei but there may be only 1, 2, or 4, as in each of the second and third divisions the nuclei divide simultaneously. The cyst wall does not stain, appearing as a colorless, hyaline, doubly outlined capsule surrounding the cyst.

The double outline of the cyst wall, the presence of 1 to 8 nuclei, the comparatively thick nuclear membrane, the large eccentrically located endosome, and the characteristic chromatoidal bodies are the important diagnostic characteristics of the cysts of this species. Also they are usually considerably larger (10 to 25 μ) than in *E. histolytica*.

Formalinized material stained with iodine shows the characteristic features described above in the same manner as in *E. histolytica*.

Metacysts: Eight minute metacystic trophozoites emerge from the cyst in the large intestine.

Entamoeba gingivalis

The cytoplasm shows little differentiation into ectoplasm and endoplasm. The whole is finely granulated or vacuolated, and food vacuoles are prominent. The spherical nucleus is vesicular, measuring about 2 to 4 μ in diameter. There is a small endosome near the center, while the peripheral chromatin granules are small and compact along the inner surface of the membrane. *E. gingivalis*, which resembles *E. histolytica* in the trophozoite, is not known to encyst but has a cystoid form in culture material. It normally measures about 12 to 30 μ in diameter, but will be much larger if treatment with antibiotics has occurred within a few days prior to fixation (Figure 3).

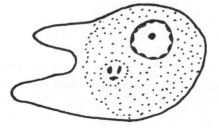

Figure 3. Trophozoite, or vegetative stage, of *Entamoeba gingivalis* from the human mouth.

Iodamoeba bütschlii

Trophozoites: The vesicular nucleus measures about 3 to 4 μ in diameter. The cytoplasm is reticulated or alveolated, depending upon the extent of degeneration of the bacteria. The endosome, which is about one-half the diameter of the nucleus, is typically surrounded by small spherules that do not take stain, so that the achromatic interspherule substance may appear as a reticulum. The endosome is found in various positions in the nucleus, and the well-defined nuclear membrane is free of chromatin granules (Figure 4).

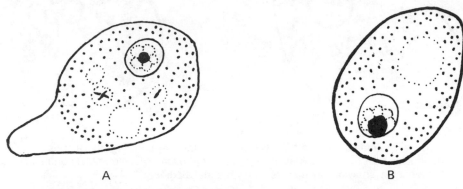

A B

Figure 4. *Iodamoeba bütschlii* from the human intestine, showing characteristic nucleus and large iodinophilous vacuole in the endoplasm. (A) Trophozoite; (B) Encysted stage.

The comparatively large nucleus, the large endosome surrounded by the peculiar spherules, the presence of bacteria in the cytoplasm, and the absence of red blood cells are important diagnostic characteristics of the trophozoites of this species.

Cysts: The cyst contents are reticulated and 1 (occasionally 2) glycogen vacuole is always present. The singe nucleus is usually situated close to the vacuole. The endosome is often attached to the nuclear membrane and may be crescentic in shape. Specimens usually measure 8 to 15 μ in diameter.

The great variation in the shape of the cyst, the presence of but 1 nucleus, the peculiar structure of the nucleus, the comparatively large endosome, and the very large glycogen vacuole are important diagnostic characteristics of cysts of this species.

Endolimax nana

Trophozoites: The body is rounded, the cytoplasm reticulated and containing bacteria. There is a vesicular nucleus, measuring about 1.5 to 3 μ in diameter. The nuclear membrane is delicate, without a distinct peripheral layer of chromatin. Usually the endosome is triangular, square, or irregularly angular in shape, and may be in the center, off-center, or attached to the nuclear membrane. In the latter case, a strand may be seen connecting it with a smaller chromatin mass on the opposite side of the nucleus (Figure 5).

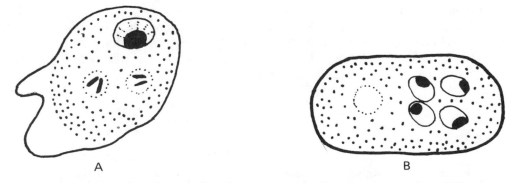

A B

Figure 5. *Endolimax nana* from human intestine, showing characteristic nucleus and cyst. (A) Trophozoite with food vacuoles containing bacteria; (B) Rectangular cyst.

The comparative small size, the peculiar variation in the endosome, and the absence of red blood cells are important diagnostic characteristics of the trophozoites of this species.

Cysts: The cytoplasm is finely reticulated. The nuclei vary in number from 1 to 4 and their structure varies as in that of the trophozoites, but appears to be characterized by an angular endosome and its variable location within the nucleus. Specimens usually measure 5 to 14 μ in diameter.

The oval shape of many of the cysts, the number of nuclei, the single or divided endosome of the nuclei, centrally situated or in contact with the nuclear membrane, and the absence of chromatoidal bodies are important characteristics of the cysts of this species.

Dientamoeba fragilis

The shape is elongate, rounded. The cytoplasm is reticulate, and the food vacuoles contain bacteria. One or 2 nuclei are present, consisting of a small vesicle measuring about 1 to 2.5 μ in diameter. The nuclear membrane is delicate, and the endosome is comparatively large, being more than half the diameter of the nucleus itself. The endosome is characterized by possessing 4 to 8 chromatin granules on its periphery and occasionally a central granule (Figure 6).

The comparatively small size, the presence of 2 nuclei, the peculiar granular structure of the centrally situated endosome, the delicacy of the nuclear membrane, the absence of chromatin granules on the inner surface of the nuclear membrane, the presence of bacteria in the cytoplasm, and the absence of red blood cells are important diagnostic characteristics. Encystment is unknown in this species.

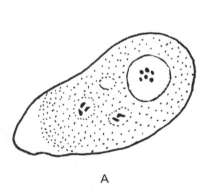
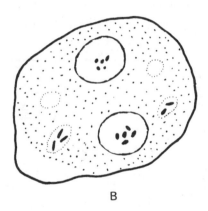

A

B

Figure 6. Trophozoite of *Dientamoeba fragilis* from human intestine. (A) Mononucleate form; (B) Binucleate form.

Superclass Mastigophora   Class Zoomastigia

The forms included in this class bear one or more whiplike flagella and lack chloroplasts. The group contains many species parasitic in vertebrates, invertebrates, and plants—as well as free-living forms.

Hemoflagellates and Related Forms

The hemoflagellates are characterized by a spindle-shaped body containing a central nucleus, basal body, from which the flagellum arises, and kinetoplast, a relatively large, deeply staining body situated near the base of the flagellum. Since the kinetoplast is a mass of DNA within the mitochondrion, it stains similar to the nucleus, and is seen with a light microscope. Some forms bear an undulating membrane. The group includes both pathogenic and nonpathogenic species.

The members of this group are parasites in the alimentary canal of insects and other invertebrates, in plant latex or in the blood or tissues of vertebrates. Life cycles may be extremely complicated, in some cases involving several different developmental stages, and will be illustrated in the laboratory.

Table I. CHARACTERISTICS OF COMMON INTESTINAL AMEBAE OF MAN

Characteristic	Entamoeba histolytica	Entamoeba coli	Endolimax nana	Iodamoeba bütschlii	Dientamoeba fragilis
TROPHOZOITES Size	18-25 μ	20-30 μ	8-10 μ	10-20 μ	3.5-12 μ
Cytoplasm	Finely granular, nonvacuolated; ingested RBC's diagnostic	Coarsely granular; vacuolated; bacteria, yeast ingested	Clear and finely granular; bacterial inclusions	Finely granular; bacterial inclusions	Endoplasm vacuolated; ectoplasm clear; bacterial inclusions
Nucleus (Stained)	Fine membrane, delicate beading; endosome central	Coarse membrane, coarse beading and large; endosome eccentric	No beading; large single, deeply stained endosome	No beading; large single, deeply stained endosome	Generally two nuclei; no beading; endosome consists of a group of chromatin grains
CYSTS Size, shape	Roundish; 7-18 μ	Roundish; 10-25 μ	Ovoidal; 5-14 μ	Irregular, 8-15 μ	No cysts
Nuclei number	Mature, 4	Mature, 8	Mature, 4	Mature, 1	No cysts
Nuclei structure	As in trophozoites	As in trophozoites	As in trophozoites	Eccentric endosome beside granular mass	No cysts
Chromatoidal bodies	Thick or slender bars; often absent	Like glass splinters or irregular	No true chromatoids	No true chromatoids	No cysts
Glycogen mass (in iodine)	Diffuse, brown. In young cysts	Large, deep brown. In young cysts	Occasionally present in young cysts	Large or small, sharply delimited, deep brown	No cysts

8

In the course of their cyclical development, flagellates of the genera *Leishmania* and *Trypanosoma* pass through stages comparable with those of the monogenetic Trypanosomatidae, and so it has become customary to refer to them by names derived from those genera (often combined with form) in which the corresponding stages are the most characteristic forms.

In the past, the following terms have been commonly used for this purpose: leptomonad, based upon the flagellated stage of *Leptomonas;* crithidial for the flagellates with a short undulating membrane attributed to *Crithidia;* leishmanial for the rounded forms of *Leishmania;* and trypanosome for the blood forms and metacyclic flagellates of *Trypanosoma*, as well as for the stages of *Herpetomonas* bearing a superficial resemblance to trypanosomes. For the latter stages the term herpetomonad has also been used for flagellates of *Herpetomonas.*

Because the existing names for the stages of development of these hemoflagellates are inappropriate and often misleading, Hoare and Wallace (1966) proposed a complete revision of the terminology. Their system is based upon characteristics of the flagellum, its arrangement in the body, as determined by its starting point (indicated by the position of the kinetoplast), and its course and point of emergence (Figure 7).

The root "-mastigote" combined with appropriate prefixes formed the following new terms:

1. Amastigote, the former "leishmanial" stage, represented by rounded forms devoid of an external flagellum (as in genus *Leishmania* et al.). Amastigote forms constitute one stage in the life cycle of all hemoflagellates. Examine a smear of a hamster infected experimentally with *Leishmania donovani* for the minute amastigotes.

2. Promastigote, the former "leptomonad" stage, represented by forms with antenuclear kinetoplast, flagellum arising near it and emerging from anterior end of the body (as in genus *Leptomonas* et al.). Promastigote forms are intermediate stages in the life cycle of all hemoflagellates. *Leptomonas ctenocephala* is an interesting parasite of the larvae of the dog flea. Examine a slide prepared from smears from cultures of *Leishmania tropical* for promastigotes.

3. Opisthomastigote, the former "trypanosome" stage, represented (in the genus *Herpetomonas* only) by forms with postnuclear kinetoplast, flagellum arising near it, then passing through the body and emerging from its anterior end. This form is one of several stages of *Herpetomonas muscarum*, occurring in the intestine of houseflies. Biflagellate forms, which are common, represent dividing individuals. Examine stained smears of the intestinal contents of houseflies for opisthomastigote forms.

4. Epimastigote, the former "crithidial" stage, represented by forms with juxtanuclear kinetoplast, flagellum arising near it and emerging from the side of the body to run along a short undulating membrane (as in genus *Blastocrithidia* and in stages of *Trypanosoma*). This form is the final stage of species of the genus *Blastocrithidia* and an intermediate one in some genera such as *Trypanosoma* and *Herpetomonas*. *Blastocrithidia gerridis* is a common parasite in the intestine of water striders. Examine stained smears of the intestinal contents of water striders for epimastigote forms.

5. Trypomastigote, the true "trypanosome" stage, represented by forms with postnuclear kinetoplast, flagellum arising near it and emerging from the side of the body to run along a long undulating membrane (as in genus *Trypanosoma*). Although trypanosomes are also of the epimastigote type, it is desirable to distinguish this important stage by a special term, which is suggestive of the stage in question but is not based upon the generic name. The trypomastigotes represent the final stage in the species of the general *Trypanosoma* and *Herpetomonas*.

Two forms of trypomastigotal bodies are recognized in the blood of vertebrate hosts. All individuals of monomorphic species are similar in structure. In a polymorphic species, both slender forms with a free flagellum and stumpy ones without a free flagellum occur simultaneously.

6. Choanomastigote, the peculiar "barleycorn" form, usually with antenuclear kinetoplast, flagellum arising from wide funnel-shaped reservoir and emerging at the anterior end of the body (typical of genus *Crithidia*).

Life Cycles

Two types of life cycles occur. In one, transmission is generally by ingestion of infective stages passed in the feces of infected animals. These include species of *Leptomonas, Crithidia, Blastocrithidia*, and *Herpetomonas*, which occur in insects and arachnids. In the other, transmission is by sucking arthro-

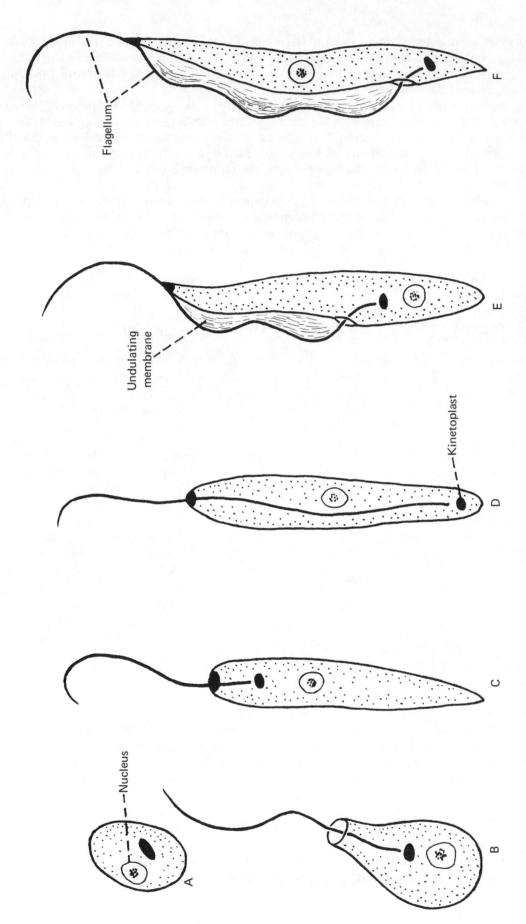

Figure 7. Types of developmental stages of Trypanosomatidae. (A) Amastigote; (B) Choanomastigote; (C) Promastigote; (D) Opisthomastigote; (E) Epimastigote; and (F) Trypomastigote.

pods and leeches. Two means of entering the vertebrate host occur here. In one, the infective stage is injected directly during the blood sucking activity of the infected invertebrate host. This is known as infection from the anterior station, i.e., from the mouth, and is inoculative in nature. In the other method, the infective form is voided in the feces while the arthropod sucks blood and is later ingested or rubbed into abrasions of the skin. This method is known as infection from the posterior station and is contaminative in nature.

Multiplication, involving 2 or more of the developmental stages, usually takes place in the invertebrate host. A few species have developed independence from the invertebrate host and are transmitted venereally, as in the case of *Trypanosoma equiperdum* of horses. In this case, all multiplication occurs in the vertebrate host.

Most species may be transmitted mechanically on the mouth parts of bloodsucking arthropods which, if interrupted while feeding on an infected animal, quickly resume feeding on an uninfected one. Surgical instruments, likewise, may serve as a means of mechanical transmission.

Trypanosoma lewisi

This species occurs commonly in wild rats. It can be reared easily in culture or in great numbers by injecting a small volume of infected blood into a young white rat.

Examine living trypanosomes in a drop of blood drawn from a heavily infected rat, viewing it with both the low and high powers of the microscope. Close the diaphragm to the point that it gives good contrast to see the otherwise transparent organisms. When available, a phase contrast microscope is preferable.

Observe the activity of the trypanosomes. Red blood cells in the vicinity of each trypanosome are moving in response to its rapid lashing action. Introduce a drop of neutral red vital stain under the cover glass to color the trypanosomes and facilitate study of their actions.

Study stained smears of blood taken from experimentally infected rats shortly after infection and at the end of the first and second weeks postinfection.

Multiplication is by binary division which at first is incomplete, resulting in rosettes of small epimastigotes still attached by their posterior ends. These rosettes are most prevalent during the early period of infection when the trypanosomes are multiplying rapidly. Look for them.

Smears made at the end of the first week show smaller trypanosomes than those made 2 weeks or longer after infection.

Note the shape of the body, the nature of the undulating membrane, and location of the internal organs, which should be identified.

Other trypanosomes that may be obtained for comparison both as living and stained specimens include *T. rotatorium* from frogs (transmitted by certain species of *Culex* mosquitoes [Desser et al., 1973]), and *T. avium* from birds. More *T. rotatorium* occur in the blood from kidneys and liver of infected frogs than in peripheral blood, and *T. avium* is more abundant in the bone marrow. A small amount of marrow from the femoral bone macerated with a toothpick in a drop of physiological solution is more likely than blood smears to reveal the trypanosomes of infected birds. A drop of neutral red will render the hemoflagellates more visible.

Life Cycle

Development of *T. lewisi* can be followed in white rats by observing stained blood smears made at frequent intervals over a period of 30 days. Note the dividing rosettes of epimastigotes and the difference in size of the trypanosomes. Observe the rise, peak, and decline of the population in the blood during this period. Do the trypanosomes disappear? What are the time relationships of these events?

In northern rat fleas *(Nosopsyllus fasciatus)*, the trypanosomes enter the epithelial cells of the stomach where binary fission occurs. Upon leaving the cells, they migrate to the rectum and transform to epimastigotes that develop into small trypanosomes known as the metacyclic form. These are voided in the feces and are infective to rats when ingested.

Passage of infective metacyclic forms from the rectum is known as infection from the posterior station—in contrast to infection from the anterior station by metacyclic forms developing in the anterior part of the alimentary tract and injected by feeding arthropods.

When northern rat fleas are available from populations of wild rats, dissections of them should be made in an effort to find the developing trypanosomes. Examine presses and dissections of the entire alimentary tract.

Examples of pathogenic trypanosomes will be studied as demonstration specimens. As you look at the different species, acquaint yourself with the important disease they cause and with other pertinent information about them. It will be impossible for you to recognize the species on the basis of slide material. Many other species occur in mammals, birds, reptiles, amphibia, and fish. Most of them appear to be harmless to their adult host. You will be given opportunity to study a slide of rat myocardium or human heart showing developmental stages of *Trypanosoma cruzi*. Recently *T. cruzi* has been reported from raccoons and other wild animals in certain southeastern states.

Study slides of amastigote types from the following demonstrations: *Leishmania donovani* from culture, cause of kala azar; *L. donovani* from spleen section; and *L. tropica* from culture, cause of "tropical sore" or Delhi boil.

If culture materials are available, *Leishmania* shows amastigote and promastigote stages which can be readily distinguished. Which is the more highly specialized, amastigote or trypomastigote stage?

Other Flagellates

The group of flagellates listed below occur as symbionts in the alimentary tract of a variety of animals, and some species are found in other parts of the host. For the most part, the species are considered to be relatively nonpathogenic. Examine the polymastigid fauna of a freshly killed frog, rat, guinea pig, chicken, and/or termite. First examine the material on a slide in the proper physiological solution, after which a vital stain should be added to show structural details. Those occurring in frogs may include *Eutrichomastix batrachorum*, *Tritrichomonas augusta*, and *T. batrachorum*; in rat or guinea pig, *T. muris* (Figure 8) and *T. caviae* respectively; in termites, a large pear-shaped form belonging to the genus *Trichonympha*.

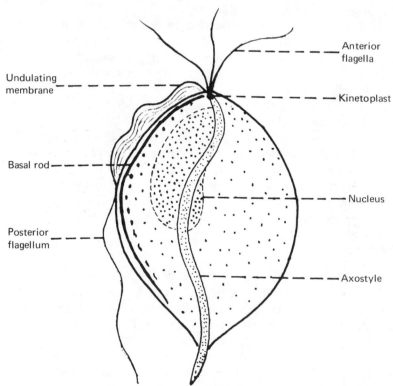

Figure 8. *Tritrichomonas muris* from mouse intestine.

Chilomastix mesnili

Trophozoites: Body outline pyriform, with the anterior end bluntly rounded and the posterior long and drawn out. The cytoplasm is reticulated and contains bacteria in food vacuoles. The spherical, vesicular nucleus is located near the anterior end of the body, has a distinct membrane, and a small eccentrically located endosome. Three flagella arise from the kinetoplast, situated slightly anterior to the nucleus; a fourth, shorter flagellum lies in the cytostomal cleft.

Cysts: Lemon-shaped, with a nipplelike structure anteriorly, surrounded by thin walls except where thickened at the nipple end. The internal structure resembles that of the trophozoite.

Trichomonas hominis (Figure 9) has a cosmopolitan distribution. There are 3 to 5 (usually 4) free flagella and a posteriorly directed flagellum attached to the body by an undulating membrane. If you cannot see these in the demonstration slide refer back to the rat or frog *Tritrichomonas* structure. Note the small cytostome at the anterior end of the base of the flagella, the axostyle, and a hyaline rodlike structure running through the center of the body. Note the kinetoplast from which the flagella originate, the nucleus, and other internal structures. As in the case of other species of *Trichomonas* and *Tritrichomonas*, cysts are unknown. Transfer to another host apparently occurs in the trophozoite stage.

Trichomonas tenax occurs in the mouth of man. Examine a scraping of tartar in physiological solution or a demonstration slide. Note the 4 anterior flagella, marginal flagellum on the undulating membrane, and axostyle.

Trichomonas vaginalis (Figure 10) occurs in the human vagina, and in the urethra and prostate of males. There is no evidence that this species causes abortion in women but *Tritrichomonas foetus*, found

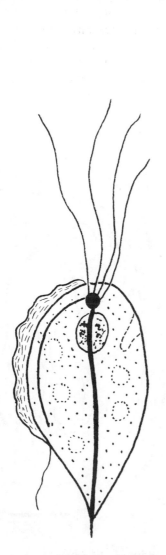

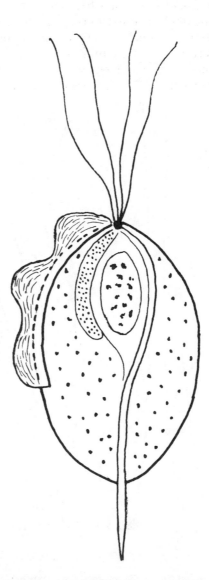

Figure 9. *Trichomonas hominis.*   **Figure 10.** *Trichomonas vaginalis.*

in the genital tract of cattle, causes abortions, stillbirths, and delayed conceptions. Living type cultures can be obtained from the American Type Culture Collection, 12301 Parklawn Drive, Rockville, Maryland 20852.

Giardia lamblia

This flagellate occurs in the human small intestine. It is common in warm moist climates, particularly in children.

Trophozoites: Pear-shaped in outline, the anterior end being rounded while the posterior tapers to a point. Along the median line there are 2 parallel rods, the axostyles, running longitudinally. Near the anterior end of the axostyles, there is a large, ovoidal nucleus on either side. Each nucleus has a large endosome. There are 8 flagella which arise as follows: one pair from the posterior end of the axostyles, a pair from the anteriorly located kinetoplast, another from the anterior part of the axostyles, and the fourth from farther back on the axostyles at a point slightly posterior to the nuclei. There is sometimes present a deeply staining rodlike body lying across the axostyles near the posterior third of the body. Examine living trophozoites from a laboratory rat.

Cysts: Oval in outline. the cyst wall is distinctly visible, and usually the cell contents shrink away a little from the posterior end, leaving a clear space. Young cysts contain only 2 nuclei, but older ones have the normal 4, usually situated at the anterior end. The axostyles, fibrils, flagella, and endosomes usually stain conspicuously. Examine formalinized stool specimen for characteristic cysts. Iodine solution should be added to render them more conspicuous (Figure 11).

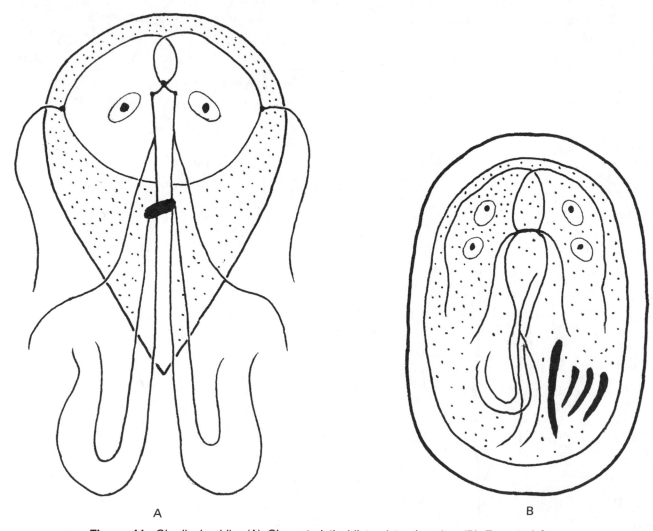

A   B

Figure 11. *Giardia lamblia.* (A) Characteristic bilateral trophozoite; (B) Encysted form.

Superclass Opalinata

Species of 2 genera occur as entocommensals in the cloaca and urinary bladder of various amphibians. The body is uniformly covered with cilia arranged in rows.

Opalina obtrigonoidea

This and other species of the genus have an oval, flattened body covered with rows of cilia and devoid of mouth, cytopyge, and contractile vacuoles. The body is uniformly covered with cilia arranged in rows. The body contains many nuclei of the same kind. Fully developed individuals exceed 300 μ in length and appear as opalescent bodies to the naked eye. Examine living specimens in physiological solution and slides of stained ones.

Life Cycle

Reproduction is by asexual and sexual means. Asexual reproduction is by plasmotomy in which the multinucleate body divides several times with cytoplasmic division occurring independent of nuclear division. Sexual reproduction is by formation of gametes which fuse to form a zygote that produces the trophozoite stage.

Repeated division in the recta of frogs in the spring produces numerous small individuals with few nuclei that encyst. They are voided with the feces. When swallowed by young tadpoles, the small nucleated forms released from the cysts are gametocytes. They divide repeatedly, eventually forming minute, fusiform gametes. Two gametes fuse to produce a zygote that develops into a new trophozoite or vegetative stage.

Mount adults in physiological solution on a slide, seal cover glass with Vaseline and observe movements, as well as living individuals in physiological solution to which some vital dye such as neutral red has been added.

Protoopalina mitotica

This species and others of the genus commonly occur in the colon of amphibians. The body is spindle-shaped and has 2 large nuclei of equal size. Examine permanent mounts of fixed and stained specimens, as well as living individuals in physiological solution to which some vital dye such as neutral red has been added.

SUBPHYLUM SPOROZOA

The Sporozoa are all parasitic and comprise a highly heterogenous group. Some have direct life cycles and others indirect ones. They include some of the most serious parasites of man, of his domestic animals, and of wild animals.

Class Telosporea   Subclass Gregarinia

The Gregarinia are parasites of the intestine, reproductive organs, celomic spaces, or other cavities of many kinds of marine, aquatic, and terrestrial invertebrates.

Morphologically, the Gregarinia are of the acephaline and cephaline types. The acephalines possess undivided bodies and inhabit the body spaces. Cephaline gregarines have a body divided into an anterior part—with a knob, hooks, or filament for attachment inside the host cell—and a posterior part which hangs outside the cell.

Cephaline gregarines commonly occur in grasshoppers, cockroaches, and mealworms. Acephaline forms are abundant, with species of *Monocystis* and *Zygocystis* in the seminal vesicles of earthworms.

Although they are relatively unimportant as parasites, the generalized life cycle of the gregarines makes them significant objects of study to aid in understanding the important sporozoan parasites of man, domestic mammals, and birds.

Monocystis lumbrici

Despite the unimportance of *M. lumbrici* as an earthworm parasite, the generalized life cycle and the close relationship to the economically important avian and mammalian species of Sporozoa make it a desirable species for study.

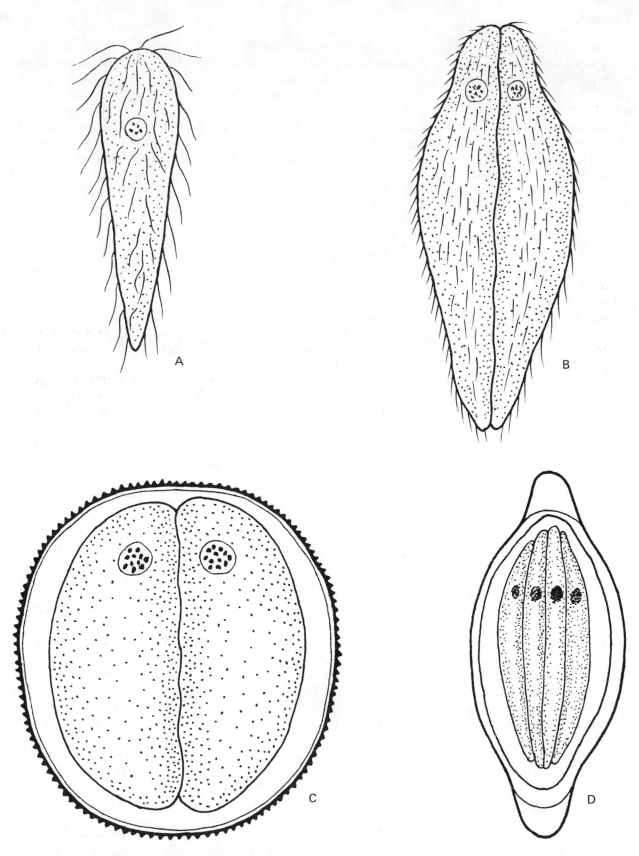

Figure 12. *Monocystis lumbrici* from seminal vesicles of earthworm, showing some stages in the life cycle. (A) Mature trophozoite covered with tails of spermatozoa of earthworm; (B) Paired trophozoites, showing remnants of sperm tails; (C) Oocyst, encysted trophozoites; (D) Boat-shaped mature sporocyst with 8 sporozoites.

To obtain living *Monocystis*, anesthetize an earthworm by placing it in 7% alcohol for about 30 minutes. After such treatment, make a cut through the dorsal body wall by inserting only the point of the scissors about midway between the head and the clitellum. Extend the cut to both sides. Beginning at the lateral points of the cut, make 2 parallel cuts through the body wall up to near the head. Then grasp the loose flap with forceps and gently pull it forward. This will expose the cream colored seminal vesicles situated alongside the esophagus. Using forceps, transfer a small portion of seminal vesicle to a clean slide and add a drop of physiological solution. Tease apart well with dissecting needles, so as to make a thin smear of the vesicle contents, add a cover glass and apply enough pressure to further spread the smear, and examine under a compound microscope.

When sporocysts are ingested by a worm, the 8 sporozoites are liberated by the digestive enzymes. The sporozoites then traverse the gut wall to reach the seminal vesicles, where they penetrate bundles of developing sperm. By this time each sporozoite has developed into a trophozoite. Since the adjacent sperm cells are deprived of their nourishment, they slowly disintegrate, their tails adhering to the trophozoite, giving it the appearance of a ciliated structure. Trophozoites now pair off and become encysted, forming oocysts, after which each undergoes numerous divisions resulting in many gametes. After the formation of the oocyst, the former trophozoites are called gametocytes by some workers. The membranes of the gametocytes disintegrate and gametes, apparently from different trophozoites, unite to form zygotes. While still within the oocyst, a wall develops around each zygote to form an elongate structure known as a sporocyst. Within each sporocyst, 3 successive divisions occur, resulting in 8 mononucleate sporozoites. Due to their large size, the stages commonly seen in seminal vesicle smears are the trophozoites and the sporocyst-filled oocysts (Figure 12).

Miles (1962) has shown experimentally that (1) the chief, if not the only, way worms are infected in nature is through ingesting mature sporocysts, and (2) sporocysts do not leave the host through openings in the body wall, but are released after death and decay of the host. Earthworm-eating birds and mammals doubtless play an important role in the dissemination of sporocysts with their feces.

Gregarina spp.

Members of this genus represent common cephaline gregarines. They occur commonly in the digestive tract of cockroaches, grasshoppers, and mealworms.

Remove the intestine from a decapitated host, tease it apart in physiological solution on a slide, add a cover glass, and examine under a microscope. Small, young stages of the parasite are inside the epithelial cells. Older and larger individuals are attached to the cells by the small anterior end of the body, with the large hind part hanging free in the lumen of the gut.

The body of the mature gregarine is divided into 2 major parts, the anterior protomerite and the posterior deutomerite. The protomerite bears anteriorly a small, simple, knoblike epimerite, which is embedded in the cytoplasm of the epithelial cell.

The pendant deutomerite detaches to form the sporadin. Two or more sporadins unite in tandem to produce a condition known as syzygy (Figure 13). Find syzygyous forms.

Eventually 2 sporadins attach along the long axis of their bodies and secrete a cyst about both. Find the conjugating sporadins known as a gametocyte. The nucleus of each gametocyte undergoes repeated division and the particles migrate to the surface where each becomes enclosed in a bit of cytoplasm. These are gametes. Union of 2 gametes produces a zygote in which develop the banana-shaped sporozoites. Find these various stages.

Subclass Coccidia

The Coccidia are cytozoic parasites of vertebrate and invertebrate hosts. They occur most commonly in the epithelial cells of the intestine but may parasitize the liver (rabbits) or kidneys (geese). The life cycle is direct, being completed without benefit of an intermediate host. The Coccidia are the cause of great economic loss among domestic and game animals in temperate climates. Serious epizootics of coccidiosis occur in poultry, rabbits, muskrats, foxes, mink, cattle, sheep, pigeons, and occasionally dogs, cats, and other animals. Common invertebrate hosts include arthropods and molluscs.

The most important genera are *Eimeria* and *Isospora*. They differ mainly in details of development within the oocysts. *Eimeria* oocysts have 4 sporocysts containing 2 sporozoites each (Figure 14), whereas *Isospora* oocysts have 2 sporocysts containing 4 sporozoites each. In both cases, the ripe oocyst contains 8 sporozoites.

The life cycle consists of 3 phases, each with its distinctive forms and functions. They are schizogony, gametogony, and sporogony. Schizogony and gametogony take place inside the cells while sporogony occurs outside the host in the soil or litter. Each of these phases will be considered in studying the Coccidia.

Because *Eimeria perforans* in the intestine and *E. stiedai* in the liver of rabbits are fairly common, either or both of them may be used as typical forms in the life cycle studies. *Eimeria stiedai* is more important as a cause of mortality among young rabbits.

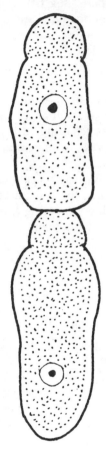

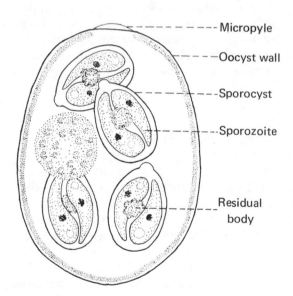

Micropyle

Oocyst wall

Sporocyst

Sporozoite

Residual body

Figure 13. *Gregarina blattarum,* a cephaline gregarine with two individuals in syzygy, showing the small anterior protomerite without the knoblike epimerite, and the large posterior deutomerite.

Figure 14. Diagrammatic sketch of ripe oocyst of *Eimeria.*

Eimeria perforans

Examine stained sections of the intestine of a heavily infected rabbit for the progressive development of the schizogenous and gametogenous stages. Observe oocysts from feces that have been incubated different lengths of time to show the major steps in sporogony.

Schizogenous stage. This is the asexual phase of multiplication inside the epithelial cells of the intestine. Four different forms appear in succession.

1. Trophozoite: These are the small single-celled, uninuclear forms inside the epithelial cells. The small nuclei of the parasite are surrounded by a halo and are easily distinguished from the large nucleus of the epithelial cells.

2. Schizont: After a period of growth, the trophozoite matures and the nucleus undergoes a series of divisions, resulting in 32 small ones. This process of nuclear division is known as schizogony and produces the schizonts. They are easily recognized by the many small nuclei scattered in the cytoplasm.

3. Segmenter: After division of the nucleus, a bit of cytoplasm surrounds each nuclear fragment, forming many tiny uninucleate individuals. The body containing these is known as a segmenter since the original uninucleate cell has segmented to form many tiny individuals.

4. Merozoites: As the individual bodies in the segmenter mature, they become spindle-shaped and eventually mature to become merozoites. The schizogenous stage terminates with the appearance of the merozoites. They are liberated into the lumen of the intestine, whence other epithelial cells are invaded by the merozoites to begin a new schizogenous cycle or to initiate the gametogenous cycle. In the latter case, the individual merozoites may be considered as being of different sexes.

Find epithelial cells filled with numerous merozoites and cells that have ruptured, liberating the merozoites.

Gametogenous stage. Some merozoites upon entering epithelial cells transform into gametocytes, or sexual cells, thereby initiating the gametogenous phase. Two types of cells are formed. They are macrogametocytes and microgametocytes.

1. Macrogametocytes: These are large cells located inside the epithelial cells, whose nucleus is crowded to one side. In sections stained with hematoxylin and eosin, the large cytoplasmic granules of the macrogametocytes stain bright pink. The cells have a thick hyaline wall and large centrally located nucleus. Upon reaching maturity, the numerous macrogametocytes become macrogametes ready for fertilization. Each one completely fills an epithelial cell. They usually appear in groups.

2. Microgametocytes: These differ from the macrogametocytes in that the nucleus breaks up into many small bits that migrate to the inner margin of the cell membrane where they align themselves in an orderly fashion. Each tiny nucleus is clothed separately in a bit of cytoplasm to form minute, slender, biflagellated microgametes. They are liberated from the cell in which development occurred. The appearance of the male and female gametes terminates the 2 intracellular phases of gametogony.

Find the characteristic microgametocytes with the many small nuclei and those with fully developed microgametes.

Fertilization occurs in the intestine to form the zygote. It develops a thick cyst wall and is known as an oocyst that is released from the epithelial cell and voided with the feces for further development outside the host.

3. Sporogony: Sporogenous development is a process of asexual multiplication in which the sporozoites are formed.

The oocyst is the beginning of the sporogenous stage. When voided with the feces, the oocysts, with equally rounded ends, and without a micropyle, measure 24 to 30 x 14 to 20 μ. Each one contains a large spherical cytoplasmic body. Sporulation, the process of forming the sporozoites by a process of division, begins and is completed in 48 hours at 33 C. The cytoplasmic mass divides twice to form 4 sporoblasts. The next step is a single division of each sporoblast, with the formation of 2 sporozoites enclosed in a sporocyst. Thus each ripe oocyst contains 4 sporocysts, each of which contains 2 sporozoites (Figure 14). The sporozoites are the infective stage. When the ripe oocyst is swallowed by a suitable host, the sporozoites liberated under the influence of digestive juices of the intestine penetrate epithelial cells and begin the cycle anew.

Sheep are host to a number of species of *Eimeria,* whose oocysts are large or small, with thick or thin walls that are smooth or rough, and with or without a micropyle. Examine wet slides prepared from a mixture of feces that have been preserved fresh in formalin and from feces that have been incubated for varying periods before preservation to show the progressive stages of sporulation. Note the different kinds of oocysts and their variations. Consult Levine, 1973, for identification of them.

Life Cycle Exercise

The life cycle of *Eimeria tenella* of chickens can be worked easily as an example of a typical coccidian (Figure 15).

Feed a massive dose of fully sporulated oocysts (Lotze and Leek, 1961) to 8 or 10 young chicks. Sacrifice 1 at each 24 hour interval. Preserve the ceca of each chick in 10% formalin for subsequent sec-

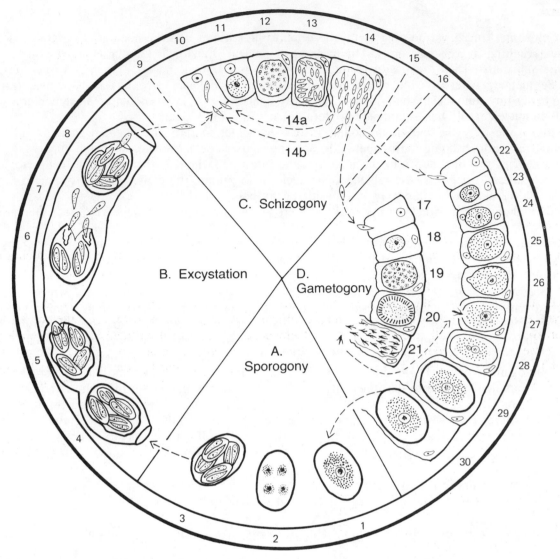

A. Sporogony—development of oocysts in litter or soil with formation of sporozoites. (1) mature oocyst; (2) sporoblast in which nucleus has undergone multiple division; (3) infective or ripe oocyst containing 4 sporocysts each with 2 sporozoites.

B. Excystation of sporozoites in crop, gizzard, and intestine. (4) oocyst in crop where action of CO_2 and enzymes causes oocyst wall to wrinkle and weaken; (5) oocyst in gizzard where mechanical action together with continued effect of CO_2 and enzymes causes many sporocysts to rupture, releasing sporozoites inside oocyst; (6) oocysts in intestine rupture, releasing intact sporocysts or individual sporozoites; (7) sporozoites free in intestine; (8) individual sporozoites escape through imperceptible opening in intact oocyst.

C. Schizogony—asexual reproduction with formation of merozoites in cecal epithelial cells. (9) sporozoite freed in intestine enters cecum and penetrates epithelial cell; (10) sporozoite inside cell; (11) growth to form trophozoite; (12) nucleus undergoes multiple division to form schizont; (13) each nucleus with a bit of cytoplasm forms a merozoite; (14) schizont and parasitized cell rupture to release numerous merozoites, completing first generation of schizonts; (14a) merozoite free in lumen of ceca enters new epithelial cell to produce a second generation of schizonts (10-14); (14b) merozoite of second generation enters new epithelial cell to produce third generation of schizonts (10-14); (15) third generation merozoite destined to produce female stage; (16) third generation merozoite destined to become male stage.

D. Gametogony—sexual reproduction with formation of gametes in cecal epithelial cells. (17-21) male cycle: (17) merozoite enters epithelial cell; (18) growth and formation of microgametocyte; (19) multiple division of nucleus; (20) formation and alignment of microgametes (sperms) on periphery of cell; (21) rupture of cell and liberation of mature, biflagellated microgametes; (22-30) female cycle: (22) merozoite enters cell; (23) formation of macrogametocyte; (24) meiotic division of nucleus with extrusion of half of chromosomes; (25) haploid macrogamete; (26) mature haploid macrogamete; (27) fertilization of macrogamete; (28) fusion of male and female nuclei to form zygote; (29) oocyst; (30) mature oocyst breaks out of epithelial cell, enters lumen of ceca, and is finally voided with feces (1).

Figure 15. Life cycle of *Eimeria tenella* of chickens.

tioning and study. Stain with hematoxylin and eosin. The first generation of merozoites appears 2.5 to 3 days postinfection and the second during the fourth and fifth days. Make smears of fresh cecal contents at these times to show the merozoites and red blood cells resulting from hemorrhage. Stain the smears with Wright's blood stain.

Gametocytes appear in the epithelial cells between the fifth and seventh days, with the formation of gametes. Oocysts appear in the feces by the eighth day.

From your stained sections and smears, recognize all of the stages described above for *Eimeria perforans* and illustrated in the drawing on the life cycle of *E. tenella* (Figure 15).

Toxoplasma gondii

The biology and relationships of this parasite have been poorly understood until recently. It is now generally conceded that *T. gondii* belongs to the Coccidia. It resembles *Isospora bigemina* of cats; in fact, possibly what has been called the small race of *I. bigemina* is indeed *T. gondii*. Its life cycle differs from that of *Isospora* in that in addition to the sexual phase, there is a secondary host in which a special type of asexual proliferation known as endodyogeny occurs in cells of the striated muscles, retina, and central nervous system of birds and mammals.

Oocysts developing in the epithelial cells of the intestine of cats and passed in the feces contain 2 sporocysts each with 4 sporozoites when mature. When these are swallowed by cats, the oocysts release the sporozoites in the small intestine. Some of them penetrate intestinal epithelial cells and undergo the sexual phase of development, as in the case of *Eimeria*, and produce more oocysts. On the other hand, some of the sporozoites do not participate in the sexual cycle in the gut but enter the general circulation and go to various parts of the body, enter cells, and transform into trophozoites that reproduce asexually.

When animals other than cats that serve only as intermediate host (mice, humans, cattle, sheep, and others) swallow mature oocysts, the sporozoites make their way into the blood and are distributed to all parts of the body. They enter various kinds of cells, mainly those of the striated muscles, retina, and brain. Following rapid asexual proliferation, the host cells rupture and liberate the trophozoites which are free for a short time in the blood and abdominal serous exudate as small banana-shaped bodies. The trophozoites enter new host cells and initiate another cycle.

As the number of trophozoites increases from repeated cycles of proliferation, the host reacts by producing specific antibodies that curtail proliferation. When this stage is reached, the infection passes into the chronic stage characterized by the formation of intracellular cysts packed with numerous trophozoites. With the retention of trophozoites in the cysts and the absence of active reproduction in new cells, immunity declines and the loaded cysts deteriorate, eventually rupture, and release the infective trophozoites. Thus a succession of proliferative and cystic phases appears in infected intermediate hosts.

Trophozoites in cysts or free in the flesh of intermediaries are the principal source of infection to carnivorous animals, including man. Ripe oocysts from fecal contamination of the environment by cats provide infection for grass-eating animals, and to persons who may swallow them. In addition to the oral route of infection, congenital infection may occur when pregnancy and the proliferative phase of toxoplasmosis with free trophozoites in the circulation occur simultaneously.

Human infections come from sand boxes and soil in yards where cats defecate and children play, from eating inadequately cooked meat, and drinking raw milk even from the human mother.

Toxoplasmosis is a zoonotic disease of man transmitted from animals either through fecal contamination of his environment by cats or in his food from the flesh and milk of domestic animals. The cat-mouse relationship is the ecological key in the life cycle. The parasite is cosmopolitan in distribution. Prenatal infections in children result in still births, abortion, hydrocephalus, mental retardation, and blindness.

Trophozoites: These are banana-shaped crescents 4 to 7 μ long by 2 to 4 μ in diameter. They possess a polar ring, conoid, toxonemes, subpellicular fibrils, nucleus, Golgi apparatus, mitochondria, and endoplasmic reticulum (Figure 16, A). Most of these are ultrastructures.

Examine free trophozoites from smears of serous exudate of infected mice and in cysts from sections of infected tissue. The latter are round to oval, intracellular bodies packed with trophozoites.

Oocysts: They are oval in shape, surrounded by a smooth double wall, and measure about 10 by 12 μ. Sporulated oocysts contain 2 oval sporocysts about 6 by 8 μ long, each with 4 elongated, curved sporozoites 2 by 7 μ in size, and a residium of cytoplasm (Figure 16, B). Instructions for the isolation of oocysts are included under Special Techniques and Further Notes, p. 273).

Schizonts: These develop in cells of the intestinal epithelium of cats and contain merozoites similar in shape to sporozoites.

Gametocytes: Microgametocytes contain a light colored, loosely formed nucleus and, later, tiny microgametes arranged in parallel fashion around the periphery of the parent cell. Macrogametocytes have a small, compact nucleus and prominently staining cytoplasmic granules.

Observe schizonts and gametocytes of *Isospora* in sections of cat intestine to see the general aspects of these stages.

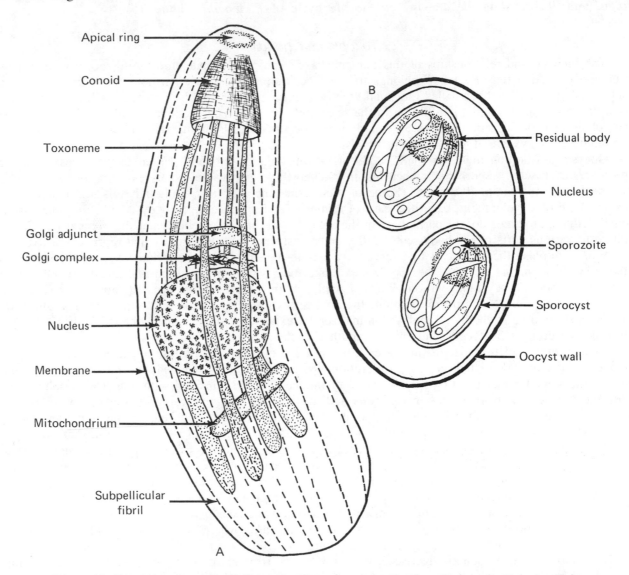

Figure 16. *Toxoplasma gondii.* (A) Reconstruction of a mature trophozoite based on electron micrographs; (B) Sporulated oocyst.

Haemosporina*

The Haemosporina are parasites of the blood of vertebrates, where schizogony occurs and gametogony begins, and of blood sucking Diptera in whose bodies gametogony is completed and sporogony occurs. Malaria is an example.

It is estimated that throughout the world 3.5 million lives are lost annually from malaria, caused by certain species of *Plasmodium*. It is difficult to estimate the debilitating effects of the disease, including

*In the Honigberg et al. (*J. Protozool.* 11:7-20, 1964) classification of the Protozoa, Haemosporina is a suborder of the order Eucoccida of the subclass Coccidia. In the older classification Haemosporidia was a subclass of the class Telosporidea.

mental lethargy, anemia, lowered resistance, etc., but some workers believe that, indirectly, malaria is responsible for over half of all human deaths in the world today. Malaria remains one of the leading causes of death among the infectious diseases of man, and it will be only through increased educational propaganda, therapeutic research, and extensive control campaigns conducted by trained workers (including doctors, technicians, engineers, conservationists, etc.) that this scourge will ultimately be eliminated as an important disease.

The student should recognize the normal constituents of the blood of vertebrates before undertaking a study of *Plasmodium* and the related forms. It is also urged that he be familiar with the preparation and staining of blood smears. For instructions consult p. 272.

Plasmodium vivax

This is the most common and widespread of the species of malaria infecting man. In vivax malaria the asexual cycle is completed in a 48-hour period and its effect on man is less severe than some other species. In the past, the disease caused by *P. vivax* was referred to as benign tertian malaria. It is now recommended that "benign tertian," referring to the clinical feature of the disease, be replaced by the unitalicized name of the causative agent and that, similarly, the names pertaining to the clinical features of the diseases caused by the other species of plasmodia causing malaria in man be replaced by the unitalicized name of the causative agent (Figure 17).

Like Coccidia, *Plasmodium* has a schizogenous, a gametogenous, and a sporogenous stage in its life cycle. The first phase of the schizogenous stage occurs in the liver, while the second phase and part of the gametogenous stage occur in the red blood cells of lizards, birds, and mammals. The concluding part of the gametogenous and all of the sporogenous stages occur in mosquitoes.

The stages in the life cycle are found in liver sections and blood smears of vertebrates, and in dissections of the alimentary canal and salivary glands of infected mosquitoes. Examine the appropriate preparations for the various stages described below.

Schizogenous cycle. The schizogenous stages consist of 2 parts, the exoerythrocytic stage in the liver and the erythrocytic stage in the red blood cells.

The exoerythrocytic phase is initiated when sporozoites are injected into the blood by infected mosquitoes while feeding. The sporozoites quickly enter liver cells where several generations of schizogenous multiplication take place, producing numerous merozoites known as metacryptozoites (they are morphologically similar to other merozoites). They enter the blood stream, come into contact with the red blood cells, penetrate them, and initiate the erythrocytic schizogenous phase of the life cycle.

Examine a section of liver for the large schizonts among the liver cells.

In *Plasmodium gallinaceum* of chickens, the exoerythrocytic schizonts occur in the endothelial cells of the brain capillaries. Examine a demonstration slide which shows the large schizonts in the capillaries. Note the numerous nuclei which will form the metacryptozoic merozoites. These enter the blood stream, penetrate the nucleated red blood cells, and initiate the erythrocytic phase.

The erythrocytic phase with its different stages will be studied from blood smears taken from man infected with P. *vivax*.

1. Trophozoites: Upon entering the red corpuscles, the metacryptozoites transform into trophozoites. The young trophozoite appears as a ring of blue cytoplasm with a single ruby red nucleus perched on one side, giving the appearance of a signet ring—hence the designation ring stage. As the ameboid trophozoites grow, they appear in various shapes, causing the infected corpuscles to increase in size and vary in shape.

The cytoplasm of infected cells becomes stippled with fine pinkish spots or granules known as Schüffner's dots. A brownish pigment known as hematin is separated from the iron-bearing hemoglobin by the parasites. It appears in the older trophozoites and subsequent stages.

Find the unicellular trophozoites in various stages of growth, together with the pigment and Schüffner's dots.

2. Schizont: Upon completion of the first division of the nucleus of the trophozoite, it becomes a schizont. Division of the nucleus continues until 12 to 24 minute nuclei appear.

Locate a schizont with many nuclei scattered throughout the uniform cytoplasm.

3. Segmenter: The cytoplasm of the schizont divides into small masses with each bit surrounding a nucleus to form a clump of small bodies. This is a segmenter.

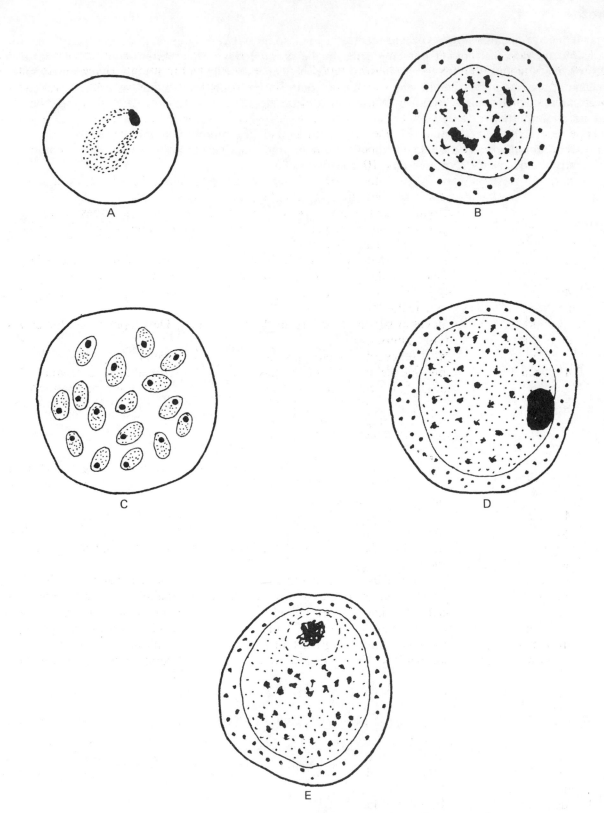

Figure 17. *Plasmodium vivax,* showing stages in the life cycle. (A) Young trophozoite in ring stage; (B) Schizont showing nuclei in parasite and Schüffner's dots in cytoplasm of blood cell; (C) Segmenter with typical number of merozoites; (D) Macrogametocyte, showing pigment in parasite and Schüffner's dots in cytoplasm of cell; (E) Microgametocyte.

Find segmenters in the red blood cells.

4. Merozoites: Each nucleus of the segmenter with its particle of cytoplasm becomes a merozoite. The mature segmenter ruptures the parasitized blood corpuscle, releasing the spindle-shaped merozoites into the blood stream with resultant chills and fever in the host. In mature segmenters, the pigment usually appears as a single mass and remains in the residuum of cytoplasm. The merozoites parasitize red blood cells and initiate a new schizogenous cycle leading to the gametogenous stage. Merozoites forming gametocytes are either male or female. The formation of merozoites occurs at intervals of 48 hours. In *P. vivax*, the prepatent period, the interval between the introduction of an organism and the demonstration of its presence in the host, is usually 10 to 12 days.

Observe the merozoites in the red blood cells. Merozoites are difficult to find in most smears. Nevertheless, one should watch for the free, minute, spindle-shaped bodies.

Gametogenous cycle. The gametogenous cycle is the beginning of the sexual stages. Large, uninucleated parasites with smooth contour that fill the red corpuscles are mature gametocytes. They consist of the male microgametocytes and the female macrogametocytes.

1. Microgametocyte: The microgametocyte has pale blue or pinkish blue cytoplasm and a large nucleus with irregularly distributed granules. It fills the red blood cell, often distorting it.

2. Macrogametocyte: This form has dense, blue cytoplasm containing a small, compact nucleus which may have a dark red nucleolus.

The gametocytes do not develop beyond maturity in the blood cells. Development continues and is completed in the stomach of mosquitoes.

Find representatives of both sexes of gametocytes in the red blood cells.

3. Microgamete: Upon entering the stomach of the mosquito, the microgametocyte produces 6 to 8 long, slender, filamentous microgametes by a process of exflagellation. They are attached to the residual mass of cytoplasm for a short time, after which they become free to seek and penetrate a macrogamete. The macrogametes are difficult to distinguish from gametocytes.

Observe the demonstration slide showing exflagellation of a microgametocyte in the stomach of a mosquito.

A class demonstration showing the process of exflagellation can be made by using blood from an infected sparrow or canary. Exflagellation will occur within 30 minutes or less in a Vaseline-sealed slide of fresh blood and can be viewed with the oil immersion objective.

4. Macrogamete: The macrogamete is similar to the macrogametocyte in appearance.

5. Zygote: Fertilization takes place in the mosquito's stomach when a microgamete enters a macrogamete to form a zygote. In plasmodia, the zygote is motile and designated as an ookinete. Formation of the zygote terminates the gametogenous cycle.

Sporogenous cycle. The sporogenous phase commences when the ookinete begins to develop.

1. Oocyst: When the migrating ookinete comes to rest between the epithelial cells and basement membrane of the mosquito's stomach, it is known as a young oocyst. Nuclear division commences and continues to the point where there are vast numbers of minute nuclear particles, known as sporoblasts.

As growth continues, the oocyst increases greatly in size. Bits of cytoplasm surround each nuclear particle to form myriads of sporozoites. The enlarged oocyst ruptures, liberating the sporozoites in the hemocel.

See demonstration side of mosquito stomach, showing oocysts of various sizes attached to the wall (Figure 18).

2. Sporozoite: When the ripe oocyst ruptures, up to 200,000 sporozoites are liberated in the hemocel of the mosquito. They travel to the salivary glands and migrate through the cells into the lumen. They are injected into the host with the contents of the salivary glands when the mosquitoes feed. Entrance of the sporozoites into the body of the vertebrate host initiates the exoerythrocytic phase of the schizogenous cycle.

See demonstration slide of sporozoites.

Three other species of plasmodia causing malaria in man are generally accepted. They are (1) *Plasmodium falciparum*, (Figure 19), the cause of falciparum malaria; (2) *P. malariae*, (Figure 20), the cause of malariae infection; and (3) *P. ovale*, the most recently accepted species, the cause of ovale malaria.

In falciparum malaria, schizogony is completed in 36 to 48 hours, usually requiring the longer period. In this species, only the ringlike trophozoites and elongated, somewhat crescent-shaped macrogame-

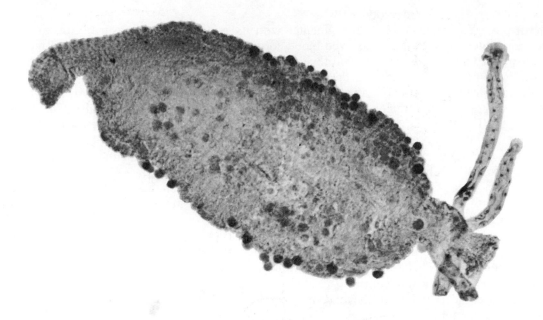

Figure 18. *Plasmodium* oocysts on stomach of infected female mosquito.

tocytes and bean-shaped microgametocytes appear in the circulating blood. The schizonts and segmenters are in the capillaries and blood sinuses of the internal organs and in the bone marrow.

This species differs conspicuously from the others in man in its crescent- or bean-shaped gametocytes and in the absence of schizonts and segmenters in the circulating blood cells.

Observe smears of human blood showing the distinctive gametocytes in the greatly distorted red corpuscles.

In the malariae infection, schizogony occurs at 72-hour intervals. Schizonts often appear as bandlike organisms, extending across the red blood cell. Segmenters contain 6 to 12, usually 8 to 10, robust merozoites instead of the 12 to 18 smaller ones in *P. vivax* and *P. falciparum*. Gametocytes are oval and similar to those of *P. vivax*; being smaller, they do not expand and distort the blood corpuscles.

Observe smears containing blood infected with *P. malariae*. Note the characteristic morphology of the schizonts, segmenters, and gametocytes of this species.

While *P. ovale*, fairly common in western Africa, has been reported from all 5 continents and from Pacific islands, the infection rate is always low. This species resembles both *P. vivax* and *P. malariae*, which may easily lead to a wrong identification of the species in the hands of inexperienced persons. As in *P. vivax*, the asexual cycle is completed in 48 hours. For further details on *P. ovale* refer to Wilcox (1960: 18).

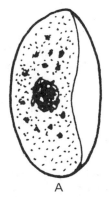
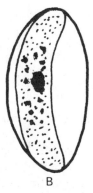

A   B

Figure 19. *Plasmodium falciparum,* showing characteristic elongated microgametocyte (A) and macrogametocyte (B).

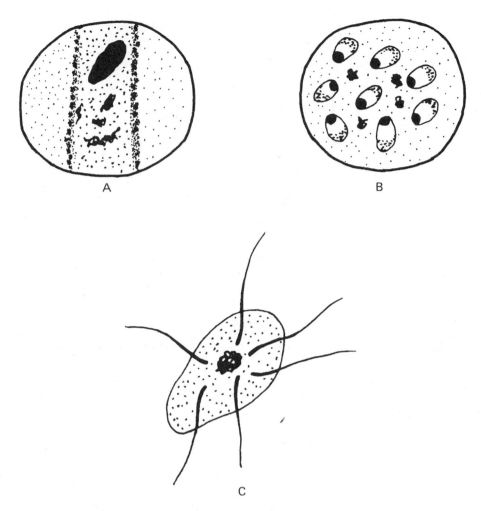

Figure 20. *Plasmodium malariae.* (A) Bandlike schizont; (B) Segmenter with characteristically small number of merozoites; (C) Exflagellating microgametocyte forming microgametes.

Plasmodium knowlesi, a species infecting monkeys, was once used extensively for malaria therapy of syphilis. It is lethal to monkeys on the 12th day but runs a mild course in man, and can be easily controlled.

The rodent-borne *P. berghei* and species of avian malaria, *P. cathemerium* et al., transmitted by anopheline and culicine mosquitoes respectively, have been used extensively in malaria research.

Various species of *Plasmodium* occur commonly in many kinds of wild birds. A study of them may be made from smears obtained from birds trapped, bled, and released unharmed. Stain the smears with Wright's or Giemsa blood stain. In studying avian plasmodia, note how the parasites are arranged in relation to the nucleus of the red blood cells (Figure 21). For further information on bird malaria consult Hewitt (1940) and Huff (1963, 1968).

Other Haemosporina Producing Malarialike Diseases in Birds and Mammals

Haemoproteus columbae occurs in the erythrocytes of pigeons only as gametocytes. The crescent-shaped gametocyte surrounds much of the nucleus of the red blood cell (Figure 22). Schizonts are found in the endothelial cells of the capillaries of the lungs. Gametogony and sporogony occur in the stomach of louse flies (*Lynchia* and related genera) similar to *Plasmodium* in mosquitoes. *Parahaemoproteus nettionis*, found in ducks and geese, is transmitted by biting midges (*Culicoides* sp.).

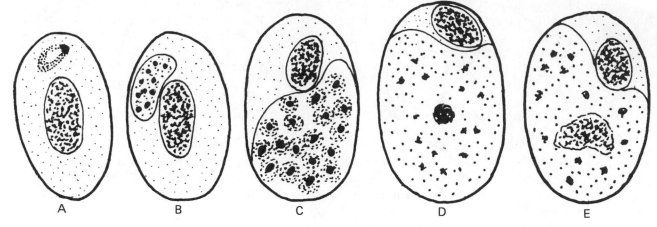

Figure 21. *Stages of Plasmodium* of birds. (A) Young trophozoite in ring stage; (B) Schizont; (C) Segmenter; (D) Microgametocyte; (E) Macrogametocyte.

Leucocytozoon simondi is in the endothelial cells of the liver and lungs of ducks as trophozoites, schizonts, and segmenters, causing great mortality. The gametocytes are in cells presumed to be macrophages. Infected blood cells may be round in shape with the parasite partially encircling the nucleus, or the cell may be greatly enlarged and distorted into a spindle shape with the nucleus compressed and pushed to one side by the large, oval gametocyte (Figure 23). The life cycle is similar to that of *Plasmodium* except that transmission is by blackflies of the genus *Simulium*. Other species of *Leucocytozoon* occur in turkeys and ruffed grouse.

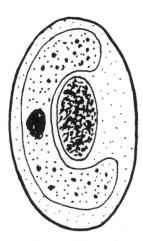

Figure 22. Halter-shaped gametocyte of *Haemoproteus* in red blood cell of bird.

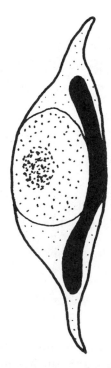

Figure 23. *Leucocytozoon* in elongated blood cell with displaced, attenuated nucleus in bird.

Babesia bigemina appears as 2 small, comma-shaped, unpigmented bodies in the red blood cells of cattle (Figure 24). It is responsible for Texas cattle fever, or hemoglobinuria. Fertilization occurs in the gut of ticks and the zygote migrates through the gut wall into a developing egg in the ovary. This is an example of transovarian transmission. As the young ticks develop, the parasites reproduce and infect various tissues of the acarine host. Sporozoites entering the salivary glands are transmitted by the larval ticks when they feed.

Observe blood smears from cattle suffering from babesiasis or cattle tick fever.

Anaplasma marginale occurs in the blood of bovines as a small deeply stained dot on the margin of the erythrocytes (Figure 25). The similarity of the clinical syndromes of anaplasmosis and piroplasmosis has suggested a protozoan character for *Anaplasma*. However, a lack of regularity associated with the development of *Anaplasma* and an undifferentiated internal structure as revealed by electron microscopy are against its being a protozoan. The presence of deoxyribonucleic and ribonucleic acid, sensitivity to broad spectrum antibiotics, and a size larger than 0.3 μ resemble rickettsial rather than viral character. There are other similarities.

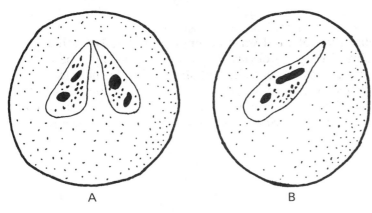

Figure 24. *Babesia bigemina* in red blood cells. (A) Cell with 2 parasites; (B) Cell with single elongated parasite.

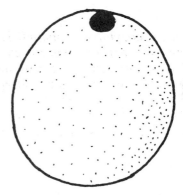

Figure 25. *Anaplasma marginale* with parasite located at margin of blood cell. In *A. centrale*, the parasite is located near the center of the blood cell.

SUBPHYLUM CNIDOSPORA

Class Myxosporidea

The members of this subphylum are characterized by their unique spores, which consist of 1 or more polar filaments and sporoplasms enclosed in a membrane consisting of a single-piece or 2 or 3 valves.

The Cnidospora are parasites of invertebrates and lower vertebrates. Infections may occur as serious epidemics in silkworms, honeybees, and fish. The life cycle is direct, being completed without benefit of an intermediate host.

Common cnidosporan parasites of fish are histozoic members of the Myxosomatidae and the histozoic and celozoic species of Myxobolidae.

Myxosoma sp.

The spore varies from pyriform, circular, or ovoid in shape. There are 2 pear-shaped polar capsules (Figure 26, B). The sporoplasm has no iodinophilous vacuole. *Myxosoma catostomi* is prevalent in the muscles and connective tissue of the common sucker. *Myxosoma cerebralis* occurs in the cartilage and perichondrium of trout, causing twist or whirling disease. The oval spores are 10 to 11 μ in length.

Observe slides containing spores of some of the Myxosomatidae. Note the pustulelike lesions on the skin of fish. Open a pustule, transfer the contents to a slide, and observe them under the high power of a microscope. Add a drop of iodine solution to determine whether there is an iodinophilous vacuole.

Myxobolus sp.

The spores are somewhat similar to those of *Myxosoma* except that there is an iodinophilous vacuole in the sporoplasm and there may be 1 or 2 polar capsules. *Myxobolus orbiculatus* occurs in the muscles of shiners and *M. intestinalis* in the intestinal wall of sunfish. The spores are small, being 9 to 13 μ in size and somewhat circular or oval in shape (Figure 26, A).

Observe smears containing spores that have been stained with iodine and sections of gut showing the lesions filled with developing forms and spores.

Henneguya sp.

Typical spores are somewhat spindle-shaped in frontal view. Each of the 2 valves of the cyst continues posteriorly as a long, slender extension. There are 2 polar capsules and an iodinophilous vacuole. *Henneguya exilis* occurs in the gills and skin of catfish and *H. microspora* in bass and sunfish. Spores proper are 18 to 20 μ long but with the valvular extensions attain lengths of 60 to 70μ (Figure 26, C).

Observe a demonstration slide of *Henneguya*, noting the spindle-shaped spore and the long extensions of the valves. Collect and examine catfish, bass, and sunfish for infections. Stain living spores with Lugol's iodine solution to demonstrate the iodinophilous vacuole.

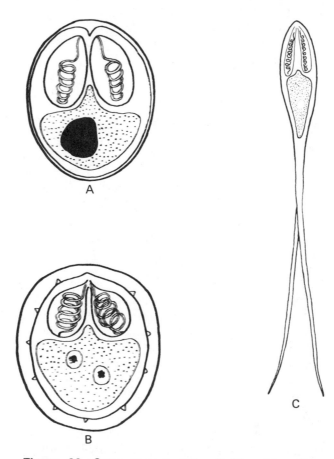

Figure 26. Some representative Cnidoporida. (A) Frontal view of *Myxobolus,* showing iodinophilous vacuole in sporoplasm; (B) Frontal view of *Myxosoma* which has no iodinophilous vacuole; (C) Frontal view of *Henneguya.*

SUBPHYLUM CILIOPHORA
Class Ciliatea

Many cilates occurring in animals are nonpathogenic commensals in their natural hosts. *Balantidium coli*, normally in swine, is an example of one that is harmless in the natural host but is pathogenic when in man.

Ichthyophthirius multifiliis

This ciliate commonly known as "ich" occurs on the skin of freshwater fish. It is especially prevalent on pond and hatchery fish, being embedded in whitish pustules. It may prove fatal by making the skin of the fish host susceptible to attack by fungi.

Fully developed ciliates are oval in shape and up to 1 mm in length. The body is covered by numerous rows of cilia and bears a circular, unciliated cytostome at one end. The large macronucleus is horseshoe-shaped and the small, spherical micronucleus lies in the concavity of the former. Numerous vacuoles occur in the cytoplasm (Figures 27, 28).

Life Cycle

Mature ciliates dropping from the pustules produce gelatinous cysts within which repeated simple transverse division takes place until up to 1,000 daughter ciliates, or swarmers, are produced within a few hours. These escape from the cyst, attach to fish, and develop.

Examine slides of fixed and stained adults and sections of infected gills and integument. If available, examine small brook trout or other species infected with "ich." Note the many small but distinctly visible white pustules. Place some adult ciliates in physiological solution on a slide, cover, and seal the cover glass with Vaseline. Watch the preparation over a period of 8 to 10 hours to see encystment, division, formation of swarmers, and finally their escape from the cyst.

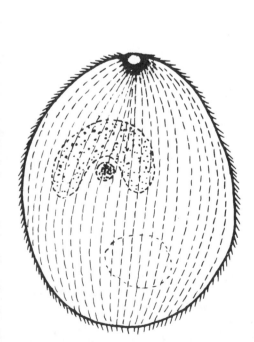

Figure 27. *Ichthyophthirius multifiliis.*

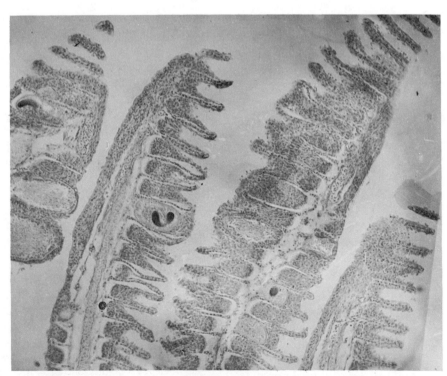

Figure 28. *Ichthyophthirius multifiliis,* photomicrograph of section of infested fish gill.

Balantidium coli

This is a common commensal of the cecum and large intestine of swine. In man, however, it is a pathogenic parasite which invades the intestinal mucosa in a manner similar to *Entamoeba histolytica*.

Trophozoites: The large, egg-shaped trophozoite measuring 50 to 100 x 40 to 60 μ is surrounded by a pellicle from which arise cilia arranged in rows. One end is more pointed than the other and at this end there is a cleft, known as the peristome, from which leads the cytostome. At the more rounded end is the cytopyge. Within the endosarc is a large, slightly curved macronucleus, usually situated near the middle of the body. Lying within the concavity of the macronucleus and in contact with it, is the micronucleus. The two contractile vacuoles, one in each end, appear as clear areas, and the food particles stain variously (Figure 29).

Cysts: Cysts stain poorly, the cyst wall remaining unstained, and the macronucleus appears to be the most conspicuous structure.

Life Cycle

Trophozoites in the intestine multiply by binary fission and feed on bacteria and particulate material. When carried into the rectum, the trophozoites round up and secrete a double wall about themselves before being voided with the feces. Infection results from swallowing the encysted stage with contaminated food or water.

Living material will ordinarily be made available from an infected guinea pig. To a drop of material in physiological solution on a clean slide, add a cover glass, and look for these large ciliates with the low-power objective. Stain with neutral red.

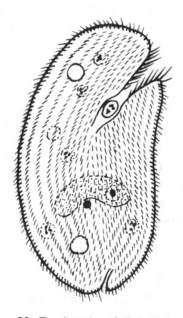

Figure 29. Trophozoite of *Balantidium coli.*

Trichodina myakkae

These barrel-shaped ciliates occur fairly commonly on the body, fins, and gills of freshwater fish. Although *T. myakkae* was originally described from largemouth bass, it has since been reported from the gills of hatchery-reared trout. Thus, it is apparent that these ciliates are not always host specific. In profile, the body is discoidal, the concave adhesive disc being attached to the host. The convex surface is smooth and bears two parallel rows of cilia, known as the adoral spiral. Marginally situated, there is a band of long cilia, known as the ciliary girdle, which enables the animal to move over the surface of the host. In addition to the usual food vacuoles and large contractile vacuole, the cytoplasm contains a large horseshoe-shaped macro- and a small, spherical micronucleus.

REFERENCES

Baker, J.R. 1969. Parasitic Protozoa. Hutchinson University Library, London, 174 p.

Kudo, R.R. 1966. Protozoology 5th ed. Charles C Thomas, Springfield, 1174 p.

Levine, N.D. 1973. Protozoan Parasites of Domestic Animals and of Man. 2nd ed. Burgess Publ. Co., Minneapolis, 406 p.

Sarcodina

Elsdon-Dew, R. 1968. The Epidemiology of Amoebiasis. *In* Advances in Parasitology, ed. B. Dawes. Academic Press, New York, vol. 6, pp. 1-62.

Neal, R.A. 1966. Experimental Studies on *Entamoeba* with Reference to Speciation. *In* Advances in Parasitology, ed. B. Dawes, Academic Press, New York, vol. 4, pp. 1-51.

WHO. 1969. Amoebiasis, Report of a WHO Expert Committee, Wld. Hlth. Org. Tech. Rept. Series, No. 421, 52 p.

Mastigophora

Adler, S.1964. Leishmania. *In* Advances in Parasitology, ed. B. Dawes, Academic Press, New York, vol. 2, pp. 35-96.

Desser, S.S., S.B. McIver, and A. Ryckman. 1973. *Culex territans* as a potential vector of *Trypanosoma rotatorium*. I. Development of the flagellate in the mosquito. J. Parasitol. 59: 353-358.

Garnham, P.C.C. 1965. The leishmanias, with special reference to the role of animal reservoirs. Amer. Zool. 5: 141-151.

Hoare, C.A. 1967. Evolutionary Trends in Mammalian Trypanosomes. *In* Advances in Parasitology, ed. B. Dawes, Academic Press, New York, vol. 5, pp. 47-91.

_____, and F.G. Wallace. 1966. Developmental stages of trypanosomatid flagellates: a new terminology. Nature 212: 1385-1386.

Jírovec, O., and M. Petroů. 1968. *Trichomonas vaginalis* and Trichomoniasis, *In* Advances in Parasitology, ed. B. Dawes, Academic Press, New York, vol. 6, pp. 117-188.

Köeberle, F. 1968. Chagas' Disease and Chagas' Syndromes: The Pathology of American Trypanosomiasis, *In* Advances in Parasitology, ed. B. Dawes, Academic Press, New York, vol. 6, pp. 63-116.

Lumsden, W.H.R. 1965 & 1970. Biological Aspects of Trypanosomiasis Research, *In* Advances in Parasitology, ed. B. Dawes, Academic Press, New York, vol. 3, pp. 1-57; vol. 8, pp. 227-249.

Sporozoa

Fallis, A.M. 1965. Protozoan life cycles. Amer. Zool. 5: 85-94.

Frenkel, J.K. 1973. *Toxoplasma* in and around us. BioScience 23: 343-351.

Garnham, P.C.C. 1966. Malaria Parasites and Other Haemosporidia, Blackwell Sci. Publ., Oxford, 1114 p.

_____. 1967 & 1973. Malaria in Mammals Excluding Man. *In* Advances In Parasitology, ed. B. Dawes, Academic Press, New York, vol. 5, pp. 139-204; vol. 11, pp. 603-630.

Hammond, D.M., and P.L. Long, eds. 1972. The Coccidia: *Eimeria, Isospora, Toxoplasma,* and Related Genera, University Park Press, Baltimore, 482 p.

Hewitt, R. 1940. Bird Malaria. Am. J. Hyg., Monog. Ser. No. 15, 228 p.

Horton-Smith, C., and P.L. Long. 1963 & 1968. Coccidia and Coccidiosis in the Domestic Fowl and Turkey. *In* Advances in Parasitology, ed. B. Dawes, Academic Press, New York, vol. 1, pp. 67-107; vol. 6, pp. 313-325.

Huff, C.G. 1963 & 1968. Experimental Research in Avian Malaria. *In* Advances in Parasitology, ed. B. Dawes, Academic Press, New York, vol. 1, pp. 1-65; vol. 6, pp. 293-311.

Jacobs, L. 1973. New Knowledge of *Toxoplasma* and Toxoplasmosis. *In* Advances in Parasitology, ed. B. Dawes, Academic Press, New York, vol. 11, pp. 631-669.

Lotze, J.C., and R.G. Leek. 1961. A practical method of culturing coccidial oocysts in tap water. J. Parasitol. 47: 588-590.

———, and ———. 1968. Excystation of sporozoites of *Eimeria tenella* in apparently unbroken oocysts in the chicken. J. Protozool. 15: 693-697.

Maegraith, B. 1968. Liver Involvement in Acute Mammalian Malaria with Special Reference to *Plasmodium knowlesi* Malaria. *In* Advances in Parasitology, ed. B. Dawes, Academic Press, New York, vol. 6, pp. 189-231.

Miles, H.B. 1962. The mode of the transmission of the acephaline gregarine parasites of earthworms. J. Protozool. 9: 303-306.

Walker, A.J. 1968. Manual for the Microscopic Diagnosis of Malaria. 2nd ed. Pan American Health Organization, Washington, D.C., Sci. Publ. No. 161. 117 p.

WHO. 1969. Parasitology of Malaria, Report of a WHO Scientific Group. Wld. Hlth. Org. Tech. Rept. Series, No. 433, 70 p.

Wilcox, Aimee. 1960. Manual for the Microscopial Diagnosis of Malaria in Man, U.S. Dept. Health, Educ., Welfare. Public Health Service Publ. No. 796, 80 p.

Ciliophora

Davis, H.S. 1946. Care and Diseases of Trout. U.S. Dept. Interior, Fish and Wildlife Service. Research Rept. No. 12, 98 p. [*Chilodon* (now *Chilodonella*), *Trichodina* and *Ichtyophthirus*, pp. 37-45].

El Mofty, M.M., and J.D. Smyth. 1964. Endocrine control of encystation in *Opalina ranarum* parasitic in *Rana temporaria*. Exptl. Parasitol. 15: 185-199.

SECTION II

HELMINTHS

Class Trematoda

The trematodes, or flukes, are all parasitic in or on animals, mostly vertebrates. The body is covered with tegument. A muscular sucker usually surrounds the mouth located at or near the anterior end of the body, except in a few species. The digestive tract consists of an esophagus and normally a pair of blind intestinal ceca, the whole being Λ-shaped. The life cycle may be direct, as in the ectoparasitic Monogenea, or indirect, with a molluscan intermediate host, as in the endoparasitic Digenea.

Subclass Monogenea

These are external parasites on the skin and gills of aquatic vertebrates, primarily fish. Some occur in the uterus of fish. A few species live in the mouth and urinary bladder of amphibians and reptiles. The adhesive organs are known as haptors; prohaptor at the anterior end of the body and opisthaptor at the rear. The life cycle is direct; the egg hatches into an oncomiracidium which is ciliated and bears numerous hooks (Figure 31, so that the larva is well adapted both for swimming and for attachment. In a few species, eggs hatch in the uterus (ovoviviparity) and the newly hatched larvae resemble the adults.

Order Monopisthocotylea

Characterized by the single opisthaptor, which may be divided into parts by septa. There are 1 to 3 pairs of large hooks and 12 to 16 marginal hooklets. The prohaptor consists of glandular areas of suckers or pseudosuckers located outside the buccal cavity. The intestinal ceca may or may not be branched or united posteriorly. Testes 1, 2, 3, or many, usually postovarian. The cirrus consists of a simple or complex cuticularized structure. A genito-intestinal canal extending from the oviduct to the intestinal ceca is lacking. Oviparous or ovoviviparous; eggs usually have a polar filament at one or both ends.

Family Gyrodactylidae

These are small flukes with poorly developed vitellaria, no eyespots or vagina, and ovoviviparous with young in the uterus. They are parasites of fish and amphibians. There are many species.

Gyrodactylus spp.

Species of this genus are parasitic on the gills of freshwater and marine bony fishes.

Description

The small body is elongate. The opisthaptor is without divisions; it bears a pair of large anchors connected by a ventral and dorsal bar; there are 16 marginal hooklets. There is a single pair of lobelike head glands. The mouth is subterminal and followed by a globular pharynx; the short esophagus divides into a pair of intestinal ceca, terminating near the posterior end of the body (Figure 30).

The testis is a small median body lying between or behind the intestinal ceca; the cirrus is armed with spines at its opening. There is a complex sclerotized accessory piece associated with the copulatory organ. The genital pore is submedian behind the pharynx. A fully developed embryo is in the uterus; the ovary is posttesticular, median; vitellaria form 2 lobes surrounding the ends of the ceca.

When born, the precocious larvae appear much like their parents. They attach to the gills of their hosts and grow directly without metamorphic changes into adults. When fully developed, the larva contains a less developed larva in its uterus; before birth a second larva appears in the uterus of the first, a third inside the second, and even a fourth inside the third.

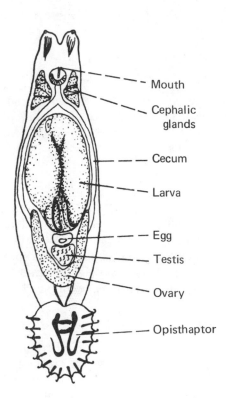

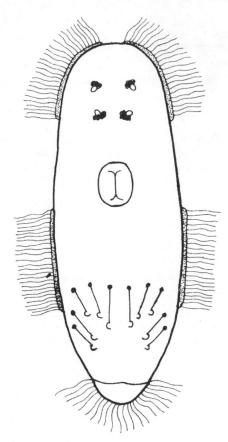

Figure 30. *Gyrodactylus medius.*

Figure 31. Oncomiracidium of *Dactylogyrus vastator.*

Family Dactylogyridae

These are small, oviparous flukes with eyes and vagina. The vitellaria are well developed. They are parasites on the gills of freshwater and marine fishes.

Dactylogyrus vastator

This fluke occurs commonly on the gills of carp. They are especially detrimental to young fish in crowded ponds.

Description

Adults are up to 1.2 mm long by 0.3 mm wide. There are 2 prominent head lobes, each with glands. Two pairs of eyespots are located anterior to the pharynx. The opisthaptor bears 1 pair of large anchors with bifurcate roots united by a single rod-shaped bar; there are 7 pairs of small marginal hooklets. There is a single, oval testis located slightly posterior to the ovary. The small oval ovary is situated near the midbody. The vaginal opening is toward the right margin of the body, slightly preequatorial.

There are 2 kinds of eggs. During warm weather, summer eggs are laid. These develop quickly and hatch, freeing the oncomiracidium (Figure 31). Several generations of worms appear during the summer. Toward the end of the summer, winter eggs are laid. These remain dormant at the bottoms of the ponds until spring when development begins as the water warms.

Family Capsalidae

Characterized by a flat, oval body with a well-developed muscular disc-shaped opisthaptor bearing 3 pairs, occasionally 2 pairs, of anchors; marginal hooklets may or may not be present. The prohaptor consists of paired suckerlike structures. There are 2 pairs of eyes. Parasitic on marine fish.

Benedenia melleni

This species occurs on the skin, gills, in nasal cavities, and around the eyes of a wide variety of marine teleosts. More than 50 species representing 18 families have been reported as hosts (Figure 32).

Description

Adults are up to 5 mm long. An ophisthaptor 1.2 mm in diameter is attached to the body by a slender stalk; there are 3 pairs of anchors with slightly forked base, and 14 small accessory branched hooks arranged radially around the margin of the sucker. The prohaptor, a pair of large unarmed suckerlike organs, lies at the anterior extremity of the body. Oral sucker is lobed; intestine bifurcates at the mouth to form 2 branched ceca that extend to near the posterior end of the body.

There are 2 roundish testes arranged transversely near the middle of the body, a voluminous, slender, pear-shaped seminal vesicle with a pair of long external, follicular prostate glands. The common genital pore opens near the left side of the mouth. The ovary is located medially near the anterior margin of the testes; there is a well-developed seminal receptacle, and a sinuous duct that terminates in a voluminous shell gland lying parallel to the cirrus pouch and opening in the common genital atrium. The vitellaria occupy most of the space between the organs. The eggs are tetrahedral in shape, usually with 1 very long and 2 much shorter filaments with hooklike tips. For life cycle details consult Jahn and Kuhn (1932).

Order Polyopisthocotylea

Characterized by an opisthaptor consisting of 2 or more suckers or clamps. The prohaptor is usually devoid of adhesive glands and the mouth surrounded by an oral sucker. A genito-intestinal canal is present. These worms are parasitic on the skin or gills, or in the oral cavity or urinary bladder, of fishes, amphibians, and reptiles.

Family Polystomatidae

The opisthaptor bears 3 pairs of cup-shaped, muscular suckers, with a number of marginal hooklets, and with or without haptoral anchors. The mouth is terminal or subterminal and surrounded by a muscular sucker. Polystomatids are parasitic in the mouth, esophagus, or urinary bladder of amphibians and reptiles.

Polystomoides oris

This and related species occur in the mouth of the painted turtle *(Chrysemys picta)* and other species of North American turtles (Figure 33).

Description

The fusiform body up to 4.2 mm long by 1.6 mm wide. The opisthaptor bears the usual 6 suckers. There are 2 pairs of large hooks between the posterior pair of suckers, and 16 larval hooks: 6 between the anterior pair of suckers, 4 between the posterior pair of suckers and 1 in each sucker. A vagina opens ventrally near each lateral margin of the body near the union of the first and second thirds of the body. The oral sucker is large and subterminal; the pharynx is as large as or slightly larger than the oral sucker; the esophagus is short, and the intestinal ceca long and diverticulate anteriorly. The spined genital pore is median and directly behind the intestinal bifurcation.

There is a median, single, large testis a short distance behind the intestinal bifurcation. The cirrus bears 24 to 27 spines. A small, comma-shaped ovary is anterior to the testis and to the left of the midventral line of the body. The vitelline follicles surround the lateral and ventral sides of the intestinal ceca from the pharynx to the opisthaptor. A uterus is absent; the ootype contains a single egg 250 x 180 μ.

Polystoma nearcticum

This species, originally regarded as a subspecies of *P. integerrimum*, the well-known European species, occurs in the urinary bladder of adult tree frogs *(Hyla versicolor* and *H. cinerea)*, and on the gills of

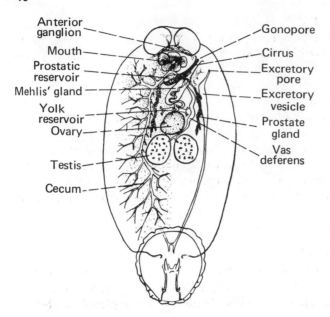

Figure 32. *Benedenia melleni.*

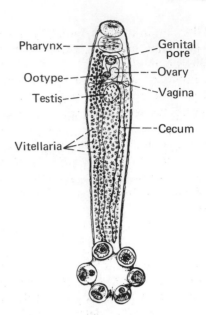

Figure 33. *Polystomoides oris.*

their tadpoles. There are 2 adult forms, the branchial form on the gills and the bladder form in the urinary bladder.

Description

The branchial forms are up to 5 mm long by 0.8 mm wide. The opisthaptor has 6 pedunculate suckers. The hooks are rudimentary or absent. The testis is spherical and the ovary elongate. The bladder forms are up to 4.5 mm long by 1.5 mm wide. The opisthaptor has 6 muscular suckers and 1 pair of large hooks. The testis is multilobate and the ovary is comma-shaped. Eggs of both forms average 300 x 150 μ.

Life Cycle

As a result of 1 form of adult worms occurring in the urinary bladder of adult tree frogs and another form of adults living on the external gills of the tadpoles, the life cycle is somewhat complicated. Eggs from bladder forms produce oncomiricidia which when attaching to the gills of young tadpoles, undergo accelerated development of the reproductive organs. But when the larvae hatched from eggs produced by the neotenic gill form, or those from eggs of the bladder form, encounter an older tadpole undergoing metamorphosis, they enter the urinary bladder via the cloacal opening and attain maturity in the spring of their third year of life, at which time the frog also reproduces for the first time. Under the influence of hormones, which appear in the host's urine at the breeding season in spring, the bladder form lays eggs which pass out when the frog enters the water to breed. For life cycle details consult Paul (1938).

Family Sphyranuridae

These are readily recognized by the bilobed opisthaptor bearing only 2 large cuplike, muscular suckers. They are parasites on the skin and gills of amphibians (Figure 34).

Sphyranura oligorchis

This polystome from the gills of the perennibranchiate mudpuppy (*Necturus maculatus*) is characterized by an opisthaptor with 2 large, cuplike, muscular suckers and 2 large hooks.

Description

Adults are up to 4 mm long by 0.7 mm wide. The bilobed opisthaptor with 2 large, muscular suckers is wider than the body; each sucker has a single hooklet inside; there is a large anchor near the posterior margin of each sucker and 14 hooklets on the margin of the opisthaptor. The terminal funnel-shaped mouth is surrounded by an oral sucker followed by a conspicuous muscular pharynx, a short esophagus, and intestinal ceca that are confluent posteriorly.

About 6 testes are arranged linearly in the posterior half of the body; a vas deferens extends anteriorly to the common genital pore just posterior to the bifurcation of the esophagus. The cirrus is armed with spines. A small pyriform ovary lies to the left of the midline and anterior to the testes; the uterus is largely anterior to the ovary; the vaginae are double but do not open to the outside. A genito-intestinal canal opens into the left cecum near the level of the ovary. The vitellaria are follicular and extend along the intestinal ceca from the ovary posteriorly. Usually only a single egg, 28 x 41 μ in size, is present in the uterus.

Life Cycle

Adults on the gills of the host lay their eggs, which settle to the bottom of the pond. Development is slow at room temperature, as hatching takes about a month. If the newly hatched larvae succeed in making contact with a host, they promptly attach and migrate to the gills. Maturity is completed in about 2 months. For life cycle details consult Alvey (1936).

Family Discocotylidae

The sclerotized clamps are symmetrically arranged with 4 in each lateral row. A pair of small suckers opens into the mouth cavity, which is followed by a muscular pharynx, esophagus, and intestinal ceca with numerous diverticula. Parasitic on fish.

1. Octomacrum lanceatum

This species is parasitic on the gills of the common sucker and chub sucker. It is a good example of clamp-bearing members of the Polyopisthocotylea (Figure 35).

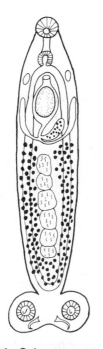

Figure 34. *Sphyranura oligorchis.*

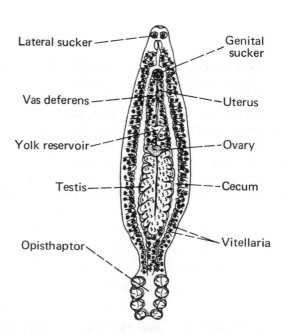

Lateral sucker

Genital sucker

Vas deferens

Uterus

Yolk reservoir

Ovary

Testis

Cecum

Vitellaria

Opisthaptor

Figure 35. *Octomacrum lanceatum.*

Description

Adults are up to 6 mm long by 2 mm wide. The opisthaptor is rectangular, and set off from the body by a slight, broad constriction. There are 2 rows of 4 suckers with edges in contact on each side of the opisthaptor; each sucker is supported by a complicated sclerotized clamp consisting of a central arched piece and 2 others, each beginning as a single part near the middle of the anterior margin of the sucker and extending around the end, where it divides into 2 branches, 1 on the posterior margin of the sucker and the other in between.

The mouth, at the anterior tip of the body, has 2 lateral suckers inside; it is followed by a muscular pharynx, and esophagus, and 2 long single, diverticulate intestinal ceca.

The male reproductive system consists of a multilobate testis and a vas deferens that ascends directly anteriorly to the genital pore, opening through a muscular, hookless genital sucker at the level of the intestinal bifurcation. The coiled ovary lies at the anterior end of the testis. A short distance from the ovary, the oviduct divides with 1 branch forming the uterus and going to the common genital pore and the other to the genito-intestinal canal. The vitellaria fill all space not occupied by the other organs.

Key to Superfamilies of Monogenea

Capital letters in parentheses refer to hosts: F-Fish, A-Amphibia, R-Reptilia, M-Mammalia, C-Crustacea, and O-Cephalopods. Small letters following each capital letter refer to the type of host within the taxonomic group: f-freshwater fish, h-holocephalians, m-marine teleosts, and e-elasmobranchs. The generic names of Figures 36 through 47 designate the relationship to the superfamilies appearing in the key.

1   Opisthaptor a single sucker or disclike organ with or without septa; with 1 to 3 pairs of anchors; prohaptor as head glands; mouth not surrounded by a sucker; genito-intestinal canal usually absent   Order MONOPISTHOCOTYLEA....2

Opisthohaptor with 2 or more suckers or rows of cuticularized clamps; prohaptor a sucker or paired suckers; head glands generally absent; genito-intestinal canal present
                                                                          Order POLYOPISTHOCOTYLEA....7

2(1) Opisthaptor with minute suckerlike structure on posterior margin (Figure 36) (Fm).........
      ....Acanthocotyloidea

No small sucker on opisthaptor; marginal hooklets usually present....................3

3(2) Opisthaptor 1- or 2-lobed, thin; 1 or 2 pairs of anchors supported by cross bars; marginal hooklets present; cirrus cuticularized, usually with accessory piece; anterior end with well-developed glands..4

Opisthaptor a muscular disc, with or without 1 to 3 pairs of anchors not supported by cross bars; marginal hooklets present or absent; intestine single or bifurcated; cirrus without accessory piece; anterior end with preoral suckers and/or glandular areas; genito-intestinal canal absent.....6

4(3) Intestine single (Figure 37) (Ffm).........................Tetraoncoidea

Intestine bifurcate..5

5(4) Opisthaptoral anchors large, with cross bars; vitellaria near posterior end of body; vagina absent; ovoviviparous (Figure 38) (Ffm, A, C, O).............................Gyrodactyloidea

Opisthaptoral anchors present but without cross bars; vitellaria lateral along full length of intestinal ceca; vagina present or absent; oviparous (Figure 39) (Ffm)..........Dactylogyroidea

6(3) Intestine single; on copepods infesting fish (Figure 40) (Fme, C).............Udonelloidea

Intestine bifurcate; on fish (Figure 41) (Feh)............................Capsaloidea

7(1) Opisthaptor with 3 pairs of anchor complexes alone or in combination with clamps; prohaptor consists of paired intrabuccal suckers (Figure 42) (Fm)...................Megaloncoidea

Opisthaptor with either muscular suckers or clamps but not both; prohaptor variable......8

8(7) Opisthaptor with 6 muscular suckers and 4 large anchors on slender stalk; head with glands opening on each side (Figure 43)(Ff).............................Avielloidea

Opisthaptor not on long slender stalk...9

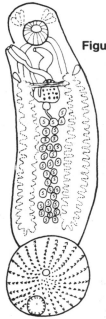

Figure 36. *Acanthocotyle williamsi.*

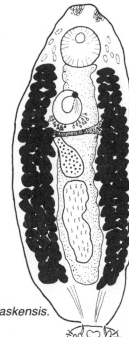

Figure 37. *Tetraonchus alaskensis.*

Figure 38. *Gyrodactylus cylindriformis.*

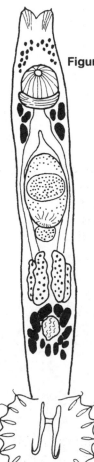

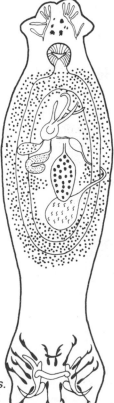

Figure 39. *Dactylogyrus albertensis.*

9(8) Opisthaptor sessile, with 6 muscular suckers (2 in Sphyranidae), appendix-like prolongation present or absent; with or without 1 to 3 pairs of posterior anchors; prohaptor a muscular oral sucker (Figure 44) (A, R, M, Fm) .... Polystomatoidea

Prohaptor not a well-developed muscular sucker surrounding mouth 10

10(9) Uterus forming many vertical or transverse loops in broader anterior portion of body, posterior part of body tapering; opisthaptor with 2 rows of simple clamps; oral sucker weak or lacking (Figure 45) (Fh) .... Chimaericoloidea

Uterus not forming such loops 11

11(10) Clamps of opisthaptor numerous, in 2 rows on the symmetrical or asymmetrical posterior end of body; prohaptor paired suckers opening in mouth cavity (Figure 46) (Fm) ...Microcotyloidea

Clamps of opisthaptor not more than 4 per side 12

12(11) Adults fused in the form of an X; intestine single with lateral branches (Figure 47) (Ff) ... . Diplozooidea

Adults not fused but independent; intestine bifurcate (Figure 35) (Ffm) ..... Diclidophoroidea

Figure 40. *Udonella caligorum.*

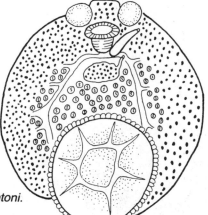

Figure 41. *Capsala lintoni.*

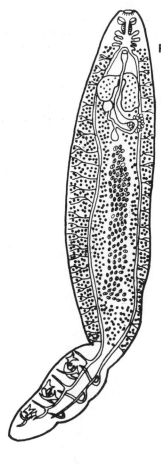

Figure 42. *Megaloncus arelisci.*

Figure 43. *Aviella baikalensis.*

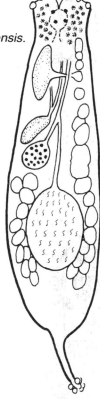

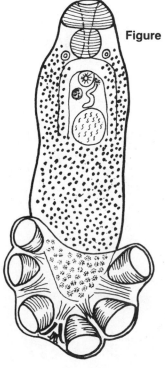

Figure 44. *Polystomoides coronatum.*

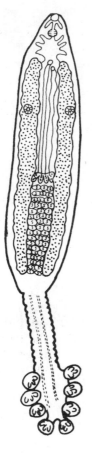

Figure 45. *Chimaericola leptogaster.*

Figure 46. *Microcotyle donavani.*

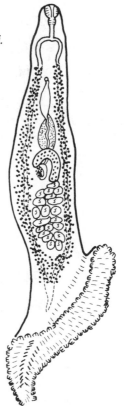

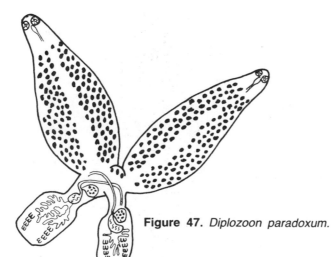

Figure 47. *Diplozoon paradoxum.*

Subclass Aspidogastrea

The Aspidogastrea (Aspidobothrea of some authors) are considered by many workers as a transitional group intermediate between the subclass Monogenea and the subclass Digenea. Anatomically, they differ from the Monogenea in lacking the opisthaptor with its sclerotized accessories and in having a simple unbranched intestine, and biologically from the Digenea in the absence of an alternation of generations in the life cycle.

This small group of trematodes is characterized anatomically by the large compartmentalized adhesive organ that encompasses almost the entire ventral surface of the body. They are endoparasites, rarely ectoparasites, of molluscs and marine turtles.

Family Aspidogasteridae

The ventral sucker is large in size, oval, or elongated in shape, and composed of numerous areoli, or compartments, arranged in 1 or several longitudinal rows. The oral sucker is rather poorly developed; a pharynx is present and the intestine is a single, simple sac. Testes are single or double. The common genital pore is median and anterior to the ventral sucker. The ovary is prestesticular; the vitellaria are paired.

Aspidogaster conchicola

This cosmopolitan fluke infects the kidney and pericardium of freshwater clams. It may occur accidentally in gastropods and some fish. Adults are up to 2.7 mm long by 1.2 mm wide. The large areolated ventral sucker occupies almost the entire ventral surface of the body; it consists of 4 rows of quadrangular sucking grooves. The narrow anterior neck, consisting of about one-fifth of the total body length, bears a terminal mouth (Figure 48).

The single testis lies in the posterior third of the body. A large cirrus sac opens into a median common genital pore on the anterior end of the ventral adhesive organ. An oval ovary lies to the right of the median line, near the equator of the sucker. The vitellaria are 2 tubular, lateral structures lying in the middle third of the ventral sucker. The operculate eggs, embryonated when laid, are 128-130 x 48-50 μ.

Life Cycle

How infection of new clam hosts occurs has not been determined. It has been suggested that the eggs hatch in the clams and the larvae develop in situ. Considering the prevalence of infection of clams,

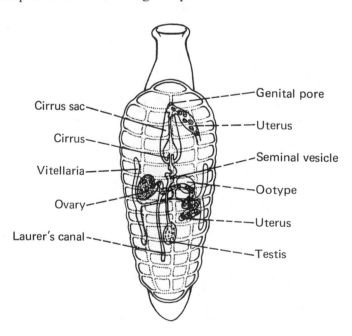

Figure 48. *Aspidogaster conchicola.*

it seems more likely that the eggs are voided by the host, hatch in water, and larvae are drawn into the gill chambers and enter the host by way of the nephridiopore. The life cycle is direct and growth of the larvae is one of gradual transformation through 4 arbitrary developmental stages. For life cycle details consult Williams (1942).

Subclass Digenea

The adults are endoparasites, occurring in all classes of vertebrates, and, in a few instances, in invertebrates. The larval stages have free-living phases necessary for getting from one host to the next. There are at least 3 parasitic stages that develop in the molluscan intermediate host. The following section includes an account of the body types of adults, and a basic description of each of the other generative stages.

Adults: The principal organ-systems generally considered in the classification are (1) the holdfast organs; (2) digestive; (3) reproductive; and (4) excretory systems. The anatomy of a generalized fluke is given in Figure 49, showing the relationship between the various organ systems. There are 7 distinct body types (Figure 50), which are: (A) gasterostomes with only the oral sucker, which is located midventrally (in all others the oral sucker is at the anterior end of the body); (B) monostomes with or without an oral sucker at the anterior end of the body and always without a ventral sucker (all others, except gasterostomes, have 2 suckers); (C) distomes (the largest group) in which the ventral sucker is close to the oral sucker in the anterior region of the body; (D) amphistomes with the oral sucker and ventral sucker at the anterior and posterior extremities, respectively, of the body; (E) echinostomes with a collar of spines around the oral sucker; (F) strigeoids with a transverse equatorial constriction that divides the

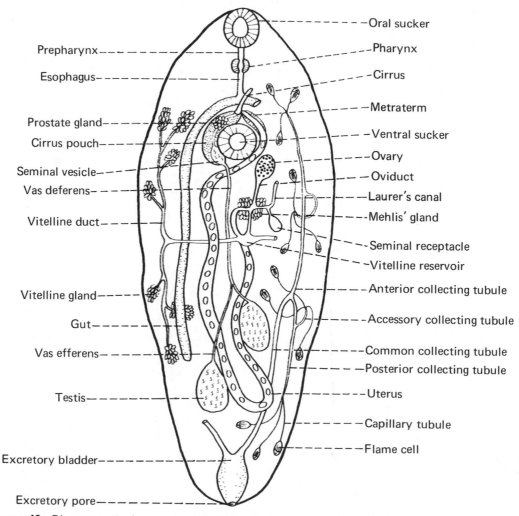

Figure 49. *Diagrammatic figure of adult digenetic trematode, showing principal parts of anatomy.*

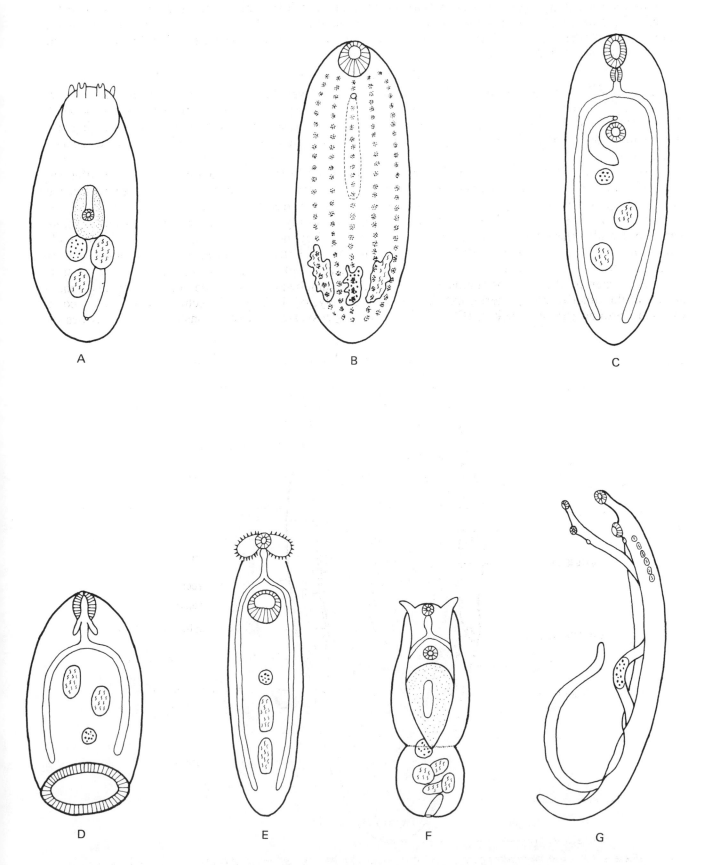

Figure 50. *Some basic body types of common adult digenetic trematodes. (A) Gasterostome; (B) Mono-stome; (C) Distome; (D) Amphistome; (E) Echinostome; (F) Strigeoid; and (G) Schistosome.*

body into a forebody with a holdfast organ and a hindbody containing the reproductive organs; and (G) schistosomes which live in the blood vessels of the final host; some are monoecious and other dioecious, in which case the males have a gynecophoric canal where the female is held. Instructions for the recovery and preparation of permanent mounts for microscopical examination are found on p. 261 et seq.

Eggs: Eggs of most species are operculate, blood flukes being a notable exception in having nonoperculate eggs. The shells are smooth and thin. The eggs of some species are undeveloped when laid but fully embryonated in others. The eggs of some species hatch in the water while others hatch in the intestine of the molluscan intermediate host.

Study eggs containing viable miracidia and, when available, eggs undergoing embryonation in water mounts. Fix eggs in hot 70% alcohol-glycerine mixture, clear by evaporating the alcohol and water in an open dish at room temperature and mount in pure glycerine, under a No. 1 cover glass, supported by cover glass slivers. Seal the cover glass with Thorne's Zut* or fingernail polish. This procedure can be utilized for mounts of eggs of all kinds of parasites, as well as miracidia and larval nematodes.

Miracidium: The miracidia are remarkably uniform throughout the trematodes. The body is more or less fusiform and ciliated. At the anterior end is a median apical gland and 1 to several bilaterally arranged pairs of penetration glands. A mass of undifferentiated cellular material known as the germ ball lies inside the miracidium (Figure 51).

Mother sporocyst: The miracidium sheds its ciliated covering inside the mollusc and transforms directly into an elongated mother sporocyst. The germ ball begins to differentiate to form either sporocysts or rediae, depending on the species of flukes, but not both in the same species.

Daughter sporocysts: Sporocysts produced by the mother sporocysts have a tubular body, usually simple, but branched in some species. There is no mouth or intestine. Reproduction in the molluscan

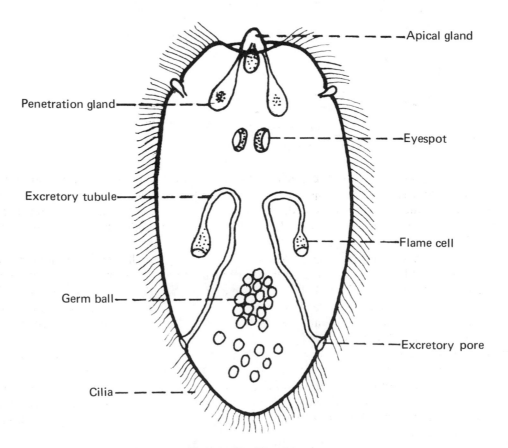

Figure 51. *Miracidium.*

*This cover glass ringing cement may be purchased from Bennett Paint and Glass, 65 West First South, Salt Lake City, Utah 84101 or made using Thorne's (1935) original formula.

host is asexual from germ balls, producing another generation of sporocysts or one of cercariae. Offspring escape through a birth pore into the tissues of the mollusc (Figure 52).

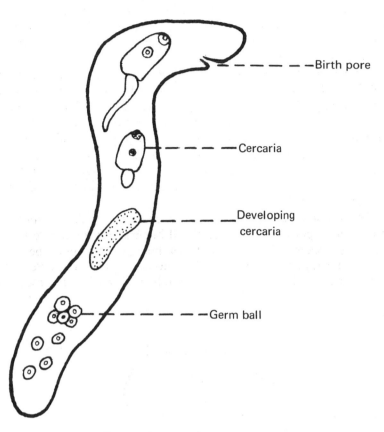

Figure 52. *Daughter sporocyst.*

Redia: The mother sporocysts of some species produce rediae instead of daughter sporocysts. Anatomically rediae differ from sporocysts in having a well-defined muscular pharynx and short, saclike gut. A second generation of rediae may be produced. The end product of asexual development of the rediae, like that of the sporocysts, is cercariae. Progeny leaving the rediae through a birth pore are located in the tissues of the molluscan host (Figure 53).

Cercaria: Cercariae are the end product of asexual reproduction in the molluscs and are tailed, immature, sexual forms, resembling adults in general body conformation. In general, they emerge from the molluscs into the water. They alternate periods of swimming by violently lashing the tail with short times of resting. The basic organ systems generally are outlined to the extent that the parts are recognizable. Some species have a set of penetration glands on each side of the anterior part of the body; their ducts open at the anterior margin of the oral sucker.

The excretory system consists of flame cells, their individual ducts, collecting ducts, all bilaterally arranged, and a medial excretory bladder, which opens externally through the excretory pore at or near the posterior extremity of the body. Extending from each side of the anterior end of the excretory bladder is a common collecting tube, which divides into an anterior and a posterior collecting tube. From each of these, there extends a definite number of accessory tubules, which in turn branch into capillaries, each terminating in a flame cell.

The flame cells of each species are distributed in a definite pattern, which may be expressed by several types of simple formulae. One type is given below for the flame cell pattern shown in Figures 49 and 54.

$$2 [(3 + 3 + 3) + (3 + 3 + 3)] = 36.$$

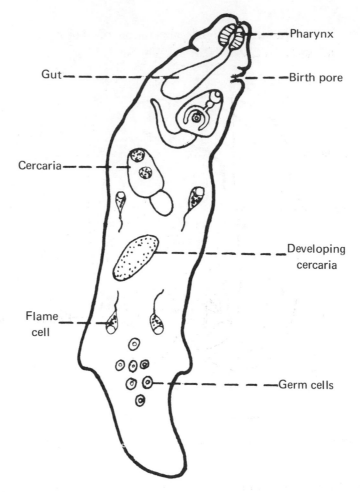

Figure 53. *Redia.*

The numeral 2 indicates that the flame cells are distributed bilaterally in the body and the brackets refer to one side. The first set of parentheses relates to the anterior collecting tube and the second to the posterior one. The 3 digits within the parentheses refer to the number of accessory tubules; the numerical value of each digit states the number of flame cells that empty through their capillaries into each accessory tubule. The second set of parentheses refers to the posterior collecting tube and its components. Thus, this diagrammatic excretory system consists of a total of 36 flame cells arranged in groups of 3 on each of 12 accessory tubules, with 3 anteriorly and 3 posteriorly on each side.

Cercariae, when fully developed, are of different body types useful in identification. The 14 basic types presented (Figure 55) are: (A) gasterostome with oral sucker on midventral surface; (B) amphistome with a sucker at each end of the body; (C) monostome with only an oral sucker which is at the anterior extremity of the body; (D) gymnocephalous without spines around or a stylet in the oral sucker and with a long tail; (E) cystocercous with a chamber in the anterior part of the massive tail into which the cercaria may be withdrawn; (F) trichocercous with long slender tail bearing numerous bristles; (G) echinostome with a spine-bearing collar around the oral sucker; (H) microcercous with short, stumpy tail and a stylet in the oral sucker; (I) xiphidiocercous with a spine in the dorsal part of the oral sucker and a long, slender tail; (J to M furcocercous types); (J) sanguinicolid with forked tail and no suckers; (K) lophocercous with forked tail and a fin over the dorsal surface of body; (L) apharyngeate schistosome with forked tail; (M) pharyngeate strigeoid with forked tail; (N) cotylocercous with short, cuplike tail; (O) rhopalocercous with tail as broad or greater than the width of the body; (P) cercariaea without tail; and (Q) rat-king consisting of a mass of cercariae attached together by the distal ends of the large tails.

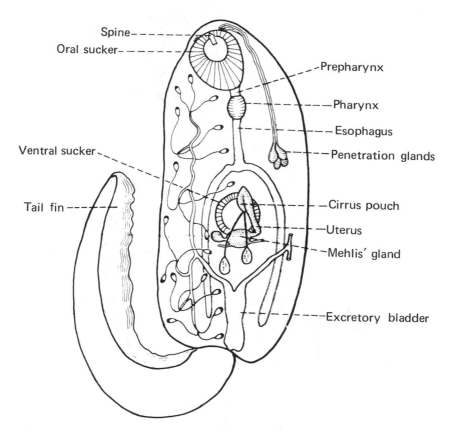

Spine
Oral sucker
Prepharynx
Pharynx
Esophagus
Penetration glands
Ventral sucker
Tail fin
Cirrus pouch
Uterus
Mehlis' gland
Excretory bladder

Figure 54. Cercaria; for labelling of excretory system see Figure 49.

In general, cercariae attach to objects in water or burrow into the bodies of aquatic animals (second intermediate hosts), drop the tail and secrete a protective cyst about themselves from material produced by cystogenous glands. The cercariae of blood flukes do not encyst but burrow into the final host and develop directly into adults. Encysted cercariae are known as metacercariae.

Key to Basic Types of Cercariae

I. No suckers
   A. Sanguincolid cercaria: fork-tailed, mouth opens subterminally (Figure 55, J)

II. Single sucker
   A. Gasterostome: oral sucker on midventral side of body, gut single (Figure 55, A)
   B. Monostome: oral sucker at anterior end of body; gut bifurcate (Figure 55, C)

III. Two suckers
   A. Tail simple
      1. Amphistome: ventral sucker at posterior extremity of body, in all others ventral sucker located otherwise (Figure 55, B)
      2. Echinostome: oral sucker surrounded by crown of spines, in all others oral sucker without crown of spines (Figure 55, G)
      3. Gymnocephalous: oral sucker without stylet; tail long, slender, simple (Figure 55, D)
      4. Xiphidiocercous: oral sucker armed with a single stylet; tail long, body with or without eyespots (Figure 55, I)
      5. Microcercous: oral sucker armed with stylet; tail stumpy (Figure 55, H)

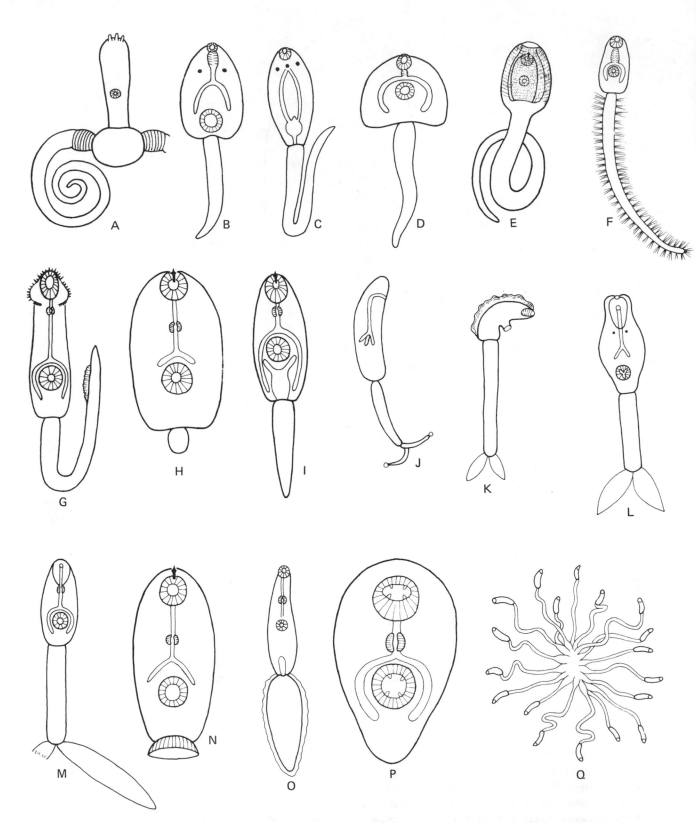

Figure 55. *Body types of some common cercariae.* (A) Gasterostome; (B) Amphistome; (C) Mono-stome; (D) Gymnocephalus; (E) Cystocercous; (F) Trichocercous; (G) Echinostome; (H) Microcercous; (I) Xiphidiocercous; (J) Sanguinicolid; (K) Lophocercous; (L) Apharyngeate furcocercous; (M) Pharyngeate furcocercous; (N) Cotylocercous; (O) Rhopalocercous; (P) Cercariaea; and (Q) Rat-king.

6. Cystocercous: tail very large with cuplike anterior chamber containing small cercaria; oral sucker with stylet (Figure 55, E)
7. Cotylocercous: tail short, cuplike; oral sucker with stylet (Figure 55, N)
8. Trichocercous: tail long, slender, and with lateral hairlike tufts; oral sucker without stylet (Figure 55, F)
9. Rhopalocercous: tail flat, broader than body; oral sucker without stylet (Figure 55, O)
10. Rattenkönig or rat-king: long tails attached at tips, forming clumps of cercariae; oral sucker without stylet (Figure 55, Q)

B. Forked tail
1. Lophocercous: with longitudinal fin over entire length of dorsal side of body (Figure 55, K)
2. Apharyngeate furcocercous cercaria (Figure 55, L)
3. Pharyngeate furcocercous cercaria (Figure 55, M)

C. Tailless
1. Cercariaea: large body (Figure 55, P)

Metacercaria: Upon encystment, metacercariae develop to the infective stage. When swallowed by the final host, they are released from the cyst by the digestive processes and the young flukes go to the respective parts of the body where development to adulthood is completed. Usually, the metacercariae are prototypes of the adults into which they develop (Figure 56). The metacercariae of the strigeoid flukes are known as the tetracotyle, neascus, and diplostomulum types, depending on the structure of the respective adults (Figure 57).

Each of the above larval stages, miracidium through metacercariae, should be studied in the living state as much as possible, with the aid of neutral red, Nile blue sulphate, or other suitable vital stain. Subsequent studies should be made on permanent mounts of fixed and stained specimens.

Parasitology laboratories should maintain several balanced aquaria in which representatives of the local species of aquatic snails are breeding. Likewise, terraria should be kept for the common species of land snails. With such breeding populations, parasitologists have readily available uninfected snails to use in life cycle studies whenever adult flukes are available.

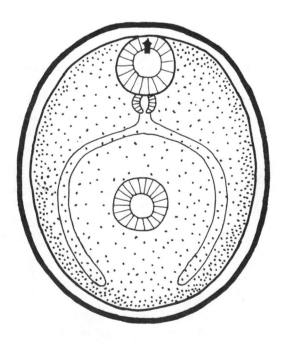

Figure 56. *Metacercaria.*

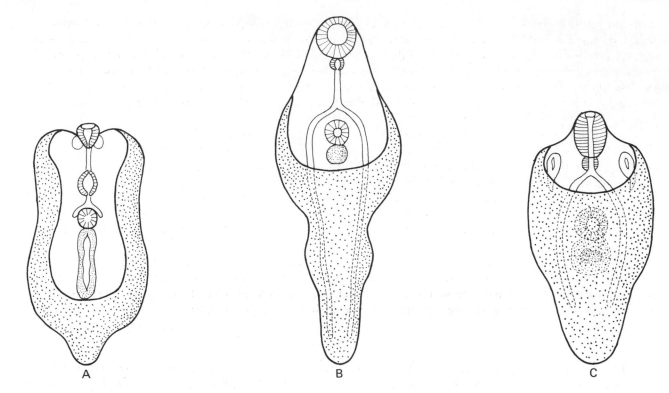

Figure 57. *Three typical unencysted metacercariae of strigeoid flukes.* (A) Tetracotyle; (B) Neascus; and (C) Diplostomulum.

Family Opisthorchiidae

Members of this family occur in the bile ducts, gallbladder, rarely in the intestine, of fish, reptiles, birds, and mammals. They are characterized by slender, flattened, clear bodies. There is a well-developed digestive system and a ventral sucker in the anterior part of the undivided body. The ovary and testes are in the posterior half of the body, and the vitellaria are lateral to the intestinal ceca and pretesticular.

Clonorchis sinensis
(*Opisthorchis sinensis* of some authors)

An important fluke in the biliary passages, occasionally the pancreatic duct, of man and fish-eating mammals of the Orient, chiefly Japan, Korea, China, Formosa, and Indo-China (Figure 58).

Description

Adults are flat, thin, transparent, and somewhat spatulate in shape, with both ends attenuated. They are up to 25 mm long by 5 mm wide. The tegument is smooth. The oral sucker is somewhat larger than the ventral sucker or acetabulum which is near the junction of the first and second quarters of the body.

The mouth opens at the bottom of the oral sucker and connects posteriorly with the small muscular pharynx; a very short cuticular-lined esophagus continues and joins the 2 epithelial-lined intestinal ceca, which extend posteriorly to the end of the body where they terminate blindly.

The large testes are deeply lobed, with 1 behind the other in the posterior third of the body. A slender vas efferens extends anteriorly from each testis to near the middle of the body where they join to form the enlarged vas deferens which continues almost directly, dorsal to the uterus, to the male genital opening at the anterior margin of the ventral sucker. There is no cirrus, cirrus pouch, or prostate gland.

The small, slightly lobed and oval ovary lies a short distance anteriorly from the forward testis. Two tubules open into the short oviduct. The first one is from the large, oval seminal receptacle located directly posterior to the ovary. A slender duct, the Laurer's canal, which arises from the oviduct, opens on the dorsal surface and may be analogous with the vagina of cestodes. Slightly beyond the opening of the seminal receptacle, the common vitelline duct enters the oviduct. A short distance from this point, the oviduct enlarges to form the ootype which is surrounded by a lightly stained glandular mass known as Mehlis' gland. The uterus arises from the ootype and continues forward in compact, intercecal coils to the genital atrium where it opens beside the male duct. Viewed ventrally, the uterus opens to the right of the vas deferens. There is no metraterm.

The vitellaria, composed of delicate follicles, are extracecal, extending from near the posterior margin of the ventral sucker to the anterior side of the front testis. A common vitelline duct conveys yolk granules from the glands to the small yolk reservoir, which empties into the oviduct.

By focusing on the ventral sucker determine if your specimen is mounted ventral or dorsal surface up. Another way to tell which surface is up is the arrangement of the female reproductive organs. If the Mehlis' gland is on the right and the seminal receptacle to the left, with the ovary between the two, the specimen is mounted ventral side up; conversely, if the relationship of these organs is reversed, the dorsal surface is up.

Ova are produced in the ovary and passed into the oviduct, where they are fertilized by sperm from the seminal receptacle and provided with yolk and shell-forming material from the vitellaria. They then go on past the Mehlis' gland, the secretions of which are involved in several aspects of egg production, and into the uterus. The eggs are described under Helminth Eggs, p. 210.

Life Cycle

The life cycle includes 2 intermediate hosts: Amnicolidae snails and fish, first and second intermediaries, respectively. The eggs hatch only when eaten by a snail. The mother sporocysts produce rediae in which gymnocephalous cercariae with a longitudinal tail fin develop. Upon coming into contact with numerous species of freshwater fish (mostly Cyprinidae) of the Orient, the cercariae drop their tails, penetrate, and encyst in the muscles. When viable metacercariae are ingested with raw fish, they excyst in the intestine and migrate to the liver by way of the common bile duct (Sun et al., 1968).

Other often available Opisthorchiidae with their hosts include *Opisthorchis tonkae,* muskrats, *Amphimerus elongatus,* ducks; *A pseudofelineus,* cats, dogs; *Metorchis conjunctus,* cats, dogs, fox, mink, raccoons, and occasionally man in Canada and the northern United States. For description and life cycle details of *M. conjunctus,* consult Cameron (1944).

Family Prosthogonimidae

In the oviducts and bursa Fabricii of birds. They are small, clear flukes with the anterior end tapering and the posterior end broadly rounded. The genital pore is near the oral sucker. The vitellaria form grapelike clusters laterally. The excretory bladder is Y-shaped.

Prosthogonimus macrorchis

This is the oviduct fluke of ducks, chickens, crows, English sparrows, ruffed grouse, and other birds (Figure 59).

Description

The body is thin, translucent, and generally oval in shape, with the anterior end somewhat narrow and the posterior end broadly rounded. The size varies, depending on the species of host, ranging from 2.5 to 2.8 mm long by 1.4 to 2 mm wide in English sparrows up to 6.3 to 8.5 mm long by 5.1 to 6 mm wide in chickens. The oral sucker is roughly half the size of the ventral one, with the latter located near the posterior end of the first third of the body. The muscular pharynx is followed by a short esophagus of about the same length that divides into 2 long slender intestinal ceca which extend to near the posterior end of the body. The common genital pore opens on the anterior right margin of the oral sucker.

The large roundish testes lie side by side near the anterior part of the second half of the body. Slender vasa efferentia unite near the posterior margin of the ventral sucker to form the vas deferens, which

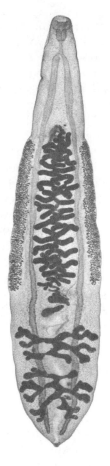

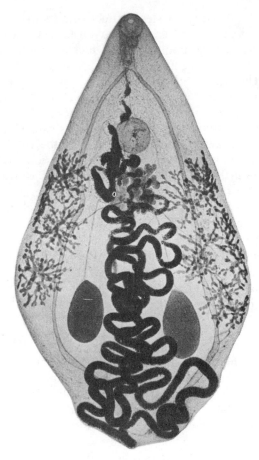

Figure 58. *Clonorchis sinensis.*　　　**Figure 59.** *Prosthogonimus macrorchis.*

continues directly to the cirrus sac. The latter extends posteriorly to near the end of the esophagus and opens through the common genital pore; it contains a convoluted seminal vesicle in the posterior end.

The multilobed ovary lies immediately behind the ventral sucker. Attached to the short oviduct directly behind the ovary is a small seminal receptacle. The highly convoluted uterus descends between the testes, fills the posttesticular part of the body, and ascends between the testes, filling most of the space between the vitellaria, the testes, and ventral sucker. It continues as a thin tube to the genital pore. The vitellaria, numbering 8 to 9 lateral clusters, begin at the level of the ventral sucker and end near the posterior margin of the testes. Eggs are operculate, with an abopercular protuberance, average 28 x 16 μ.

Life Cycle

The eggs, unembryonated when voided with the feces, develop in the water and hatch in the intestine of the snail intermediate host *Amnicola limosa porata*. The freed xiphidiocercariae which developed in sporocysts are drawn into the rectum of dragonfly naiads by the respiratory currents, burrow through the wall, and encyst in the abdomen. Species of *Leucorrhinia*, *Tetragoneuria*, *Epicordulia*, and *Mesothemis* dragonflies serve as second intermediate hosts. Birds become infected when they eat dragonflies, either as naiads or adults, containing metacercariae. For life cycle details consult Macy (1934). Because of the difficulty of obtaining adults, the life cycle is not recommended as a class project.

Family Dicrocoeliidae

Parasites of the liver, bile ducts, gallbladder, and pancreas of reptiles, birds, and mammals. In addition to the location in the final host, the family is characterized by the posttesticular position of the ovary and the small, brown, embryonated eggs whose operculum sets in a rimlike thickening.

Dicrocoelium dendriticum

This species, the lancet fluke, occurs in the liver and gallbladder of sheep, cattle, deer, rabbits, and marmots. It will also develop in guinea pigs (Figure 75).

Description

The translucent, dorsoventrally flattened body measures up to 12 mm long by 2.5 mm wide. The oral sucker is subterminal; the ventral one, in the anterior quarter of the body, is slightly larger. The small, round pharynx lies next to the oral sucker; it is followed by the slender esophagus that extends about midway between the pharynx and ventral sucker, and bifurcates, sending 2 long, slender intestinal ceca to the beginning of the last quarter of the body. The common genital pore is just posterior to the intestinal bifurcation. The slender excretory bladder reaches anterior to the ovary. The slightly lobed testes are close together, obliquely arranged in the body with the anterior one near the posterior margin of the ventral sucker. The vasa efferentia unite at the level of the ventral sucker to form a short vas deferens which enters the cirrus pouch, forming a sinuous seminal vesicle. The cirrus pouch reaches almost to the ventral sucker.

The somewhat oval ovary is located slightly posterior to the hind testis, a rather unusual position. A small seminal receptacle and Mehlis' gland are close behind. The uterus consists of a descending portion with many loops and an ascending one also with many loops, which cross each other below the gonads. The vitellaria consist of numerous, small follicles located extracecally and extend from the posterior margin of the hind testis to or slightly caudad from the middle fifth of the body. Eggs operculate, 38-45 x 22-30 μ.

Life Cycle

The land snail, *Cionella lubrica*, becomes infected by eating eggs containing miracidia. Sporocysts in the digestive gland of the snail become distended with the cercariae. The cercariae escape from the sporocyst periodically and accumulate in masses of mucus called slimeballs. These are sticky and adhere to vegetation and debris. The slimeballs, in reality, are packages of numerous cercariae, in a covering which protects them against desiccation. They are carried away by the ant *Formica fusca*, which uses them as food and becomes infected by ingesting the cercariae. Encysted metacercariae accumulate in the abdomen and head of ants and await a possible transfer to 1 of the several final hosts. The final host becomes parasitized by accidentally ingesting the infected ants while grazing.

Life Cycle Exercises

In enzootic areas, such as the northeastern region of the United States and eastern Canada, this is a good species to study a life cycle in which adult terrestrial insects and land snails are utilized.

Remove from an infected host the gallbladder intact together with the flukes and bile. Both contain eggs. Wash the contents of the gallbladder into a dish of water and strain out the flukes, keeping the bile, which contains many eggs. Clean the fluid by repeated sedimentation and decantation, either by gravity or centrifugation. The cleaned sediment will contain innumerable eggs. Mature flukes placed in a dish of water freely shed their eggs, which are fully embryonated.

The snail host, which is circumboreal in distribution, occurs on shady, moist soil, on dead raspberry canes, under stones, boards, and other objects. They can be attracted by placing wet sacks on the ground in suitable areas.

Small terraria prepared from petri dishes are suitable for maintaining the snails. To prepare, put a circle of blotting paper in the top and bottom of the dish and wet until the one in the top adheres and the one in the bottom glistens. Too much as well as too little water is detrimental. Provide food in the form of 2 pieces of ailanthus (tree of heaven) leaf torn, not cut, into small pieces about one-fourth inch square and placed with under side uppermost to provide soft tissues to eat. Torn unbruised pieces last several days without spoiling. Two pieces of large curled maple leaf provide both food and places for the snails to crawl, attach, and defecate. Add to these a piece of cured raspberry cane. A quarter to a half spoonful of dry soil in a little heap completes the "petri terrarium." Introduce 8 to 10 snails together with several hundred fluke eggs. Give constant surveillance to the dishes and promptly remove dead snails.

The embryonated eggs swallowed by foraging snails soon hatch in the intestine. The miracidia burrow through the gut wall and transform into simple, unbranched mother sporocysts.

The daughter sporocysts escape from the mother sporocysts and occur most commonly in the digestive gland, the so-called liver. These are unbranched, saclike bodies with the anterior end somewhat attenuated. The birth canal passes through this slender portion, opening at the end. Mature daughter sporocysts, which require up to 50 days to develop, are readily recognized by the presence of developing and mature cercariae inside them. Examine snails from pastures where sheep occur in an attempt to obtain both generations of sporocysts.

Being in land snails, the delicate cercariae cannot survive without being safeguarded against drying. Protection is provided by the snails by enveloping groups of cercariae in a jellylike matrix of mucus as they are expelled from the respiratory chamber. These globular masses of mucus, called slimeballs, are up to 2 mm in diameter and contain variable numbers of cercariae, ranging up to several hundred each. As the slimeballs dry, a tough covering is formed which retains the moisture, thus providing the cercariae with an aquatic environment having nutritive and isotonic properties. The slimeballs are deposited on vegetation and objects over which the infected snails crawl.

The cercariae are of the distome xiphidiocercous type, having a stylet in the oral sucker and a long slender tail without a finlike structure (Figure 55, I). The suckers are nearly equal in size, with the ventral one near the middle of the body. The body contains 2 sets of glands: large ones, of which there are 24 in the posterior half of the body, and 4 to 6 small ones just posterior to the oral sucker. The digestive tract consists of a small pharynx, a short, slender esophagus whose length is about half the distance from the oral sucker to the ventral one, and the very short intestinal ceca which extend laterally. The excretory bladder is saclike and extends anteriorly almost to the ventral sucker. The flame cell pattern is

$$2 \; [(\; 2 \; + \; 2 \; + \; 2) \; + \; (2 \; + \; 2 \; + \; 2)].$$

The time required to produce cercariae after ingestion of eggs by the snails is unknown but it is at least 60 days and possibly more. Watch for slimeballs deposited by naturally infected snails kept in the terraria. They are expelled after cool, rainy periods. Slimeballs resemble egg masses of the snails in appearance, but naturally differ in contents.

Adult ants carry the slimeballs into the nest. Whether infection of ants occurs in the adults or in the larvae and pupae which are fed by the brown workers is unknown. Up to 36% of the ants in some areas are naturally infected with as many as 128 cysts but the average number is 39.

Examine the gasters of ants from pastures occupied by infected sheep for metacercariae. Tear the gasters apart in physiological solution to release the encysted flukes.

The metacercariae in the gasters are enclosed in firm, resilient, colorless cysts having the texture of thick gelatin. A mass of small glands appears at the anterior margin of the ventral sucker. The intestinal ceca have increased greatly in length. Metacercariae excyst in the small intestine and are extremely active. They migrate quickly to the opening of the common bile duct on Vater's papilla, enter, and creep up the duct. In white mice, the metacercariae are in the liver within 1 hour after ingestion. The prepatent period is about 43 days in sheep.

If metacercariae can be obtained from naturally infected ants, feed them to 2 white mice. Examine 1 mouse 30 minutes after and the other 60 minutes after ingesting the cysts. With the aid of a dissecting microscope, note where the excysted flukes are in the gut 30 minutes after swallowing them. In the other mouse, watch the flukes entering the opening of the common bile duct and migrating up it.

Other Dicrocoeliidae utilizing land snails as the first intermediate host are *Eurytrema procyonis* of raccoons, cats, foxes, and species of *Brachylecithum* and *Tanaisia* of birds.

Family Notocotylidae

Parasitic in the intestine and ceca of birds and mammals. They are monostome flukes with the ovary lying between the 2 transversely arranged testes. All 3 gonads are in the posterior extremity of the body. The genital pore is median and near the posterior margin of the oral sucker (Figure 68).

Quinqueserialis quinqueserialis

This species is common in the cecum of muskrats and meadow voles, throughout the range of both hosts in North America. It is an excellent model for a trematode life cycle.

Description

The body is up to 5 mm long by nearly 2 mm wide and somewhat oval, with the posterior end slightly broader than the anterior end. The ventral surface of the body has 5 longitudinal rows of adhesive glands. Each row has 16 to 18 of the distinct wartlike projections. The esophagus is extremely short and the slender intestinal ceca terminate near the posterior end of the body.

The much-lobed testes are opposite each other near the posterior end of the body. The short vasa efferentia join to form a large vas deferens, which extends to the equatorial region of the body and joins the long cirrus pouch, which empties through the common genital pore at the level of the intestinal bifurcation.

The lobed ovary lies between the testes. The uterus consists of many transverse folds between the ceca from the ovary to the posterior end of the cirrus pouch. The terminal part of the uterus is surrounded by strong muscle fibers, forming a metraterm about two-thirds the length of the cirrus pouch. If the specimen is mounted ventral surface up, the metraterm is to the right of the cirrus pouch. The vitellaria are weakly developed, located extracecally, and extend from the posterior end of the cirrus pouch to the anterior margin of the testes. A Mehlis' gland and seminal receptacle are median and just anterior to the ovary. Eggs operculate, 22-32 x 11-18 μ.

Life Cycle

The eggs are embryonated when laid, but do not hatch until eaten by the snail *Gyraulus parvus*. The miracidia penetrate the intestinal wall of the snail and transform into sporocysts containing 4 mother rediae in the adjacent tissue. They leave the sporocyst, migrate to the liver, and by the 20th day postinfection free daughter rediae are present. The cercariae escape through the birth pore, and are mature by the 26th day. Upon striking an object, they encyst within 5 minutes, at which time they are infective. Final hosts become infected when they eat aquatic plants bearing the metacercariae. Prepatency takes about 16 days.

Life Cycle Exercise

Inasmuch as these flukes are common, it is not difficult to obtain them from the intestine of voles and from muskrats. Living flukes transferred to a dish of physiological solution extrude eggs in the form of ropy masses. The eggs contain fully developed miracidia when laid.

Place parasite-free snails in a dish with eggs. Snails readily eat the eggs, which hatch in the alimentary canal, and the miracidia migrate to the surrounding tissues. The intramolluscan stages can be found by dissecting infected snails at appropriate times.

Examine exposed snails about 10 days after they have ingested eggs. Young oval, round, or saclike mother sporocysts are embedded in the mantle lining. Continue examinations every third or fourth day for 4 weeks or until rediae appear in the mother sporocysts. Mature sporocysts contain 4 mother rediae.

When snails are kept at room temperature, small mother rediae appear embedded in tissues surrounding the intestine as early as 14 days postinfection. The rediae possess a terminal muscular pharynx and a short, saclike intestine. By the 19th day, mother rediae are in the digestive gland. The body is saclike and contains germ balls and daughter rediae. Mature rediae produce only daughter rediae. Young mother rediae cannot be distinguished from young daughter rediae when only germ balls are present.

Daughter rediae appear in the digestive gland by the 20th day postinfection. They are similar to mother rediae in anatomy but differ physiologically in producing cercariae with eyespots. Note the birth pore.

The monostome cercariae (Figure 55, C) are immature at the time they leave the daughter rediae. A short period of maturation is spent in the digestive gland. By the 26th day, cercariae begin emerging from the snails between 9 and 11 A.M. The ovoid body, with 3 eyespots, has a tail about twice the body length. The body is opaque, due to the great number of cystogenous glands under the tegument. Upon emerging from the snails, the cercariae swim about for a short time and then encyst on vegetation or other objects in the water. The process of encystment is rapid.

Infect 6 or 7 meadow voles with 30 to 40 metacercariae each. The cysts may be ingested with lettuce to which they are attached or put in the back of the mouth by means of a bulbed pipette.

Examine 1 vole on the second day postinfection and 1 every other day thereafter until the 10th day. On the 10th day, begin examining the feces of the remaining voles for eggs to determine prepatency. This should occur about the 16th day. Two weeks after the appearance of eggs in the last vole, collect the worms which will be fully developed and of known age, fix, stain, and mount as representatives of the series.

Other Notocotylidae are *Notocotylus stagnicolae* of birds, primarily ducks, and *N. urbanensis* from muskrats. Species of *Notocotylus* have 3 rows of glands on the ventral side of the body instead of 5 as in *Quinqueserialis*.

The life cycles of species of *Notocotylus* are basically similar to that described for *Quinqueserialis quinqueserialis* except that a different molluscan intermediate host is involved. The snail *Stagnicola emarginata* and its varieties are intermediaries for *Notocotylus stagnicolae* and *Physa gyrina* for *N. urbanensis*.

Family Plagiorchiidae

The members of this family are parasites of all classes of vertebrates. Primarily they are parasites of the alimentary canal, but representatives occur in the biliary and respiratory systems. While difficult to characterize simply, members of the family have a pretesticular ovary, the uterus descends and ascends between the testes, and the excretory bladder is usually Y-shaped. Many species utilize immature aquatic insects as intermediate hosts. Considering the large number of insectivorous vertebrates, it is no wonder that this is a large family (Figures 49, 96).

Haematoloechus medioplexus

This is one of the common lung flukes of adult *Rana pipiens* and *Bufo americanus* in North America.

Description

These are generally elongate worms, up to 8 mm long and 1.2 mm wide. The tegument is densely spined. Ventral sucker small, in first third of body, anterior to ovary. Ovary oval to round, situated submedially anterior to testes. Seminal receptacle posterior to and slightly overlying the ovary. Rounded testes posterior to ovary and seminal receptacle, diagonally situated in third quarter of body. Vitellaria consist of about 10 groups of 6 or more follicles, on either side, extending along the ceca from the level of the posterior testis to near the anterior end. Uterus continues posteriorly in short transverse folds between testes, filling the posttesticular region; ascending limb fills anterior intercecal region with short, transverse folds. Genital pore situated slightly posterior to pharynx. Eggs operculate, 22-29 x 13-17 μ.

Life Cycle

Eggs pass up the lungs into the mouth of the frog, are swallowed and escape with the feces. When voided, the miracidia are mature, but, unless the eggs are eaten by the flat snail *Planorbula armigera*, hatching does not occur. The sporocyst gives rise to cercariae, which remain suspended in the water. As a result of this aimless swimming, they are passively drawn by respiratory currents into the branchial basket of *Sympetrum* dragonfly naiads. Once in the second intermediate host, the cercariae enter and encyst in the gill lamellae. As the naiad metamorphoses, the metacercariae collect in the posterior body where further development is contingent upon the adult dragonfly being eaten by a frog *(Rana pipiens)* or a toad *(Bufo americanus)*. Metacercarial excystment occurs in the stomach of the final host, after which the worms ascend the esophagus and enter through the glottis to the lungs.

Life Cycle Exercise

The study will be pursued as a field project rather than a series of controlled laboratory experiments. They should include (1) finding adult worms in the lungs of frogs; (2) obtaining and observing eggs; (3) collecting snail intermediate hosts from the same habitat where infected frogs occur; (4) obtaining the intramolluscan stages of the fluke from naturally infected snails; (5) getting cercariae from them; (6) observing cercariae being drawn through the anus into the respiratory chamber of dragonfly naiads; and (7) collection of metacercariae from naturally infected naiads.

When live gravid worms are placed in water, they void many eggs. Hatching does not occur normally until they are eaten by the snail intermediate host. Try artificial means of inducing them to hatch, such as described for *Halipegus occidualis* (p. 80).

From the area where infected frogs occur, which is widespread, collect snails and place several in each of a number of small widemouthed jars containing pond water. Observe the jars each morning by holding them so as to have a dark background and with plenty of light on them for seeing cercariae tumbling in the water. The cercariae are shed mostly during the night. When infections are found, isolate the snails in individual jars in order to find the infected ones.

The cercariae bear a stylet in the dorsal wall of the oral sucker and a finlike structure over the dorsal side, tip, and ventral side of the relatively short tail. The stylet and fin place them in the ornate xiphidiocercaria group (Figure 55, I).

Dissect an infected snail and find the sporocysts. There are no rediae in the Plagiorchiidae. The sporocysts contain developing and fully formed cercariae. What kind of sporocysts are they? Should you happen to find a sporocyst producing only sporocysts, what would it be? Examine living sporocysts and locate the birth pore near or at the end of the body through which its progeny escapes.

Place a dragonfly naiad in a small dish of water with many cercariae. While observing under a dissecting microscope, note how the cercariae are swept through the anus into the branchial basket by the respiratory currents created by the expansion and contraction of the abdomen. Also observe how the naiad reacts to the presence of the cercariae in the branchial basket where they are penetrating the gills. Keep the naiads for several days to allow metacercariae to develop. Open the naiads and observe the metacercariae in delicate cysts in the gills.

If metacercariae are available, feed them to a frog at intervals of 2 days for a week or thereabouts. Examine the frog 1 day after the last worms are fed to it. Note the differences in size of the worms. Are they in different locations?

Other commonly encountered Plagiorchiidae with their hosts include: other species of *Haematoloechus*, frogs, and toads; *Glypthelmins quieta*, frogs, *Plagiorchis proximus*, muskrats.

Family Echinostomatidae

The members of the family, common in aquatic reptiles, birds, and mammals, are characterized by a head collar armed with a single or double row of spines.

Echinostoma revolutum

This is a cosmopolitan parasite of ducks and occasionally mammals, especially muskrats, and sometimes man (Figure 84). It shows great biological versatility in being able to parasitize so many species of animals in the course of development. The miracidia penetrate and develop in species of 4 genera of snails from 3 families. Cercariae encyst in many species of pulmonate snails, often those in which they developed. They also encyst and develop in fingernail clams, tadpoles, and bullheads.

Description

The body is about 15 mm long and with parallel sides, having a ratio of length to width of 15:1 in fully relaxed specimens. The size varies greatly depending on the final host. The prominent head collar bears 37 spines, arranged in 3 groups, 2 of which are paired. They include 2 corner groups of 5 spines each, 2 lateral groups of 6 each, and a dorsal group containing 15 alternating spines. There are transverse rows of cuticular spines variable in number and size which cover the body posteriorly to the ventral sucker. The ventral sucker is larger than the oral sucker and varies in position from the anterior end of the body. The muscular pharynx is close to the oral sucker, followed by a short esophagus that branches at the anterior margin of the ventral sucker to form 2 slender, simple intestinal ceca extending to the posterior end of the body. The excretory bladder is a long, slender, saclike vesicle reaching to the posterior margin of the hind testis.

The gonads are arranged one behind the other, with the paired testes hindmost. Between the anterior testis and the ovary, which is foremost, is the Mehlis' gland. The oviduct extends posteriorly for a short distance and then turns forward as a long, convoluted uterus that opens through the common genital pore. A short cirrus pouch lies dorsal to the ventral sucker. The vitellaria extend laterally from the oral sucker to the posterior end of the body, being confluent behind the testes. Eggs operculate, 92-145 x 66-83 μ.

Life Cycle

The eggs hatch after about 3 weeks at room temperature, and the miracidia enter a suitable snail (*Helisoma trivolvis*, *H. antrosa*, and *Physa occidentalis*). The miracidia transform into mother sporocysts which give rise to 2 generations of rediae, the second of which produces cercariae in 9 to 10 weeks after infection of the snails. Cercariae enter and encyst in several species of pulmonate snails and tadpoles of frogs. *Physa gyrina*, *P. occidentalis*, *Helisoma trivolvis*, *Fossaria modicella*, and *Pseudosuccinea columella* are common molluscan second intermediaries. The final host becomes infected by ingesting the hosts with metacercariae.

Life Cycle Exercise

A plentiful supply of gravid worms is available to students who have access to wild ducks during the hunting season or muskrats during the trapping season. Flukes from hosts dead for some time are likely to be dead themselves, but the eggs are viable. Eggs are best obtained by dissecting them from the uterus of worms.

In eggs that have been incubated at room temperature for 6 days, the miracidia are nearly spherical, ciliated, and have 2 flame cells. During the next few days, the eyespots appear and the cuticular plates become evident. After 10 to 12 days, they have the appearance of a hatched individual except for being somewhat smaller. Hatching occurs in 18 to 30 days. The miracidia are covered with unicellular epidermal plates arranged in 4 rows of 6, 6, 4, and 2 from anterior to posterior. Eyespots are in the form of dark crescents. There are 2 flame cells on each side of the body, opening laterally near the junction of the third and last tiers of epidermal plates. An apical gland is present.

Young laboratory-reared snails are preferable for experimental infections. Place a large number of recently hatched miracidia with the young snails in a small dish of water, leaving them together for several hours.

Upon entering the snails, the miracidia shed the ciliated covering and transform into sporocysts. By 40 days postinfection they contain a small number of mother rediae.

There are 3 generations of rediae. The first can be recognized during its sojourn in the mother sporocyst and the third by the presence of cercariae. Examine the liver of snails 70 to 80 days postinfection for rediae containing rediae and cercariae. The rediae vary greatly in size and, of course, have the pharynx and primitive gut. In addition to rediae or cercariae in the lumen, the rediae contain masses of germ cells.

Watch for cercariae, beginning about 40 days postinfection, when snails are kept at room temperature. Echinostome cercariae are unmistakable because of the head collar and the collar spines on it. They are arranged in the same pattern as on the adult worms. Tegumental spines are seen best when cercariae are stained lightly with fast green or methylene blue. A slender prepharynx precedes the oval muscular pharynx. The esophagus bifurcates at the anterior margin of the ventral sucker and the 2 ceca extend to the end of the body. The short, broad, saccular excretory bladder sends common collecting tubules to the anterior end of the body, and turn caudad again, sending off 6 groups of 3 flame cells each along the side of the body from the hind end to the front end. From the posterior end of the bladder, a tubule extends caudad into the tail a short distance, divides, and empties laterally on each side. The formula of the flame cell pattern is

$$2 \ [(3 + 3) + (3 + 3) + (3 + 3)].$$

Infect snails and tadpoles by placing them in a dish with cercariae. Tadpoles and physids are killed by too heavy infections, but helisome snails tolerate many more cercariae. The cercariae enter the anus of tadpoles and migrate up the ureters to the kidneys, or go through the respiratory duct to the gill chamber and encyst within a few hours. Cercariae enter the anus of small bullheads and encyst in the kidneys. The same specimen of snail which served as the first intermediate host, producing cercariae, may likewise serve as the second intermediary, for the same cercariae may reenter and encyst in it.

Metacercariae become infective within a day after entering the second intermediate hosts. Feed several infected snails to 5 pigeons, chicks, or rabbits. Examine a host every other day for 8 days to obtain stages of developing flukes. Prepatency is about 12 days.

Family Schistosomatidae

The Schistosomatidae occur in the blood vessels of birds and mammals and are known collectively together with the Spirorchiidae, of turtles, and Sanguinicolidae, of fish, as blood flukes. The Schistosomatidae are dioecious with males being the larger. Usually, there is a ventral groove (gynecophoric canal) in which the long, slender female is carried. The apharyngeate, fork-tailed cercariae (Figure 55, L) penetrate directly into the final host and develop to maturity, thereby eliminating the metacercarial stage. In this respect, this family of blood flukes is unique biologically.

Schistosoma mansoni

This species occurs commonly in the mesenteric venules of the lower part of the ileum and large intestine of man across equatorial Africa, parts of South America, including the coastal areas of Venezuela and Brazil, and in Puerto Rico (Figure 63).

Description

The characteristic of prime significance is the occurrence of separate male and female individuals.

Male: the body is up to 12 mm long with many conspicuous tuberculations over the entire surface caudad from the ventral sucker. The lateral margins of the body extend as flaps to form a ventral gynecophoric canal in which the female is held. The oral sucker is terminal, with the ventral sucker a short distance posterior. The gynecophoric canal extends from the ventral sucker to the posterior end of the body. The digestive tract consists of the apharyngeate esophagus and the ceca which bifurcate at the level of the ventral sucker, reuniting preequatorially and continuing to the caudal tip of the body. A cluster of 7 (3 to 13) small testes just behind the anterior bifurcation of the intestine opens by means of a short duct into the genital pore in the anterior part of the gynecophoric canal.

Female: the slender body is up to 17 mm long. The oral sucker is terminal and the ventral sucker only a short distance caudad. The intestine is similar to that of the male, dividing at the level of the ventral sucker and reuniting preequatorially. The elongate ovary lies between the ceca of the intestine; there is a small crooked seminal receptacle near its posterior end; the oviduct extends forward, soon enlarges to form the ootype, which is surrounded by Mehlis' gland, and then merges with the uterus, which continues as a rather straight tube to the genital pore immediately posterior to the ventral sucker. Usually 1 to 4 eggs appear at a time in the short uterus. The vitelline gland extends as a cylindrical structure with lateral extensions to the posterior end of the body.

Other important human blood flukes include S. japonicum from the Orient, particularly China, Formosa, Japan, and the Philippines, and S. haematobium which is widespread throughout Africa and on Malagasy Republic (formerly Madagascar), as well as in some parts of the Near East. Males of S. japonicum have 7 (6 to 8) testes arranged in a line. The intestinal ceca unite postequatorially. The body covering is smooth. Males of S. haematobium have a cluster of 4 to 5 large testes. The integument is finely tuberculate and the intestinal ceca unite postequatorially. The eggs of these 3 species, which are nonoperculate and contain fully embryonated miracidia, are described under Helminth Eggs p. 210.

Life Cycle

As stated above, the life cycle of the Schistosomatidae and related families of blood flukes is unique in that the cercariae penetrate the final host and develop directly into the adult without going through the metacercarial stage.

Upon being deposited in the venules of the intestinal wall, the eggs work their way into the lumen and are voided with the feces. Once in the water, the miracidia escape through a rent in the shell and swim away. In the western hemisphere, the important snail intermediate hosts are the discoidal snails *Australorbis glabratus* and species of *Tropicorbis*.

The miracidia penetrate the snails, shedding the ciliated epithelium in the process. Inside the snail, they transform into mother sporocysts which produce numerous elongate daughter sporocysts. Periodically, these give rise to vast numbers of apharyngeate, fork-tailed cercariae (Figure 55, L). Upon coming into contact with the human skin, the cercariae quickly burrow with the aid of the penetration glands and body movements through the skin. Inside the body, they migrate via the blood stream through the heart, lungs, heart, and out to the mesenteric vessels and intestinal wall and liver. Prepatency takes 5 to 7 weeks.

The life cycles of *S. japonicum* and *S. haematobium* are basically similar. The snail hosts of *S. japonicum* are species of the dextrally spired snail *Oncomelania* while those of *S. haematobium* are species of the sinistrally spired snails *Bulinus* and *Physopsis*.

Schistosomatium douthitti

This species is naturally parasitic in the blood vessels of the mesenteries, liver, and small intestine of voles and muskrats in North America. It is an excellent model for the life cycle of a schistosome parasite of a mammalian host to study in the laboratory. The size of worms of the same age varies greatly, especially in heavy infections.

Description

Male: Mature males from experimentally infected white rats measure 1.9 to 4.5 mm (average 3 mm) in length, from *Peromyscus maniculatus* 2.9 to 5.3 mm, and from *Mus musculus* 2.5 to 5.3 mm. Specimens from naturally infected *Microtus* measure up to 6.3 mm.

The body is divided into 2 parts, a longer, narrow forebody and a shorter wide hindbody. The lateral parts of the hindbody fold over to form the gynecophoric canal. The oral sucker is smaller than the ventral sucker. Spines are distributed over much of the surface of the body, including the sucker.

The digestive tract consists of a mouth opening inside the oral sucker, an esophagus surrounded by numerous unicellular glands, and 2 irregular ceca that divide at the anterior margin of the ventral sucker and fuse near the end of the body to form a short single tube, preceded immediately by several transverse connections.

The reproductive system is well defined. The testes consist of 15 to 36 follicles located at the anterior end of the gynecophoric canal. A short vas deferens empties into an arcuate, posteriorly directed cirrus pouch that opens through the genital pore located near the anterior field of the testes.

Female: Adults are filiform. Specimens resulting from experimental infections in white rats measure 1.6 to 3.6 mm long, from *Peromyscus maniculatus* 2.5 to 4.3 mm, from *Microtus pennsylvanicus* 1.3 to 4.8 mm, and *Mus musculus* 1.3 to 5.4 mm.

The oral sucker is smaller than the ventral one, which is borne on a stalk. Spines are distributed over the entire oral sucker but limited on the ventral sucker; they cover the lateral edges of the dorsal and ventral surfaces from the anterior end of the body to the level of the ovary. The digestive system is similar to that of the male.

The reproductive system consists of an ovary, oviduct, seminal receptacle, ootype, Mehlis' gland, uterus, and vitellaria with vitelline duct. The ovary is spiral and located near the posterior end of the anterior half of the body.

The oviduct, beginning at the posterior end of the ovary, forms a small seminal receptacle that turns forward. At the anterior end of the ovary, the duct enlarges to form the ootype, which is surrounded by the follicles of Mehlis' gland. A long uterus follows, extending anteriorly to the genital pore located at the posterior margin of the ventral sucker. The anterior end of the uterus is muscular, forming a metraterm.

Vitelline follicles fill the body posterior to the ovary. Two vitelline collecting ducts join medially at the anterior end of the vitelline gland to form the common vitelline duct. It may be distended with vitelline granules before emptying into the oviduct at the posterior margin of the ootype.

The nonoperculate eggs are unembryonated when laid but are fully embryonated by the time they reach the lumen of the intestine and appear in the feces. They measure 94-122 (average 108) x 74-98 (average 82) μ.

Life Cycle

Adults occur in meadow voles *(Microtus pennsylvanicus)* and muskrats *(Ondatra zibethicus).* The eggs pass through the intestinal wall, and are voided with the feces. The eggs reaching water hatch in a short time and the miracidia enter snails and transform into mother sporocysts which in turn produce daughter sporocysts. The lymnaeid snails suitable as hosts include the common dextral pond snails *Lymnaea stagnalis, L. reflexa, L. palustris* and the sinistral physids *Physa ancillaria* and *P. gyrina.* Daughter sporocysts give rise to cercariae. The cercariae penetrate through the skin of the final host and make their way along the blood stream to the hepatic portal veins where they mature.

Life Cycle Exercise

The natural sources of these flukes are voles and muskrats because of their close association with water where the cercariae occur. Laboratories maintaining cultures of these flukes are usually willing to provide living material for experimental studies. How would the live flukes be sent to your laboratory? Examine the venules of the liver, mesenteries, and intestinal wall of the hosts for the worms. Remove them to dishes of physiological solution.

Embryonated eggs are in the feces of infected mice. Macerate the feces in water and sediment and decant in cylinders. The heavy eggs precipitate quickly and the fine detritus and coloring matter are removed in the supernatant water. In water, the eggs containing mature miracidia swell immediately. The miracidia are clearly visible and surrounded by a layer of clear liquid and a vitelline membrane.

Prepare a water mount of eggs on a depression microscope slide or one on which the cover glass is supported so as not to exert pressure on them. As the egg swells due to absorption of the water, the miracidium becomes increasingly active. In a short time, a longitudinal split appears in the shell, the edges roll back, and the miracidium escapes, swimming away.

The newly hatched miracidium is pyriform and completely clothed with cilia. These are borne on 21 flattened epidermal cells arranged in 4 transverse rows of 6, 8, 4, and 3 cells each. A lateral process in the form of a small knob is located between the first and second rows of epidermal cells.

Conspicuous internal organs apparent in fixed and stained specimens include an apical gland (formerly called the primitive gut) and a pair of penetration glands, one flanking each side of the apical gland.

The excretory system consists of 2 pairs of flame cells. One flame cell of each pair is located lateral to a penetration gland and its mate at the level of the third tier of epidermal plates. Each pair opens separately on the side of the body at the level of the third tier of epidermal plates. The excretory system can be seen in living miracidia slightly pressed by the cover glass in water mounts and under high magnification. Germ cells fill the greater part of the body behind the apical gland.

The progeny of a single miracidium will be either male or female flukes but not both. Hence, all cercariae resulting from the infection of a snail with one miracidium will produce adults of the same sex. This is also true of the other dioecious blood flukes.

Place a young laboratory-reared snail of a suitable species in a shell vial with a few miracidia. They attach to the soft parts of the snail and quickly penetrate the epithelium. Mother sporocysts, daughter sporocysts, and cercariae develop in the body of the snail host.

Upon entering the snail, the miracidia shed the ciliated epithelium and transform into mother sporocysts. They are located in the head and neck region. The body is a long, thin-walled, tubelike structure with a smooth cuticle. Germ cells attached to the body wall give rise to germ balls. These develop into the daughter sporocysts. When the daughter sporocysts are produced, the mother sporocysts disintegrate. Dissect snails 10 to 15 days after infection in search of the elongate mother sporocysts.

The daughter sporocysts migrate to the liver where development continues. By 36 days after infection, young daughter sporocysts are present. The anterior end is covered with relatively long spines and the remainder of the body with short ones. The body cavity contains numerous germ balls. Mature sporocysts are long, slender, and tortuous.

Appearance of cercariae in the water is evidence of infection of snails. Forty to 54 days postinfection, put snails in a shell vial of pond water and check for cercariae by viewing the vial in a bright light against a dark background. The cercariae appear as tiny white spots moving rapidly in the water. Another procedure is to stopper the vial and examine it in good light under a dissecting microscope.

The cercariae rest by hanging with the body parallel to the underside of the surface film and the tail at an angle. When swimming, they progress tail first by rapid vibrations of the tail. They may crawl, measuring worm fashion, over the substratum.

Snails known to serve as hosts collected in habitats frequented by voles and muskrats should be checked for cercariae shed by them. Cercariae have eyespots, a short, forked tail, and an esophagus without a pharynx (Figure 55, L). The ventral sucker is slightly postequatorial. Small spines cover the entire body. The anterior end of the cercarial body forms the head organ that develops into the oral sucker of the adult. The mouth is ventral and subterminal, opening into an apharyngeate esophagus that divides to form 2 short laterally directed intestinal ceca located midway between the eyespots and ventral sucker.

Penetration glands consist of 3 pairs anterior to the ventral sucker and 3 pairs posterior to it in cercariae still inside the snails. One pair of glands is used up in emerging from the snail, leaving 5 pairs in free cercariae. Large spines at the anterior end of the body in the concavity of the head organ aid the cercariae in penetrating the vertebrate host.

The genital system appears as a globular mass of cells located immediately behind the ventral sucker. The excretory system consists of an excretory bladder in the posterior end of the body and 6 pairs of flame cells, 5 pairs of which are in the body and 1 pair in the base of the tail stem. Two common collecting tubules arise from the anterior end of the excretory bladder and extend forward to the level of the ventral sucker, divide and send out an anterior and a posterior collecting tubule, each with 3 flame cells. The excretory bladder discharges through 2 ducts that go around the Island of Cort located in the base of the tail and then unite to form a single duct that extends posteriorly through the stem of the tail, and then divides to empty through a pore at the tip of each furca. The flame cell pattern is

$$2 \; [(3) \; + \; (3)].$$

Infect 25 young white mice by placing them individually or by twos or threes in a dish of shallow water containing numerous cercariae. Allow 20 to 30 minutes of exposure, after which return them to normal conditions in the cages.

Because of the minute size of the larval flukes during the early period special procedures are necessary to find them after penetrating the mice. Two methods might be used: serial sectioning of certain parts of the hosts or pressing parts of tissue thin between glass plates, and making permanent mounts to show the almost microscopic worms *in situ*.

Examine mice according to the protocol given below and search for migrating larval flukes and adult worms. Development in the mice includes 5 phases: (1) penetration of the skin; (2) migration to the lungs; (3) migration to the liver; (4) entry into the liver; and (5) entry into the portal veins of the intestine.

1. Penetration of skin: Within 1 hour, cercariae have reached the dermis and by 12 hours some are in the subcutaneous tissue where the bulk of them occur on the third day. Some have reached the pleural cavity by 24 hours postinfection. Make preparations of skin and subcutaneous tissue to demonstrate larvae. Flush the body cavities with physiological solution into a dish and examine sediment for larvae.

These cercariae are able to penetrate the skin of humans but are unable to enter the circulation and develop to maturity. In the skin as a foreign protein, they cause inflamed wheals that produce dermatitis of cercarial origin. It affects bathers and others who come in contact with infested waters and is known commonly as ''swimmers' itch.'' The effects may be serious when large numbers of cercariae penetrate the skin.

2. Route of migration from skin to lungs: Larval flukes probably enter the lymphatic vessels and are carried to the heart and pulmonary artery within 24 hours. They penetrate the cardiac tissue. The maximum number of larvae are in the posterior lobes of the lungs during the third to sixth days. Prepare heart muscle, the pulmonary artery, and lungs by making presses of pieces 24 to 36 hours after infection to demonstrate the larvae.

3. Migration from lungs to liver: Larval flukes in the lungs appear in the larger vessels of the pulmonary artery, the tissue, and alveoli. They migrate to the posterior lobes, going toward the serous membrane. They escape from the lungs into the pleural cavities as early as 24 hours after infection but the maximum number is present by the sixth day. Most of them have left by the ninth day. On the sixth day after infection, sacrifice a mouse and flush the pleural cavities with physiological solution, catching the fluid in a dish. Examine the sediment for larvae.

By the fourth day after infection, larvae appear in the diaphragm and ligament connecting the diaphragm with the liver, and in the peritoneal cavity. They are most numerous in the body cavity by the ninth to tenth days. Examine another mouse on the tenth day. Flush the celomic cavity with physiological solution into a dish and examine the sediment for larvae. Make presses of the diaphragm from around the esophagus, aorta, and other parts; fix; stain; and mount for study.

4. Penetration into the liver: The first larvae invade the liver from the peritoneal cavity by the 36th hour and on the third day, the van has already entered the larger blood vessels. The number in the liver increases rapidly after the fourth day, with most of them inside after the 10th to 12th days. Make presses of pieces of liver and look for live worms in good light under a dissecting microscope.

5. Migration to portal veins of intestine: By the eighth day, the much larger larvae begin to arrive in the intestinal veins and by the 13th day, most of the flukes are in them. The sexes are recognizable as early as the eighth day postinfection and mature worms are present by the 17th day. Prepare presses and sections of the small intestine of infected mice between the 15th and 20th days.

Other nonhuman blood flukes whose anatomy and life cycles might be studied include species of *Trichobilharzia*, the adults of which occur in ducks and related shore birds. When the cercariae of these freshwater species penetrate the human skin, they cause a mild to severe schistosome dermatitis known as "swimmers' itch." The cercariae of *Austrobilharzia variglandis* of ducks and *Gigantobilharzia huttoni* are some that develop in marine snails and cause "clam diggers' itch." A number of sea birds, including some of the passerine species, that spend considerable time around water, harbor blood flukes whose cercariae cause dermatitis in man.

Family Spirorchiidae

Parasites of the blood vessels of the small intestine of turtles. The flukes are monoecious, lack a ventral sucker, and possess 1 or more testes. The esophagus is surrounded by gland cells.

Spirorchis parvus

This species occurs in the arteries of the wall of the pyloric stomach and small intestine of the painted turtle, *Chrysemys picta*.

Description

Adults are up to 2 mm long, thin, translucent, and lack a pharynx. Needlelike bristles cover the anterior end of the body for the length of the oral sucker. There are 4 to 5 testes, and a seminal vesicle that opens through a laterally placed common genital pore caudad from the testes. The ovary is slightly anterior to the genital pore. Vitellaria fill the body from the bifurcation of the intestinal ceca to near the posterior end of the body. Eggs, nonoperculate, average 54 x 38 μ.

Life Cycle

When the eggs reach water, the miracidia become active, and hatching occurs within 4 to 6 days. Miracidia attack and enter ram's horn snails (*Helisoma trivolvis, H. campanulata*); only young snails are susceptible to infection. The miracidia develop into mother sporocysts. Development is completed in about 18 days, at which time there are daughter sporocysts. These escape and migrate to the digestive gland, where cercariae develop and escape. Upon coming into contact with turtles, especially young ones, they penetrate the soft membranes and enter.

Life Cycle Exercise

Mature worms must be dissected from the large venules of the intestinal wall. In naturally infected turtles, masses of eggs accumulate in the intestinal wall, embryonate, and eventually break into the lumen of the gut. They are voided unhatched with the feces. Since hatching occurs soon after the eggs reach the water, feces should be collected daily upon being passed, comminuted and processed by sedimentation and decantation. The eggs are readily recognized, being, along with *S. elephantis*, the only embryonated, nonoperculate fluke eggs in painted turtles.

Sediment and decant freshly voided feces to remove fine debris and coloring matter. Pour the cleaned sediment into a small dish of water and change several times daily. At room temperature, hatching occurs between 2 and 5 A.M. as soon as 4 days after entering the water.

The miracidia have a large apical gland flanked by a penetration gland of about the same length on each side. Immediately behind the apical gland is a pair of crescentic, brown eyespots. The germinal mass lies immediately caudad from the eyespots. A lateral process extends from each side of the body opposite the posterior end of the apical gland. A pair of flame cells on each side of the body opens through excretory pores near the middle of the body. There are 4 tiers of epidermal plates. Beginning anteriorly, the tiers contain 6, 6, 4, and 2 plates, respectively.

To assure parasite-free snails for infection, only laboratory-reared ones should be used. Young snails between 4 and 25 days of age are most favorable for infection. Expose them to a large number of newly hatched miracida. Observe the reaction of the miracidia in the presence of the snails. Watch them attach to and burrow into the snails. Do the latter show evidence of annoyance?

Upon gaining entrance to the snails, the miracidia transform to mother sporocysts which are readily found in the free lateral margin of the mantle in 5 to 8 days. By 18 days, they attain a length of 2 mm and contain young daughter sporocysts. In sporocysts 5 days old, no body cavity or embryos are visible; at 7 to 8 days germ cells are numerous and a few embryos appear.

Upon leaving the mother sporocyst, daughter sporocysts appear in the lymph spaces of the mantle and in the digestive glands where they develop.

Dissect sporocysts from the digestive glands, beginning 18 to 20 days after infection of the snails. Continue dissections at intervals of 4 to 5 days until cercariae appear.

A birth pore develops at one end. Embryos of cercariae appear when the daughter sporocysts enter the digestive gland. The body wall is thin and the cavity large. It contains germ balls and cercariae.

Study living sporocysts dissected from the liver. Note the birth pore at the anterior end, the germ balls, and cercariae in various stages of development.

The furcocercous cercaria is apharyngeate, brevifurcate, distomate, and bears a cuticular crest along the humped dorsal side (Figure 55, K). They are very active upon escaping from the snails, swimming and resting intermittently as is characteristic of the furcocercous cercariae. They creep geometrid-fashion over the bottom of the dish. Watch for the appearance of cercariae issuing from the experimentally infected snails, beginning about the third week after infection. Hold a vial containing the infected snail so that light falls on it and view toward a dark background. The swimming cercariae appear as small, white specks moving rapidly in the water.

The protrusible head organ is pyriform. The ventral sucker is smaller than the head organ and located near the union of the second and last thirds of the body. There are 6 pairs of penetration glands in cercariae that emerge from snails. The ducts open at the anterior end of the head organ. There are 7 pairs of sharp penetration spines on the anterior end of the head organ. The mouth is subterminal and ventral, opens into a long slender esophagus that bifurcates into 2 short ceca just anterior to the ventral sucker. A prominent mass of germ cells located behind the ventral sucker represents the primordium of the reproductive systems.

The excretory system consists of a V-shaped excretory bladder without an Island of Cort. Two collecting tubes arise from the anterior end of the bladder and a caudal one from the hind end extends to the end of the tail, bifurcates, and empties at the tips of the furcae. The flame cell formula is

$$2 \left[(1 + 1 + 1) + (1 + 1 + 1) \right].$$

Collect *Helisoma* snails from ponds where painted turtles live and isolate individuals in vials in pond water. Observe them daily for the characteristic and easily identified cercariae of *S. parvus*. Cercariae from naturally infected snails are suitable for infecting turtles.

Young turtles are most suitable for infection experiments. The cercariae attack the soft tissues between the toes, in the flanks, around the eyes, the anus, and in the nasal passages. Turtles being attacked by cercariae react as if trying to escape from them.

After exposing young turtles to infection, keep them in aquaria and examine the feces to determine the prepatent period.

Family Fasciolidae

Parasites of the liver and intestine of mammals. These are large, flat distomes, with suckers close to each other. Intestinal ceca are highly branched or unbranched. Testes usually branched but may be smooth or lobed; ovary branched or smooth. The uterus has few coils and is anterior to the ovary. The vitellaria are profusely developed and distributed.

Fasciola hepatica

This is the common liver fluke of sheep, cattle, goats, and rabbits. It is widespread throughout the world and is of great economic importance to the livestock industry. It was the first digenetic trematode to have its life cycle worked. Simultaneously and independently in 1881, Leuckart in Germany and Thomas in Great Britain discovered the life cycle and described for the first time the larval development of digenetic trematodes. *Fasciola hepatica* has a life cycle differing from the intestinal form of the family in that the metacercariae migrate from the small intestine and through the celom to the liver. The liver capsule is penetrated by the young flukes which burrow through the parenchyma, eventually reaching the bile ducts where they mature (Figures 60, 82).

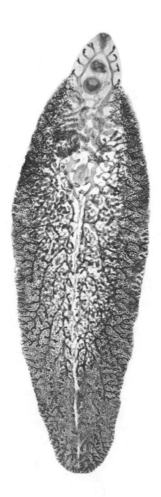

Figure 60. *Fasciola hepatica.*

Description

Fully grown specimens are up to 30 mm long by 13 mm wide. They are flat and leaf-shaped, with a small cephalic cone on a broad anterior part. The anterior portion of the body is beset with scalelike spines. The oral sucker is smaller than the nearby ventral sucker. There is a well-developed pharynx; the highly dendritic intestinal ceca occupy most of the body.

Two profusely branched testes occupy the posterior half of the body, with the vasa efferentia extending anteriorly to the cirrus pouch at the anterior margin of the ventral sucker. The common genital pore is median and opens at the level of the bifurcation of the intestinal ceca.

The dendritic ovary lies to the right of the body midline, anterior to the testes. A short oviduct enlarges to form the ootype surrounded by Mehlis' gland. The uterus loops anteriorly to the common genital pore. The vitellaria are profuse, extending the full length of the body. The eggs are described under Helminth Eggs, p. 210.

Life Cycle

The eggs are unembryonated when expelled in the feces. They require 9 to 10 days to mature at summer temperature. Upon hatching, the miracidia invade a suitable snail host and transform into sporocysts. Within 5 to 7 weeks, 2 generations of rediae and the cercariae have been produced. The common snail intermediate hosts in the United States are *Stagnicola bulimoides,* including its various subspecies, *Fossaria modicella*, and *Pseudosuccinea columella*. Elsewhere, other species of dextral lymnaeids serve as intermediaries. *Limnaea truncatula* is the common host in Europe and *Lymnaea tomentosa* in Australia. The cercariae encyst upon moist or submerged vegetation, where they remain until eaten by the final host. When viable cysts ingested by suitable hosts reach the duodenum, they excyst and the metacercariae migrate through the intestinal wall into the celom and to the liver, which they enter.

Life Cycle Exercise

In endemic areas, such as the southern states, Pacific Coast, parts of the mountain and Great Lakes states, adults can usually be collected in numbers at abattoirs. The complete life cycle can be studied in the snail intermediate host and a mammalian final host such as mice, rats, or rabbits.

Adults can be gotten from the condemned livers of cattle and sheep in abattoirs where these animals are slaughtered. By cutting across infected livers and squeezing them, flukes and eggs emerge by the handfuls from the open ends of the biliary ducts.

Many eggs can be obtained from the gall bladder or washed from the bile obtained in collecting the flukes. Also, gravid worms discharge eggs when placed in a dish of water. Because of their large size, these eggs can be seen easily with a dissecting microscope.

When incubated in light at room temperature, miracidia develop and begin hatching in about 8 days, with mass emergence by the 9th to 10th days. Eggs kept in darkness do not hatch as readily. When fully embryonated, the miracidium with its semilunar, dark eyespots, and ciliated covering can be seen through the egg shell. The anterior end, with the unciliated papillalike prolongation, or rostellum, is at the operculated end of the egg. Germ balls appear inside the miracidium. Alongside it in the egg are granules of residual yolk material.

Upon escaping from the eggs, the miracidia swim rapidly in an aimless fashion. The anterior end is broader than the posterior and bears the naked rostellum. There are 5 tiers of epidermal plates, beginning at the anterior end with 6, 6, 3, 4, and 2 plates each. These plates bear the cilia. Several sensory papillae appear between the first and second rings. There is a large primitive gland at the anterior end of the body, flanked by a pair of small penetration glands. The excretory system consists of large flame cells in the anterior part of the body; each long tubule opens to the outside laterally in the posterior half of the body.

Because of their amphibious nature, the snails require an area of exposed mud. Prepare a large moist chamber dish or small aquarium with the bottom covered with mud so that it will be exposed on one side. Maintain a small pool of water on the low side. Keep the dish or aquarium covered to provide a saturated atmosphere inside. Preferably the mud and water should come from a habitat where the snails thrive. Lettuce or dried maple or ailanthus leaves provide suitable food. An excellent diet is easily prepared by putting chopped grain feed in a pail of water and allowing it to ferment. The sediment is readily eaten by the snails and does not foul the water in the aquarium.

Very young laboratory-reared snails or specimens collected from areas not frequented by the vertebrate hosts can be infected by putting the appropriate species in a small container with miracidia. Prevent the snails from crawling out of the water during exposure to the miracidia.

Watch the miracidia attack the snails and the reactions of the latter. Put a Physidae or other kind of snail in the dish and notice how the miracidia shun it. Put some mucus from one of the acceptable snail hosts in a dish containing miracidia and note how they congregate around it.

The miracidia attack and enter the snail over the entire exposed part of the body. Upon entering the epithelium, the miracidia shed their ciliated covering. The eyespots persist but become detached from each other and lose their crescent shape.

After 24 hours, the miracidia are present in large numbers in the foot, having undergone very little change. In 48 hours, they are roughly circular. After 4 days, the sporocysts are elliptical. By the eighth day, sporocysts may attain full development and contain motile rediae. At this time, the sporocysts are numerous in the pulmonary chamber. They are readily recognized as elongate, saclike bodies without a pharynx or primitive gut and they contain rediae.

There are 3 generations of rediae. The first 2 produce rediae and are difficult to differentiate but the third is readily recognized because it contains cercariae. By the ninth day, the first rediae appear, moving about in the liver. Obviously, these are mother rediae since they are the first to appear. They are sac-shaped, with a muscular pharynx at one end followed by a short, thin-walled primitive gut. A pair of lateral outpocketings is present near the posterior end of the body. A collar appears near the anterior end. Development is rapid in the liver. By the 19th day, third generation rediae with developing cercariae are present. In 24 days, the cercariae have developed tails and exhibit motility. At 31 days, cercariae are well developed, show much pigmentation, and are active.

Cercariae (Figure 55, D) are shed naturally through the branchial pore of snails around 49 days after infection. As many as 800 or more may be produced in 1 day by a single heavily infected snail.

The body of the gymnocephalus cercaria is elliptical to almost circular, with a notch where the tail attaches. The suckers are about equal in size. The digestive tract consists of a muscular pharynx, short esophagus, and long simple ceca without a lumen. Cystogenous glands fill the body, tending to obscure all internal organs. The bilobed excretory bladder is thin-walled at first but is surrounded later by layers of muscle. The stout tail is up to twice the length of the body. Encystment may occur within a few minutes after escaping from the snail.

Upon coming into contact with a solid object in the water—even the sides and bottom of the dish containing the snails—a swimming cercaria attaches by means of the suckers, rounds up, detaches the tail, and secretes a flow of mucus that completely surrounds it. The mucus hardens and forms a 2-layered cyst that adheres to the grass or other objects. The metacercariae are somewhat smaller than the cercariae, having discharged the mucus for forming the cyst. While they still have characters common to the cercariae, they are essentially young flukes.

A plate of glass set upright in the water of a dish containing snails shedding cercaria will accumulate a large number of metacercariae for experimental infection. Scrape them off with a razor blade and store in a vial of water for infection experiments.

Rats, mice, and rabbits serve well as final hosts. Due to their small size and lower cost, mice are considered best for this exercise. Moreover, the flukes grow rapidly in them. However, mice are susceptible to damage by flukes migrating in the liver and may die as a result.

Cysts containing metacercariae may be introduced into the stomach by means of a small flexible tube attached to a syringe or mixed with moist food presented to hungry animals. For short term experiments, use 100 to 150 cysts per mouse, for longer terms about 50, and for observation on prepatency and longevity of flukes in the liver use 2 to 5 cysts. Infect 15 to 20 mice for examination at appropriate times to determine the route of migration and progress of development in them. This number will allow for deaths resulting from liver damage by the flukes.

Metacercariae excyst in the intestine within 2 to 3 hours. Within 24 hours after entering the intestine, the metacercariae have migrated through the intestinal wall and are in the celom. There is no noticeable change in development of these individuals over the time of excystment.

Necropsy a mouse at about 30 hours after infection. Open the body cavity and flush it and the viscera with physiological solution, catching all of the fluid in a tray. Examine the sediment for metacercariae that were migrating through the body cavity to the liver.

By the third day, the flukes begin burrowing into the liver. They have increased in size, as growth is rapid in mice. The transverse vitelline ducts are visible and the testicular rudiment is enlarged. The flukes are in the liver parenchyma.

Examine a heavily infected mouse on the fourth day by flushing the celom with physiological solution into a dish. Examine the sediment for young flukes and compare the number found with that recovered on the second day. Macerate the liver in a separate container so as not to destroy the metacercariae, allow the material to stand in physiological solution for an hour or 2 and then clean it by sedimentation and decantation. Examine the sediment under a dissecting microscope for the small flukes. What percentage of the number administered to the mouse was found in the celom and in the liver tissue?

By the 8th day, the body has doubled in size and about 13 lateral sacculations appear on each side of the intestinal ceca. The rudiment of the ovary is slightly bulged. Examine the celom and liver for young flukes. Determine the number of them in each place.

By the 11th day, the anterior body cone characteristic of the adult has appeared. The cecal outgrowths have secondary and tertiary branches. The testicular rudiments are separate and slightly lobed; and the rudimentary ovary is growing out of the shell gland complex. A rudimentary cirrus is visible.

By the 15th day, there is greater branching of the cecal outgrowths. The rudimentary cirrus is prominent inside the cirrus pouch and the testes branch laterally. The posterior end of the body has changed from a rounded to a broadly pointed shape.

By the 21st day, the cephalic cone and body shoulders are well demarcated. The ceca show short outgrowths on the median sides. The cirrus and pars prostatica are S-shaped.

By the 28th day, the characteristic body shape of the adult is established. The lateral fields of the body are filled with vitellaria. Some flukes have entered the bile ducts and the others are approaching them. How they gain access to the ducts is unknown. Carefully dissect the liver and look for flukes in the bile ducts.

By the 37th day, the flukes have attained sexual maturity and eggs begin to appear in the feces. Demonstrate them by fecal examinations. Prepare the sexually mature flukes from mice as stained permanent mounts and compare them for size with those obtained from cattle and sheep.

Other important Fasciolidae of the liver include *Fasciola gigantica*, common in cattle in Hawaii, and *Fascioloides magna* in deer, elk, and moose, as well as cattle, in North America. Their life cycles are similar to that of *Fasciola hepatica*. *Fasciolopsis buski* occurs in the intestine of man and swine in the Orient. This species does not have the migratory phase in its cycle as do *F. hepatica* and the other liver-dwelling forms.

Family Clinostomatidae

Parasites of reptiles and birds, such as herons and others eating freshwater fish. They are related to the blood flukes and strigeoids in having cercariae with forked tails. These flukes have flat bodies of medium to large size and the oral sucker may or may not be surrounded by a collarlike fold. The testes are tandem, with the ovary between them.

Clinostomum complanatum

This species, the metacercarial stage of which is known as the yellow grub, occurs in flesh of food fish, so is of special interest to the sports fisherman. Parasitic in the mouth of herons, egrets, nightherons, cormorants, gulls, and other fish-eating birds (Figure 77).

Description

The fairly thick, stout, spinose body measures up to 4.6 mm long by 1.7 mm wide, being widest at the level of the gonads. The anterior end is truncate, with a large circumoral, collarlike fold, and the small oral sucker located centrally. The larger ventral sucker is in the posterior part of the first quarter of the body. A very short esophagus divides to form the long intestinal ceca, with numerous, small outpocketings. The common genital pore is to the right of the anterior testis.

The lobed, almost triangular, tandemly arranged testes are in the anterior part of the second half of the body; they are wider than long. The cirrus pouch is oval, and contains a coiled cirrus. The small, oval ovary lies between the testes but toward the right side and behind the cirrus pouch. A large Mehlis' gland surrounds the ootype, which continues as the uterus with a few intertesticular loops, continues around the left and anterior margins of the front testis, and bends posteriorly to the genital pore. A Laurer's canal is present. The vitellaria, consisting of small follicles, fill the intercecal space between the intestinal bifurcation and the uterus and are extracecal from there to near the posterior end of the body, becoming confluent behind the second testis. Eggs operculate, 114-120 x 72-78 μ.

Life Cycle

The eggs upon reaching water hatch after about a week. Upon coming into contact with ram's horn snails (*Helisoma trivolvis* and *H. antrosa*), the miracidia enter and develop into sporocysts. Inside them rediae develop. The first generation rediae escape from the mother sporocyst and move to the digestive gland or liver, where mother rediae produce daughter rediae. The mature cercariae escape from the daughter rediae into the water. Upon coming into contact with a fish, the cercariae attach to the skin, enter, and develop into metacercariae. When the infected fish is easten by a suitable final host, the worms develop to maturity in about 3 days (Plate I).

Life Cycle Exercise

If herons, mergansers, and other large fish-eating birds are available, living hosts can be examined for parasites by looking in the mouth without harm or discomfort to them. Nestlings are especially good sources of adult worms. As in many other trematodes, adults void eggs when placed in water. The eggs are undeveloped when laid. They may be recovered from the feces found below nests of the avian hosts. When kept at room temperature in water, eggs develop and hatch in 8 days.

The newly hatched, fully ciliated miracidia are pyriform and very active. There are 3 brown eyespots arranged in a triangle with the apex pointed posteriorly. The body covering consists of 4 tiers of epidermal plates with 6, 8, 4, and 3 per row, beginning anteriorly. The apical gland has a rounded posterior end and a protrusible narrow apex surmounted by a small stylet. Three penetration glands lie on each side of the apical gland. One pair of flame cells is located on each side of the body. A large germ ball is in the narrow posterior end of the body.

Place young snails with newly hatched miracidia in a small dish kept at room temperature. These will be the source of the intramolluscan stages to be studied.

Mother sporocysts occur in the digestive gland by the eighth day. There are 3 eyespots, as in the miracidia, and up to 9 germ balls. There appears to be no birth pore. Dissect snails at intervals of about 4 days between the time of infection and the third week postinfection to obtain developing mother sporocysts and rediae.

Mother rediae appear by the 29th day after infection of the snails. The broad intestine extends to the posterior end of the body. Each mother redia contains 3 to 15 developing daughter rediae. The daughter rediae are recognizable by the tegumental folds, presence of developing cercariae, and a birth pore in the anterior quarter of the body.

The pharyngeate furcocerous cercariae (Figure 55, M) are small, narrow, and have a long slender tail. The body bears on its lateral margins numerous minute spines and 6 pairs of long, slender, protoplasmic hairlike structures and 1 on the outer margin of each furca. The body is curved, with a convex dorsal and concave ventral surface. There is a median dorsal fin extending from the level of the eyespots to the caudal end of the body. The paired eyespots are slightly preequatorial on the body. The anterior extremity of the body forms an elongate, oval penetration organ and bears 8 to 9 transverse rows of minute spines on the anterior end. The mouth is located medially, slightly posterior to the equator of the penetration organ, and is followed by a long prepharynx that extends slightly beyond the eyespots; a small pharynx, an extremely short esophagus, and a broad saclike cecum follow. A spherical mass of cells overlapping the posterior part of the cecum represents the developing ventral sucker. Four pairs of penetration glands are located at the level of the intestinal cecum; their ducts extend over the dorsal side of the penetration organ, opening at its anterior extremity. The genital primordium partially overlaps the ventral sucker. There are 5 pairs of flame cells, the last pair being located in the base of the tail; the

Plate I. *Life cycle of Clinostomum complanatum:* (A) Free-swimming miracidium before entering snail; (B) Sporocyst with rediae; (C) Redia with daughter rediae; (D) Cercaria produced by daughter rediae; (E) Same, having escaped from snail; (F) Metacercaria in flesh of fish; (G) Adult in cormorant. Numbers 1 and 2 indicate first and second intermediate hosts and 3 final host. After Hunter and Hunter (1935) from Meyer (1954).

excretory bladder is small, sends out a common collecting duct from each anterior lateral margin and a median posterior duct that extends to the end of the tail, dividing so that a branch opens at the tip of each furca.

At least 18 species of native fish are infected naturally with the metacercariae. They include bass, yellow perch, pickerel, trout, and catfish. Guppies are ideal for experimental studies for 2 reasons: first, they are easily infected and the metacercariae develop more rapidly than in other fish, and second, they have the advantage of being cheap, easily maintained, and virtually free from infection. Put the fish in a small aquarium with snails that are shedding cercariae. Vary the period of exposure so that all of the fish will not acquire a lethal infection. The first evidence of formation of cysts in guppies appears as swellings under the skin in about 3 weeks. By 7 weeks, the cysts are well developed and large, light colored swellings. Examine fish at weekly intervals, beginning with the first appearance of the cysts under the skin.

The metacercariae are enclosed in an inner, thin, hyaline cyst formed by the parasites and a thick outer one produced by the host. When mature, the metacercariae are precocious and resemble the adult worms in size, shape, and structure but lack eggs in the uterus.

Since domestic birds are refractory to infection, this aspect of the study presents difficulties. Where rookeries of gulls, terns, cormorants, or herons of different kinds are available, hatching eggs can be obtained and young birds used for experimental infection. Feed them fish that have mature metacercariae, either experimentally or naturally infected ones. Upon being released from the cysts in the stomach, the young but large flukes migrate up the esophagus to the mouth and mature within 3 days after being swallowed. They attach tightly by means of the anterior end to the buccal membranes of the host. They voluntarily release their hold in about 2 weeks and are lost.

Family Paramphistomatidae

Parasites of the cloaca of amphibians and rumen of ruminants. The group is characterized by the ventral sucker being at the posterior end of the body (Figure 69).

Megalodiscus temperatus

The prevalence of this species in frogs and the ease with which the life cycle can be studied make it an ideal model for class use. Specimens occur in the cloaca of various species of adult *Rana* and their tadpoles.

Description

The body is cone-shaped with the ventral side somewhat flattened. They attain a length of 6 mm, a width of 2.5 mm, and a thickness of 2 mm. The oral sucker is well developed with large, buccal pouches at the posterior end. A large ventral sucker which may have papillae in the center is located at the posterior end of the body. A short prepharynx is followed by a globular pharynx. The intestinal ceca begin at the pharynx and extend along the sides of the body as simple tubes to near the posterior end of the body. A large excretory bladder just anterior and dorsal to the ventral sucker opens on the dorsal side of the body near the caudal end.

The testes are arranged transversely near the posterior part of the anterior half of the body. A short vas deferens with a muscular end extends directly to the genital pore located at the beginning of the ceca. There is no cirrus pouch. The spherical ovary is located dorsally in the body on the median line near the ventral sucker. The oviduct extends caudad a short distance, enlarges to form the ootype surrounded by Mehlis' gland, and extends forward in broad loops to the common genital pore. Laurer's canal extends from the oviduct to the dorsal side of the body above the ovary. The sparse vitellaria consist of 2 groups on each side of the body between the testes and ventral sucker. Eggs operculate, 120-128 x 60-70 μ.

Life Cycle

The eggs hatch promptly upon entering the water, and the miracidia penetrate the young snails of *Helisoma* spp., where they transform into mother sporocysts. Three generations of rediae are produced.

Within the final generation of redia, cercariae develop; these emerge and encyst on the surface of various species of frogs and their tadpoles. Ingestion by adult frogs of their own stratum corneum, on which the cercariae have encysted, results in infection. Tadpoles become infected by eating metacercariae in pond ooze, or by taking in cercariae during respiration. Prepatency may be as short as a month, but it is usually longer.

Life Cycle Exercise

This life cycle is unique among the amphistomes in that the metacercariae are attached to the dark spots of the skin of the frog final host instead of on vegetation or inanimate objects in the water. Frogs are so commonly infected that living flukes are readily available. Adults removed from frogs to water promptly void fully embryonated eggs. Hatching occurs readily at room temperature. Observe the hatching of eggs.

Miracidia are pyriform in shape, ciliated, and rapid swimmers. The body is covered by 4 tiers of 20 epidermal cells with 6 in the first row, 8 in the second, 4 in the third, and 2 in the fourth. A lateral papillalike protrusion appears on each side between the first and second tiers of epidermal plates. A long apical gland extends from the anterior extremity to beyond the equator. There are 2 pairs of small penetration glands. Three flame cells appear on each side and their ducts open laterally between the third and fourth tiers of epidermal plates.

Young laboratory-reared *Helisoma trivolvis* 4 to 6 weeks old are readily infected, even when exposed to a single miracidium. Observe miracidia attacking and penetrating snails. Twenty-four hours after entering the snails, the miracidia have shed the ciliated epithelium to become mother sporocysts, and migrated to the stomach wall. Here they produce rediae.

At least 3 redial generations are produced, the first of which is the so-called mother redia. By the 14th day, the mother rediae contain the next generation in the form of germ balls and rediae. Between the 17th and 19th days, the first generation rediae escape from the mother sporocyst. Second generation rediae begin to escape from the mother rediae by the 24th to 26th days and are plentiful by the 50th to 60th days. The pharynx, gut, and lateral body appendages are well developed. The rediae migrate to the tip of the liver, where development continues. The gut of the older ones is dark brown. These rediae contain the next generation of rediae. By the 64th to 80th day after infection, the apical whorl of the liver is filled with third generation rediae. In older infections, the entire liver is involved. These rediae are distinguished by the presence of cercariae in the body cavity.

Dissect infected snails at the appropriate times to find (1) mother sporocysts without pharynx or gut located on the stomach wall and containing a single redia; (2) first generation rediae with pharynx and gut and containing rediae; (3) second generation rediae in the tip of the liver and containing rediae; and (4) third generation rediae filled with amphistome cercariae (Figure 55, B).

The cercariae leave the rediae in an immature condition in which the oral sucker, intestine, and germinal cells are segregated into 3 distinct groups. Development continues in the liver tissue, where the eyespots appear, the tail elongates, the suckers and intestine develop, and the reproductive system appears.

Collect cercariae from experimentally infected snails, or from naturally infected ones from marshes and swamps where frogs are infected with *Megalodiscus temperatus*. Isolate the snails in vials or small jars provided with a piece of fresh lettuce for food. Examine the container frequently by holding it before a dark background in a position so that light falls on it. The large conspicuous amphistome cercariae with 2 eyespots are easily seen and readily recognized.

Transfer infected snails to a small aquarium, the bottom of which is covered with pebbles. Supply the snails with pieces of lettuce for food and powdered calcium carbonate. They will live a long time under these conditions, shedding cercariae all the while. More cercariae will be shed on sunny than on cloudy days. Put an adult frog in a bottle with cercariae. The jar should contain enough water to cover the frog. The cercariae quickly settle on the dark spots of the skin, detach the tail, and each secretes a cyst about itself that adheres to the skin. Cercariae encyst also on tadpoles.

Adult frogs become infected when the stratum corneum bearing the metacercariae is shed and subsequently eaten, as is commonly done. The cysts pass through the intestine unaffected by the digestive juices. Cysts with slits through which the young flukes have escaped appear in the cloaca. Tadpoles become infected upon eating metacercariae in pond ooze, or by taking in active cercariae during respiratory movements. Cercariae eaten by tadpoles quickly encyst in the mouth, but later detach and are swallowed. Excystment in tadpoles occurs in the small intestine. Young flukes in tadpoles remain in the small intestine during the process of metamorphosis and go to the rectum when it is formed. Often the flukes may be lost during metamorphosis.

Introduce cercariae directly into the cloaca via the anus to determine if they can develop into adult flukes by dispensing with the metacercarial stage. In a similar experiment, place metacercariae in the cloaca to ascertain if they are able to excyst and develop without benefit of going through the stomach and intestine.

Some commonly available amphistomes include *Zygocotyle lunata* of the cloaca of ducks, with *Helisoma antrosa* as the snail intermediary; *Stichorchis subtriquetrus* of the cecum of beavers, with *Fossaria parvus* as the snail host; *Wardius zibethicus* of the cecum of muskrats, with *Helisoma antrosa* as the snail intermediary; *Paramphistomum cervi* in the rumen of cattle, deer, sheep, and goats, with *Stagnicola bulimoides techella* as a common intermediary.

Family Halipegidae

These flukes are listed by some workers as Halipegidae and by others as Halipeginae in the Hemiuridae. In either case, they are parasitic in the mouth, esophagus, stomach, intestine, and sometimes in the eustachian tubes of amphibians and alimentary canal of fish.

They are plump or elongate flukes with a smooth body. The ventral sucker is large and equatorial or postequatorial in position. The testes are in the posterior half of the body with the compact, bilobed ovary behind them. There is no cirrus pouch (Figure 74).

Halipegus occidualis

Adults occur in the esophagus and anterior part of the stomach of frogs, the Pacific giant salamander *(Dicamptodon ensatus)*, and rough-skinned newt *(Taricha granulosa)*.

Description

Body up to 5.8 mm long by 1.7 mm wide in the testicular region. The subterminal oral sucker is smaller than the nearly equatorial ventral sucker. The tegument is thick and smooth. A muscular pharynx is followed by a short esophagus which divides into narrow, sinuous intestinal ceca extending to near the posterior end of the body.

The large oval testes are obliquely placed, extracecally and postacetabular. There is no cirrus pouch and the common genital pore opens medially near the posterior margin of the oral sucker. The small ovary is located medially near the posterior end of the body. The uterus proceeds anteriorly in broad transverse folds, emptying through the common genital pore. The vitelline gland consists of 2 clusters of 4 to 6 oval follicles that fill the end of the body posterior to ovary. Eggs operculate, with a long aboper-cular filament, average 50 x 21 μ (fixed), 61 x 26 μ (live).

Life Cycle

The eggs do not hatch unless eaten by snails. Molluscan intermediaries include *Helisoma antrosa* in the eastern part of the United States and *H. subcrenatum* in the Pacific Northwest. The nonciliated miracidia have a spiny tegument with 12 to 14 spines at the anterior end. After burrowing through the intestinal wall, the miracidia slowly change into mother sporocysts. They produce elongate, cylindrical rediae which contain developing cercariae, there being no second generation of rediae. When the discharged cercariae are ingested by the copepods *Cyclops vernalis* and *C. serrulatus* or the ostracod *Cypridopsis vidua*, they become metacercariae. Infection of final hosts occurs when they eat the infected crustaceans.

Life Cycle Exercise

The embryonated eggs hatch when eaten by the snail intermediate host. The unique cercariae develop without encystment when eaten by ostracod or copepod second intermediate hosts.

Adults are obtainable from the esophagus of adult frogs. While the species is widely distributed, the occurrence is not necessarily continuous. Eggs may be obtained by placing gravid worms in water or by dissecting them. Examination of feces of final hosts will reveal those infected with species of *Halipegus* by the presence of eggs with the long abopercular filament. No other frog parasite has similar eggs. While fully embryonated when laid, the eggs do not hatch until eaten by the snail first intermediate hosts.

Since they do not hatch spontaneously in water, miracidia for observation must be obtained by treating the eggs to induce hatching. Place eggs on a slide in a 0.9% NaCl solution. Just as the liquid has evaporated, add more of the same solution. Within half a minute, the miracidia become active, the operculum loosens, and the larvae emerge.

Eggs placed in physiological saline, preferably Helix physiological solution, on a slide with a cover glass, hatch when gentle pressure is applied by a dissecting needle. When uninjured, such miracidia live for up to half an hour. If stained with 1% Nile blue sulphate or cresyl blue, what appears to be internal glandular structures and a cluster of germ cells can be seen. The epithelium is shed. The body is covered with minute body spines. There is a crown of about 12 to 14 large spines at one end of the body.

As already pointed out, balanced aquaria with populations of laboratory-reared snails common to the region should be maintained for experimental work. When laboratory-reared snails are not available, specimens should be sought where frogs do not occur, if possible. Examine 100 or more individuals to determine the percentage naturally infected, to compare with the percentage that become infected under experimental exposure.

Young snails put in a small covered glass dish containing water, some powdered calcium carbonate, and masses of ova of *H. occidualis*, ingest the eggs and calcium carbonate. Examine the feces for empty egg shells, the presence of which is evidence of hatching in the gut. The miracidia penetrate the snail from the lumen of the gut by migrating through the wall. Individuals that fail to burrow through the gut wall may be voided with the feces. Do you see miracidia in them?

Young sporocysts along the digestive tract are similar to the miracidia; they have the glandular structures at the anterior end of the body together with a germinal mass inside. By the end of 2 weeks, rediae appear in the sporocysts which are in the digestive gland of the snail hosts. The sporocysts continue to develop and grow in size for much longer. Examine infected snails by careful dissection at intervals of 3 to 4 days to obtain a series of developing sporocysts containing fully developed rediae. There appears to be but a single generation of rediae. Therefore, any mature rediae in the digestive gland contain only cercariae in various stages of development. They are light orange in color and tend to be constricted at various points by small folds. Rediae appear in large numbers in the digestive gland by the 50th day, almost destroying the entire liver. Rediae produce many thousands of cercariae; they contain up to 100 at a time.

The cercariae are of the cystophorous type and are designated as cercariocysts. The short, broad cercaria is surrounded by the cystic tail, which has a long, slender tube called the delivery tube folded inside. When the delivery tube is extended, the cercaria escapes through it. Place a cercariocyst on a slide in water, add a cover glass, and apply gentle pressure. The delivery tube extends and the cercaria proper is discharged through it with explosive force. Stimulation of the delivery tube by touching it with a teasing needle may provoke extension and discharge of the cercaria.

While cercariocysts are shed more or less continuously, they appear most abundantly between 2 and 10 P.M. daily. Upon escaping from the snails, the cercariocysts sink to the bottom of the container, are relatively inactive and capable of surviving up to 2 weeks. During this period the cyst imbibes water and the cercarial body becomes opaque and eventually dies.

In areas where infected frogs occur, snails naturally infected with *H. occidualis* may be found and provide a source of all the intramolluscan stages for study.

To infect the crustacean second intermediaries, place them in a small dish with 6 to 7 day old cercariocysts. Copepod nauplii are more susceptible than adults. In the intestine, the delivery tube is stimulated and the cercaria released. Up to a dozen or more cercariae occur in the intestine at 1 time. It appears that the delivery tube with structural modification and explosive manner of discharge is instrumental in aiding the cercaria to pass through the intestinal wall into the body cavity. After allowing young crustaceans to eat cercariocysts, examine them promptly for cercariae or cercariocysts in the intestine and cercariae in the body cavity.

The metacercaria never encysts in the crustacean hosts, but continues to be active in the body cavity. Growth to infectivity is slow. As development proceeds, large villi extend to the outside through the excretory pore. Maximum size is reached in about 30 days. Metacercariae also occur in naturally infected dragonfly *(Libellula incesta)* and damselfly *(Lestes rectangularis)* naiads which happen to eat infected ostracods or copepods, but are unnecessary for the completion of the life cycle.

When metacercariae in the crustaceans or odonate naiads are fed to frogs or tadpoles, they migrate to the mouth and develop. Prepatency takes about 60 days during summer weather.

Family Troglotrematidae

Flukes with a plump or rounded body, the ventral sucker near the midbody, and the tegument spined. They are parasites of the kidney and skin of birds and the intestine, frontal sinuses, and lungs of mammals. The family is best known for *Paragonimus westermani,* the Oriental lung fluke, which infects man.

Paragonimus westermani

Adults occur in cysts either singly or in pairs in the human lungs (Figure 65). The common animal hosts belong to the Felidae (including tigers, lions, leopards, panthers, and wild cats), swine, and dogs. The most important endemic areas are in Korea, Japan, Taiwan, central China, and the Philippines.

Description

Adults are plump, spinose worms resembling a coffee bean in size and shape. Often they are quite brown due to the profuse brownish vitellaria and loops of the uterus filled with brown eggs. Worms are up to 12 mm long, 6 mm wide, and 5 mm thick. The oral sucker is slightly larger than the ventral sucker, which is located slightly anterior to the midbody. A muscular pharynx close to or in contact with the oral sucker is followed by an extremely short esophagus. Simple, sinuous intestinal ceca extend to the posterior end of the body. The excretory bladder extends forward as a narrow, irregular sac to the level of the posterior margin of the pharynx.

The highly lobed testes are arranged transversely about midway between the ventral sucker and hind end of the body. There is no cirrus or cirrus pouch. The common genital pore is near the posterior margin of the ventral sucker. The ovary is a large irregularly shaped organ to the left and slightly behind the ventral sucker, but anterior to the testes. A looped uterus lies opposite the ovary on the other side of the body. The vitellaria consist of extensively branched, small follicles filling the lateral fields of the body and overlapping the intestinal ceca the full length of the body. The eggs are described under Helminth Eggs, p. 210.

Life Cycle

The eggs require 16 days or more to develop and hatch. The miracidia penetrate Pleuroceridae and Thiaridae gastropods. The mother sporocysts produce rediae which in turn spawn characteristic cercariae. They are small and of the microcercous type, with a small tail and bearing a well-developed stylet in the anterior part of the oral sucker. The cercariae penetrate crayfish or crabs, enter the viscera and muscles, and encyst. The final host becomes infected by eating crustaceans containing living metacercariae. Upon being released from the cyst in the intestine, the young flukes burrow through the gut wall into the celom, migrate anteriorly, passing through the diaphragm and into the pleural cavities whence they burrow into the lungs.

Paragonimus kellicotti

This is the North American representative of lung fluke. It appears to be a normal parasite of mink but occurs in cats, dogs, muskrats, swine, goats, and even man. In surveys, it has been found in 15% of the mink examined in Michigan and 7% in Minnesota, and in 12% of the muskrats in Michigan. It is widespread in the United States.

Similarity between *Paragonimus westermani* and *P. kellicotti* has led some workers to consider them as synonyms. Because of the similarity of the 2 species, the basic anatomical description of *P. wester-mani* given above will suffice for *P. kellicotti*. The life cycle is basically similar to that of *P. wester-mani*. It is a long one and not well suited for a short term study.

Life Cycle Exercise

Adults are not easy to get unless one has access to the carcasses of mink and muskrats during trapping seasons in areas where the flukes are enzootic. The adult worms appear singly or more often as pairs in cysts in the lungs. Adults are best obtained by collecting metacercariae from crayfish, as described below, and feeding them to final hosts.

Living worms void eggs when placed in water. They can be dissected from flukes or rinsed from the cysts in the lungs. Some of them develop and hatch in about 2 weeks at room temperature but most do not hatch for much longer. Eggs operculate, 75-118 x 48-65 μ.

The miracidia are active inside the shell before hatching. They are normally pyriform, completely ciliated except on the spaces between the epidermal plates and points where the anterior and lateral papillae appear. The ciliated epidermal plates are arranged in 3 tiers of 6, 6, and 3 each, beginning anteriorly. The body is filled with germ cells, as revealed by vital stains. The excretory system consists of a pair of flame cells on each side with the tubules opening laterally between the second and third tiers of epidermal plates.

The first intermediate host is the operculate Pomatiopsidae snail, *Pomatiopsis lapidaria*. They are amphibious snails of nocturnal habits. They thrive in captivity when kept in open shallow dishes containing damp soil and decomposing leaves. Young snails about 1 mm long from the natural habitat are readily infected when placed with miracidia. Old snails are refractory.

The sporocysts occur in practically every part of the body but especially along the esophagus, stomach, and intestine. The sporocysts are elliptical when young, but become elongated and of various shapes, mostly sacculate, as they mature—a process requiring about 2 weeks at room temperature. They contain 10 to 12 stumpy mother rediae. Dissect infected snails, beginning 1 week postinfection and every week thereafter until the sporocysts mature, to obtain them in the various stages of development.

Young first-generation rediae appear in about 1 month after infection. They are located in the lymph space adjacent to the stomach and liver. The pharynx is conspicuously large and the gut rarely longer than the pharynx. A cuticular collar may be present.

Mature rediae containing fully developed second generation rediae appear about 2 months postinfection. There is no collar, but a birth pore is present near the pharynx. About a dozen daughter rediae are present inside, only about half of which are developed at one time.

Second generation rediae containing the stylet-bearing, microcercous cercariae (Figure 55, H) appear in the liver about 70 days after infection. They are similar to the preceding generation. Up to 30 cercariae are present in the body cavity.

Dissect snails to obtain mother rediae containing daughter rediae and specimens of the latter filled with cercariae.

The period of time for cercariae to develop varies. They appear as early as 78 days after infection in some snails. Mature cercariae have a microcercous type tail almost spherical in shape. Both the body and tail are spinose. The subterminal oral sucker bears a stylet and is larger than the ventral sucker located near the midventral part of the body. There are 14 penetration glands, seen best in living specimens stained with one of the vital dyes.

Species of *Cambarus* crayfish serve as the second intermediate host. The cercariae attach to and penetrate the soft, thin, chitin between the segments, principally on the under side of the tail. Infect young crayfish by placing them in a small dish containing a large number of cercariae.

Upon entering the crayfish, the cercariae drop the tails and migrate to the heart, attach to it in a transverse dorsal band, and secrete a thick, dark brown wall about themselves. In crayfish kept at room temperature, the migrating cercariae reach the heart 12 hours after entering the crayfish and are encysted by the 24th hour.

Growth is slow. The intestinal ceca appear by the 17th day and the excretory bladder reaches the pharynx by the 22nd day. At 35 days, the organs and worms are distinctly larger. Metacercariae 46 days old are full size and infective. Dissect infected crayfish at the various periods suggested above and observe the developing metacercariae.

Feed mature metacercariae to a young cat. Migrating worms are in the body cavity within 5 hours after swallowing them, in the pleural cavities within 9 hours, and in the lungs by the 14th day. Feed other metacercariae to white rats and examine the body cavity by flushing it out with physiological solution to find the worms in it and over the viscera. Did you find larvae in the pleural cavities? In the lungs? Beginning the 4th week postinfection of the cat, make daily examinations of the feces for eggs. Prepatency takes between 5.5 and 6 weeks.

While mink and muskrats are usually unavailable for examination, crayfish can usually be obtained in large numbers. They are an especially good indicator of the presence of lung flukes in an area, as a large percentage of them will show the conspicuous metacercarial cysts lined up over the dorsal surface of the heart. Crayfish purchased from supply houses for use in general zoology laboratories often provide an excellent source of preserved metacercariae.

Living cysts collected in surveys may be fed to cats, as mentioned above. Feed them to white rats and to mice and follow the infections by daily examinations of animals for the first week to determine their action in these hosts.

Nanophyetes salmincola

This, the salmon poisoning fluke, occurs in the small intestine of a number of fish-eating mammals in the Pacific Northwest. It is infected with *Neorickettsia helminthoeca*, the etiologic agent of salmon poisoning disease, which is transmitted transovarially by the fluke, and is usually fatal to Canidae.

Description

The pyriform body is slightly flattened and measures up to 1.1 mm long by 0.5 mm wide. The oral sucker is followed by a pharynx and an esophagus of about equal length. The intestinal ceca bifurcate at the anterior margin of the ventral sucker and extend to the posterior end of the testes. The ventral sucker is located about midbody. The common genital pore is median and directly behind the ventral sucker.

The 2 large, oval testes are symmetrically located at the sides in the posterior half of the body. The cirrus pouch is pyriform and contains a large seminal vesicle divided into 2 parts by a transverse constriction. The round ovary is located adjacent to the ventral sucker on the right side and in the intercecal space. The uterus consists of 2 coils posterior to the ovary. The vitellaria are formed by large irregular follicles and extend from the pharynx to the posterior end of the body, filling all the space unoccupied by the other organs. The eggs are operculate; unembryonated when laid; size 64-80 x 34-50 μ.

Life Cycle

Eggs develop slowly, at room temperature requiring at least 85 days to hatch; many take much longer. Upon coming into contact with the snail intermediate host *(Oxytrema silicula)*, the miracidia enter. Cercariae are liberated in great numbers, in long strands of mucus. When they contact Salmonidae and certain nonsalmonid fish, the cercariae enter and are carried throughout the body in the vascular system. The mammalian final hosts become infected when they eat fish harboring the metacercariae, which are older than 10 days of age. Hamsters serve well as final hosts. For life cycle details consult Bennington and Pratt (1960), and Millemann and Knapp (1970).

Family Allocreadiidae

Common parasites of the alimentary canal of fish, both freshwater and marine, the world over. They are small to medium-sized, elongate flukes. The oral sucker is terminal or subterminal and followed by a prepharynx, pharynx, and 2 long, intestinal ceca. The testes are in the hind part of the body and a cirrus pouch is present. The ovary is usually pretesticular and the follicular vitellaria are extensive. The excretory bladder is tubular or saccular (Figure 85).

Crepidostomum cooperi

This species occurs in the digestive tract of a number of species of fish, among which are: bass, yellow perch, sunfish, trout, pickerel, and bullheads.

Description

The species of *Crepidostomum* are characterized by 6 fleshy papillae on the anterior margin of the oral sucker. Eyespots are often present. Adults are up to 1.5 mm long by about one-seventh as wide in relaxed specimens. The papillae on the oral sucker are of about equal size, with 2 pairs projecting anteriorly and the ventral pair laterally. The ventral sucker is near the posterior end of the anterior half of the body and is slightly smaller than the oral sucker. The oval pharynx is followed by a short esophagus, which bifurcates at the anterior margin of the ventral sucker to form 2 long intestinal ceca. The excretory bladder is sacculate, extending to the anterior margin of the front testis. The testes, of variable shape but usually round to oval, are tandem in the posterior half of the body. The long, slender cirrus sac extends from the common genital pore at the anterior margin of the ventral sucker to the posterior margin of the ovary. The pear-shaped ovary is close behind and usually overlapping the ventral sucker. A seminal receptacle of variable size is close to the posterior margin of the ovary and a Laurer's canal opens dorsal to the intestinal cecum on the opposite side. The convolutions of the uterus extend posteriorly to the anterior testis usually, but may reach the posterior one. The vitelline follicles extend from about midway between the pharynx and ventral sucker along the sides of the body, uniting behind the posterior testis. Eggs operculate, 50-70 x 30-55 μ (fixed) or 62-75 x 42-55 (live).

Life Cycle

The unembryonated eggs, upon reaching water at summer temperatures, mature in 7 days and hatching begins at 10 days. Miracidia penetrate various species of *Sphaerium* and *Musculium*, the fingernail clams. They penetrate between the inner and outer layers of gills and the mantle where they transform, presumably into mother sporocysts. There appear to be 2 generations of rediae. Daughter rediae contain cercariae, a distinguishing feature. The oculate xiphidiocercariae are active swimmers, and are positively phototactic. They attack and penetrate the naiads of species of *Hexagenia*, the burrowing mayflies, encysting in the muscles. The metacercariae are fully formed in 2 to 3 weeks. When infected mayfly naiads are eaten by susceptible fish, the metacercariae escape from the cysts and develop to maturity in 3 to 4 weeks. For life cycle details refer to Awachie (1968), and Choquette (1954). Other species vary in the kinds of intermediate hosts utilized. *Crepidostomum metoecus* of trout utilizes the gastropod *Lymnaea peregra* as the first intermediate host, and the scud *Gammarus pulex* as the second intermediary; *C. farionis* develops in the fingernail clam *Pisidium casertanum* and the amphipod *Gammarus pulex*; and *C. cornutum* develops in the fingernail clam *Musculium transversum*, with metacercariae in crayfish.

Family Heterophyidae

The members of this family are small to tiny flukes covered with scalelike spines, common in the intestine of birds and mammals, including man, throughout the world. The ventral sucker is usually well developed and may or may not be enclosed by a large suckerlike genital atrium.

Heterophyes heterophyes

This small, scaly fluke (Figure 73) occurs frequently in man in the Nile Delta and the Orient. The second intermediate host is a food fish and infection is acquired when raw fish containing metacercariae are eaten. The flukes also occur in dogs, cats, foxes, and other piscivorous mammals, thus intensifying the infection in the endemic areas.

Description

These small flukes measure up to 1.7 mm long by 0.4 mm wide. The entire body is covered with scales which have lateral projections and are more numerous anteriorly. The oral sucker is small, followed by a short prepharynx, tiny pharynx, and short esophagus which bifurcates into slender intestinal ceca reaching to the posterior end of the first half of the body. A large genital sucker, on the left margin of the acetabulum, bears numerous multidigitate spines. The excretory bladder is elongate and saclike, reaching almost to the ovary.

The small, oval testes, located side by side intercecally, are in the posterior end of the body. The vasa efferentia unite slightly anterior to the ovary to form a voluminous U- or retort-shaped seminal vesicle which opens into the genital sucker through a muscular ejaculatory duct without benefit of a cirrus pouch or cirrus.

A small oval or round ovary lies medially a short distance anterior to the testes, with the ootype, Mehlis' gland, and seminal receptacle close by. The uterus loops posteriorly on the right side of the body, traverses to the left, goes posteriorly between the testes, then runs anteriorly along the right side to the level of the ventral sucker and coils across the body to the left side, emptying through a metraterm into the genital atrium. About 14 coarse vitelline follicles lie extracecally on each side in the posterior third of the body. The eggs are described under Helminth Eggs, p. 210.

Life Cycle

The eggs hatch in the intestine of the brackish water snail *Pirenella conica* in Egypt and *Cerithidia cingulata* in the Orient. The intramolluscan stages include the mother sporocyst and 2 generations of rediae. The cercariae are of the lophocercous type, i.e., those with a long tail bearing a finlike structure and an oral sucker without armature such as a stylet. The second intermediate host is a fish, commonly a brackish water mullet, *Mugil* sp. Infection of the final hosts occurs when viable cysts are ingested with raw or incompletely cooked fish.

Cryptocotyle lingua

This species occurs in the intestine in fish-eating birds such as gulls, terns, puffins, murres, guillemots, cormorants, jaegers, eiders, loons, grebes, and herons in the boreal regions.

Description

The body, up to 2 mm long by about 0.5 mm wide, is covered with scales. The small oral sucker is followed by a much smaller pharynx and an esophagus of about equal length; the simple slender, intestinal ceca extend along the side of the body to the hind end. The ventral sucker forms a part of the anterior margin of the large genital sucker located near the middle of the body. The slender excretory bladder extends to the anterior margin of the testes, branches, sends slender arms anteriorly to the level of the pharynx, loops back, and extends posteriorly again.

The slightly lobed testes are in the posterior quarter of the body; the vasa efferentia join to form an elongate, voluminous seminal vesicle which empties in the depth of the genital sinus.

A clover leaf-shaped ovary is located on the right side and separated from the testis by the large, oval, seminal receptacle. The uterus fills the intercecal space with transverse loops between the ovary and the posterior margin of the genital sucker. The vitelline follicles extend along the lateral sides from about midway between the pharynx and genital sucker to the posterior end of the body, uniting medially anterior to the margin of the genital sucker. Eggs operculate, 40-56 x 18-30 μ.

Life Cycle

The adults live deep between the villi of the small intestine of the host, where they produce eggs in great numbers. These embryonate in seawater in about 10 days. The edible periwinkle snail, *Littorina littorea*, becomes infected by eating the fully embryonated eggs. After being eaten, the eggs hatch, the miracidia invade the tissues of the snail, and pass through the usual sporocyst and redial stages. The released cercariae readily penetrate the skin and especially the fins of fish. Most fish of the shore zone, particularly sluggish swimmers such as the cunner, *Tautogolabrus adspersus*, are naturally infected with these metacercariae. The final host becomes infected when it eats a fish with the metacercariae.

Life Cycle Exercise

Along coastal regions where the marine intermediate hosts are available, *C. lingua* offers an excellent opportunity for conducting the life cycle of a fluke whose cercariae encyst in marine snails. Mature flukes are almost always present in terns and gulls and are most abundant in their fledglings. The heavily infected fledglings pass enormous numbers of eggs in their feces. The fecal material may be removed by straining through several layers of cheesecloth, followed by repeated sedimentation and decantation in seawater until clear. The eggs are partially embryonated when voided in the feces. Several days of incubation are required to complete development.

Collect snails from rocks below nesting cliffs of gulls and terns, or on mud flats frequented by the birds at low tide. Infected *Littorina littorea*, as Willey and Gross (1957) have shown, can be detected by observing snails on the side of the glass vessel. In infected specimens, the expanded foot is a distinct brown, in contrast to the whitish color of the foot in normal individuals. Dissect infected snails to find, collect, and prepare temporary mounts of the developing larval stages.

Fully developed miracidia are ovate, more pointed posteriorly, and with the anterior end directed toward the operculum. The miracidium is covered with cilia except at the anterior tip. There is a pair of unicellular glands in the central portion. The opaqueness and dark color of the egg shell make detailed study of the enclosed miracidia difficult.

The small ovoidal to vermiform-shaped sporocysts occur in the lymph spaces of the snail's digestive gland. Rediae, the only stage normally present in naturally infected snails, occur in such great numbers that the digestive gland is discolored as a result of the mass infection.

The rediae are colorless, thin-walled, sausage-shaped bodies. The body is filled with germ cells and germ balls in the posterior end and immature cercariae in the anterior part. The excretory system is double, with excretory pores lateral and near the junction of the middle and posterior thirds of the body.

Being incompletely developed when leaving the rediae, the cercariae remain in the lymph spaces of the digestive gland to mature. After completion of growth, the pleurolophocercous cercariae leave the snails by way of the gills. The cercarial tail bears a dorsoventral fin which begins near the body on the dorsal side, extends around the end and about midway to the base on the ventral side. A small pharynx lies about midway between the oral sucker and a mass of penetration glands. The esophagus and intestinal ceca are not apparent. There is no ventral sucker. A pair of eyespots is at the level of the pharynx. About 18 penetration glands are in the third quarter of the body; their ducts extend forward from the glands as a single, broad group, but divide into 4 groups just behind the oral sucker. One group bends around each lateral margin of the oral sucker and 2 pass separately over the dorsal surface, all opening at the anterior margin of the sucker. A mass of cells directly behind the penetration glands represents the genital primordium. The excretory bladder is Y-shaped and has a thick wall.

The cercariae penetrate the skin of cunners, especially the fins and tail where, fortunately, it is easy to see them. The oval cysts are clear, unstainable, and tough. The metacercariae grow faster than the cyst increases in size; this requires them to assume a U-shape. The fish forms a strong connective tissue sheath around the metacercarial cyst of worm origin.

Metacercariae may be induced to excyst in vitro by placing the cysts in 0.5% sodium carbonate with pancreatin for 4 hours and then transferring them to distilled water. The cysts rupture promptly, freeing the young flukes. When placed in Ringer's solution, they will live 2 to 3 days. In dextrose-salt solution they will live up to 14 days at room temperature.

To observe cercariae penetrate the cunners, cut out a piece of the tail and put it in a dish containing cercariae. Watch the process of penetration under a dissecting microscope. It requires from 1 to 2 hours. Note that tails are discarded as the body of the cercariae enters the skin of the tail. The recurved spines on the body are of great mechanical advantage to the cercariae in burrowing into and migrating through the skin, as doubtless are the histolytic effects of the secretions of the penetration glands.

Feed cysts in cunners to 6 to 8 white rats. Examine a rat at 5, 8, and 15 hours after swallowing the cysts by carefully looking at the contents of the stomach and small intestine. The connective tissue part of the cyst of host origin is digested off in the stomach within 5 hours but there are no cysts in it after 7 to 8 hours. Encystment occurs in the upper part of the small intestine 6 to 12 hours after the cysts are swallowed. Since rats are unfavorable hosts, the worms rarely mature in them. Dogs are favorable hosts in which the worms mature but are soon lost in a "self-cure" phenomenon.

Family Strigeidae

This family, together with the Diplostomatidae and several others, constitute the superfamily Strigeoidea, whose adults are characterized by a body divided into an anterior and posterior part, by the presence of a tribocytic organ (adhesive gland) in the forebody, and the genital pore at the posterior end of the body (Figure 71). Their cercariae are pharyngeate, forktailed, and develop in elongate sporocysts. The entire group is parasitic in the intestine of birds and mammals.

Cotylurus flabelliformis

This species occurs in the small intestine of ducks in North America. The pharyngeate, furcocercous cercariae (Figure 55, M) penetrate snails and develop to infective metacercariae of the tetracotyle type (Figure 57, A).

Description

The body is divided by a deep transverse constriction into a cup-shaped forebody and a cylindrical hindbody about twice the length of the former.

The body is strongly flexed dorsally at the junction of the 2 regions. The holdfast organ consists of 2 transverse, partly cleft lips which often protrude from the forebody. Directly beneath the holdfast organ is a disc-shaped adhesive gland whose processes extend through the holdfast organ to the tips of the lips, or lappets. The oral sucker is located dorsally at the anterior margin of the forebody; the oval ventral sucker is behind the oral sucker and in the bottom is the cup-shaped forebody. The digestive tract consists of a well-developed oval pharynx, an esophagus of about equal length, and intestinal ceca that terminate somewhat short of the posterior end of the hindbody.

The testes occupy the second and third quarters of the hindbody. From the dorsal view, they are bean-shaped and from the side heart-shaped. The vasa efferentia unite to form a tubular seminal vesicle that enlarges into a globular ejaculatory pouch located at the posterior margin of the hind testis. The common genital pore is large, funnel-shaped and located subdorsally near the posterior end of the hindbody.

The oval ovary lies on the dorsal side of the anterior testis. The oviduct, after giving off Laurer's canal which opens dorsally, extends caudad to Mehlis' gland. From here the anterior limb of the uterus extends forward to the ovary, bends ventrad and continues posteriorly to the common genital pore. The vitellaria are in a single field in the ventral half of the hindbody, extending the full length of it. Eggs operculate, 100-112 x 68-76 μ.

Life Cycle

The significance of this life cycle is the tetracotyle type of metacercariae (Figure 57, A) found only in the Strigeidae. The unembryonated eggs hatch in water, and the miracidia attack and enter several species of snails, among which are *Physa sayi*, *P. parkeri*, *Lymnaea emarginata*, *L. stagnalis*, and *L. perampla*. The miracidia transform into mother sporocysts, which in turn produce daughter sporocysts. These give birth to cercariae. After leaving the first snail, the cercariae enter other snails *(L. emarginata, L. stagnicola, L. reflexa, L. palustris,* and *Fossaria abrussa)* and develop into the type of metacercariae known as tetracotyles. The final host becomes infected upon eating the snails containing the tetracotyles.

Life Cycle Exercise

Gravid worms may be obtained from the intestines of wild ducks available during the hunting season. An alternative approach to getting them is to find infected snails and feed the tetracotyles to baby chicks, ducklings, or goslings. Prepatency takes about a week. Eggs are voided by live, gravid worms in water. They may also be recovered from the feces of experimentally infected birds. The unembryonated eggs hatch in about 3 weeks at room temperature.

Expose young, preferably laboratory-reared snails to miracidia. They penetrate rapidly, and soon transform into mother sporocysts. Mature mother sporocysts up to 5 cm long containing numerous germ balls, embryos of daughter sporocysts, and fully developed metacercariae (tetracotyles) occur in the region of the digestive gland of naturally infected *Lymnaea stagnalis*. About 6 weeks after infection, daughter sporocysts give birth to cercariae. The daughter sporocysts are large, numerous and entwined in the digestive gland.

These cercariae are among the largest of the pharyngeate, forktailed forms (Figure 55, M). Upon escaping from the snail host, they swim back and forth almost constantly, turning in a spiral manner. They soon lose their tails, sink to the bottom and creep over the substratum. The furcae are longer than the tail stem. The body is covered by small backward pointing spines, except for a bare circumoral area on which is a clump of about 18 forward pointing spines located dorsal to the mouth. The tail stem bears 6 pairs of long, lateral, hairlike processes. The ventral sucker in the anterior part of the second half of the body is heavily spined.

There are 2 unpigmented eyespots a short distance anterior to the ventral sucker. The digestive system consists of a short prepharynx, small pharynx, long esophagus reaching almost to the ventral sucker, where it branches to form intestinal ceca which extend to near the caudal end of the body. Two pairs of penetration glands lie near the anterior margin of the ventral sucker. Their ducts pass dorsal to the penetration organ and open lateral to the circular mouth. The reproductive systems are represented by a spherical cluster of cells near the posterior end of the body. The last 2 cells on each side are in the base of the tail. The excretory bladder sends a long tubule through the tail stem to open at the tip of each furca. The flame cell pattern is

$$2 \ [(2 \ + \ 2) \ + \ (2 \ + \ 2 \ + \ 2)].$$

Upon escaping from the snail first intermediate host, the cercariae penetrate other snails and develop into metacercariae of the tetracotyle type. Upon entering certain lymnaeid snails, they develop in the digestive gland and reproductive organs. In physid and planorbid snails, the metacercariae develop as hyperparasites in the rediae and sporocysts of other flukes. They even occur in their own mother sporocysts.

The developing and mature metacercariae occur in the digestive and reproductive organs of the normal snail hosts. About 20 to 30 days at room temperature are required for the cercariae to develop to mature metacercariae and assume the characteristics of the tetracotyles. Young metacercariae are free in the tissue of the snails but as maturity approaches they always become encysted.

The development of the metacercariae in the snails is one of a striking and complete metamorphosis by which the cercaria transforms into a tetracotyle. The body of a fully developed tetracotyle is pear-shaped with the forebody broad. A large semicircular fold forming the ventral part of the cup-shaped forebody extends from each side of the oral sucker around the posterior margin of the acetabulum. The large holdfast organ is located between the fold and the ventral sucker. The lateral suckers, or cotylae with medial openings, form long clefts extending posterior from the level of the mouth to slightly beyond the acetabulum.

The digestive system consists of a well-developed pharynx, a much shorter esophagus, and very short intestinal ceca that bifurcate at the anterior margin of the acetabulum.

The reproductive system is represented by a large, somewhat bean-shaped cellular mass in the hindbody directly posterior to the holdfast body. The excretory bladder is represented by 2 vesicles, 1 at the end of each intestinal cecum. An excretory pore is located at the posterior extremity of the body and opposite an opening in the cyst.

While ducks are the natural final hosts, young chicks 2 weeks old and ducklings become infected when fed mature tetracotyles. Prepatency takes about a week.

Family Diplostomatidae

Similar to the Strigeidae, but the forebody is more flattened, the hindbody is conelike, cylindrical or flat (Figure 72). Metacercariae either of the neascus or diplostomulum type (Figure 57, B, C).

Uvulifer ambloplitis

This species occurs in the intestine of the belted kingfisher (Megaceryle alcyon) as adults and under the skin of fish as metacercariae whose cysts contain black pigment; because of this, the infection is known as black spot.

Description

The body is divided into an anterior and a posterior portion. The bowl-shaped forebody, located on a necklike part of the hindbody, contains the ventroterminal oral sucker, acetabulum, and circular holdfast (tribocytic) organ. The holdfast organ, just behind the acetabulum, bears a T-shaped opening at the apex. The hindbody is sharply distended dorsally and laterally by the uterus and large testes. The pharynx is followed by a short esophagus that bifurcates just before the ventral sucker into 2 intestinal ceca extending to the posterior end of the body.

The large testes lie in the posterior two-thirds of the hindbody. The vasa efferentia join ventral to the second testis to form a voluminous seminal vesicle, which continues as a muscular ejaculatory duct opening into the genital cone at the posterior end of the body. There is no cirrus pouch.

The small, oval ovary lies anterior and to the left of the forward testis. The uterus arches posteriorly along the left side of the front testis, forms an S-shaped loop between the testes, extends anteriorly along the ventral side of the anterior testis to the union of the 2 body parts, enlarges, and bends sharply backward to the anterior margin of the hind testis, and constricts to form a narrow oviduct that empties through the genital cone. There is no seminal receptacle. Lateral and ventral vitellaria extend from the anterior margin of the genital atrium to the base of the necklike portion of the hindbody. Eggs operculate, 90-99 x 56-66 μ.

Life Cycle

The eggs hatch in about 3 weeks at room temperature, and the miracidia attack and enter ram's horn snails, *Helisoma trivolvis*. Here the miracidia transform into mother sporocysts, which give rise to daughter sporocysts, which in turn produce cercariae. The cercariae enter and encyst in bass and related host species, where they develop into metacercariae of the neascus type (Figure 57, B). The kingfisher final hosts become infected when they consume fish containing the mature encysted neascus.

Life Cycle Exercise

When kingfishers are available for examination, adult worms are likely to be found in large numbers in the small intestine. In the absence of kingfishers, feed metacercariae from fish to pigeons and young chicks in an effort to obtain gravid worms.

Eggs may be obtained from adult worms when available. They are likely to occur in the droppings of nestlings and almost surely those of adult birds. Live adult worms void eggs when placed in water. If obtained from feces, remove the detritus and coloring matter by repeated washing and decantation in water. At room temperature, hatching takes about 3 weeks.

The pyriform, ciliated miracidia are covered by epidermal plates of 6, 6, 6, and 3 per row, beginning anteriorly. The pyriform apical gland is about one-third the length of the body. There is 1 pair of flame cells on each side of the body. A large, spherical germ ball is in the posterior extremity of the body.

Place young snails in a dish with newly hatched miracidia and examine them at intervals of several days in an effort to find the developing intramolluscan stages.

Mother sporocysts are characterized by the simple, unbranched, saclike body with 1 pair of eyespots. Fully developed ones attain lengths up to 0.3 mm and contain germ balls and daughter sporocysts. Mature specimens are easily found in naturally infected snails collected from areas frequented by kingfishers.

Daughter sporocysts occur in the digestive gland. They are unbranched saclike bodies. Older ones appear similar to a string of beads with enlargements formed by the individual cercariae. A birth pore is on one side or the tip of the body. Because of the manner in which the long daughter sporocysts are entangled in the tissues of the digestive gland, it is difficult to remove them intact.

The forktailed, pharyngeate cercariae (Figure 55, M) in water hang with the body downward and folded back upon itself. They slowly sink, then dart upwards a short distance, rest again, and descend again. Upon touching the substratum, they swim upwards.

A circumoral spineless area bears a clump of about 12 forward-pointing spines on the dorsal side. Small spines are present over the surface of the anterior third of the body and on the furcae. There are 7 pairs of long, protoplasmic hairlike processes on each side of the tail stem. The undeveloped ventral sucker is represented by a small mass of cells between the anterior pair of penetration glands. The mouth opens in the spineless area and is followed by a short prepharynx, a small spherical pharynx, and short, single saclike gut slightly longer than the diameter of the pharynx. A pair of unpigmented eyespots is near the equator of the body. The large penetration glands are in 2 rows in the posterior half of the body.

Their ducts run forward and open on each side of the mouth. The rather large genital primordium is at the posterior end of the body.

The small excretory bladder sends 2 common collecting tubules lateral and forward, and 1 median posterior tubule, which extends the length of the tail stem, bifurcates at the end and empties at the tip of each furca. The flame cell pattern is

$$2[(2) + (3 + 2)].$$

Place small bass, sunfish, rock bass, or minnows with cercariae in a small dish. Beware of too heavy infections lest the fish be killed. The black spots caused by this and related species of metacercariae occurring in fish, especially bass and sunfish, are well known to anglers and parasitologists. The cyst comprises an outer pigmented layer of host origin and an inner tough, hyaline layer produced by the parasite. It becomes pigmented in about 22 days.

The larva is a neascus (Figure 57, B), 1 of 3 types of metacercariae characteristic of the superfamily Strigeoidea. The other 2 are the tetracotyle already discussed under the Strigeidae, and the diplostomulum to follow. The forebody is very thin, leaflike, and spoon-shaped, without lateral sucking cups. The entire ventral surface, except the ventral sucker, is spinous. The hindbody is well developed, conical in shape, and distinctly set off by a transverse constriction. The ventral sucker is near the union of the middle and last thirds of the forebody, and the prominent cone-shaped holdfast organ lies directly behind the ventral sucker; a muscular pharynx is followed by a very short, slender esophagus which bifurcates into 2 slender intestinal ceca reaching the posterior end of the body. A small copulatory bursa may be everted from the posterior end of the body.

Dissect infected fish at intervals of 3 to 4 days for a period of 4 weeks to follow the development of the cercariae to the neascus type of metacercaria characteristic of this and related species of Diplostomatidae. Feed mature, encysted neascus to pigeons and young chicks. Determine if they are susceptible and whether worms mature sexually by fecal examinations beginning a week postinfection. If no eggs appear, necropsy the birds and examine for worms.

Crassiphiala bulboglossa is another strigeoid, with a neascus type of metacercaria, that produces black spots similar to that of *Uvulifer ambloplitis*, in Cyprinidae, Cyprinodontidae, Esocidae, Etheostomatidae, Percidae, and Umbridae fish. Adults occur in kingfishers, but differ markedly from those of *U. ambloplitis* from the same host. For a description and life cycle details consult Van Haitsma (1935) and Hoffman (1956), respectively.

Diplostomum baeri eucaliae

This species occurs in the intestine of ducks as adults (Figure 72 is representative of the genus) and in the central nervous system and eyes of sticklebacks as unencysted metacercariae of the diplostomulum type (Figure 57, C).

Description

The body is distinctly 2-segmented, with the parts about equal in length and the hindbody bent dorsally at the junction of the 2 parts. The length of fixed specimens is up to 1.4 mm; the flattened forebody is 0.9 mm long by 0.4 mm wide, and the cylindrical hindbody is 0.6 mm long by 0.4 mm in diameter. A small marginal pseudosucker, an identifying characteristic, is located on each side of the body opposite the oral sucker. The ventral sucker is in the anterior portion of the second half of the forebody. A prominent oval holdfast organ lies directly caudad to the ventral sucker. The digestive system consists of a prepharynx, a pharynx, an esophagus about as long as the pharynx, and intestinal ceca which terminate at the posterior end of the hindbody. The common genital opening is subdorsal near the posterior end of the hindbody.

The large testes, arranged in tandem, occupy most of the hindbody. A large curved seminal vesicle lying behind the second testis opens into the genital atrim. There is no cirrus pouch. The small transversely oval ovary is located medially at the anterior margin of the forward testis. The ootype and Mehlis' gland lie between the testes. The uterus opens in the common genital atrium. Vitelline follicles extend throughout the width of the body from slightly posterior to the intestinal bifurcation to the hind end of the body. Eggs operculate, average 102 x 59 μ (live).

Life Cycle

The eggs, unembryonated when voided, develop and hatch in 12 days at room temperature. The miracidia penetrate the snails *Stagnicola palustris* and *S. p. elodes* and transform into mother sporocysts. At 3 weeks, they contain daughter sporocysts, which escape and enter the liver. The sporocysts have matured by the 30th day and are releasing cercariae. Cercariae penetrate the skin of brook sticklebacks (*Eucalia inconstans*), where they develop into metacercariae of the diplostomulum type (Figure 57, C) in the brain. The metacercariae become infective between the 13th and 23rd days. Ducks acquire the infection by eating sticklebacks harboring infective diplostomulae.

Life Cycle Exercise

When brook sticklebacks infected with *D. baeri eucaliae* are available, feed 200 to 400 diplostomulae to newly hatched, unfed chicks. Prepatency takes 3 days. Feed the chicks commercial dog chow or other suitable food.

Eggs for experimental purposes are obtained best from the feces of heavily infected chicks. Separate the eggs from the fecal matter by repeated sedimentation in water and decantation. At room temperature, active miracidia are present in 10 days and hatching occurs 2 days later. Observe eggs in water during development.

The miracidium is somewhat fusiform in shape. Two pigmented eyespots close together are in the anterior third of the body. The primitive gland at the anterior end of the body is nearly cylindrical. Expose young, parasite-free snails, preferably laboratory-reared ones, to newly hatched miracidia. Keep snails at room temperature. Mother sporocysts develop in the mantle of the snail, and by the 14th day they contain daughter sporocysts. Cercariae are shed by experimentally infected snails 30 days after exposure to miracidia. Dissect snails at 15, 20, 25, and 30 days postinfection to obtain developing stages. Press pieces of mantle between glass plates, fix and process as permanent mounts.

The pharyngeate furcocerous cercariae (Figure 55, M) are covered by spines arranged differently on separate parts of the body. One band of minute irregularly distributed spines is on the anterior half of the penetration organ; following these spines are 9 to 12 evenly spaced rows of spines encircling the body from the middle of the penetration organ to the equator of the acetabulum. Minute spines are scattered irregularly over the remainder of the body. The ventral sucker is in the anterior part of the second half of the body. The oval penetration organ covers half the distance between the anterior tip of the body and the anterior margin of the acetabulum. The mouth opens at the anterior extremity of the body. The prepharynx and esophagus are shorter than the muscular pharynx; the intestinal ceca extend to near the caudal end of the body. There are 4 large penetration glands behind the ventral sucker; their ducts pass dorsally over the penetration organ where they expand to form an enlarged reservoir, and then constrict in diameter to small ducts opening beside the mouth.

Expose 12 sticklebacks to about 100 cercariae each by placing individual fish and the cercariae together in a small container. Follow the development of the metacercariae. On the first day, many larvae are in the blood and optic lobes. Hemorrhage occurs behind the eyes and in the brain. By the third day, larvae are all in the brain. The body divisions are not yet evident. By the fifth day, most larvae are in the optic lobes; some are invading the retina. Division of the body into 2 parts is beginning. By the ninth day, the hindbody, lateral (pseudo-) suckers, and holdfast organ are well formed. After about 15 days the metacercariae are infective. Observe the development of the larvae. The diplostomulum type metacercariae are characterized by the large forebody and the marginal pseudosuckers, 1 on each side of the oral sucker. These may be everted earlike. In these various respects, they differ from the tetracotyle and neascus larvae discussed above.

Newly hatched, unfed chicks serve as satisfactory experimental hosts. Using a bulbed pipette, place 100 mature diplostomulae in the throat of each of 3 baby chicks. Prepatency is 3 days. Determine patency in this unnatural host.

Key to Common Families of Digenea
Capital letters in parentheses refer to hosts: F-Fish, A-Amphibia, R-Reptiles, B-Birds, M-Mammals.

1   Mouth near midventral surface of body (Figure 61) .... (F) Bucephalidae
     Mouth near anterior end of body 2

2(1) In blood vessels 3
      Not in blood vessels 4

3(2) Monoecious (Figure 62) .... (F) Sanguinicolidae
      Dioecious; male usually with gynecophoric canal or flaps to embrace female (Figure 63)
      .... (B, M) Schistosomatidae

4(2) Encysted in tissues 5
      In locations other than blood vessels or cysts in tissues 6

5(4) Body slender anteriorly, sometimes expanded posteriorly (Figure 64) ....(F) Didymozoidae
      Body oval, fleshy, spiny (Figure 65) .... (R, B, M)[1] Troglotrematidae

6(4) Ventral sucker absent 7
      Ventral sucker present 9

7(6) Testes median to ceca, which unite posteriorly to form ringlike intestine; may lack oral sucker
      (Figure 66) .... (B) Cyclocoelidae
      Testes lateral to ceca 8

8(7) Testes near midbody; cirrus pouch absent; ventral surface of body without rows of glands
      (Figure 67) .... (B) Eucotylidae
      Testes near posterior extremity of body; cirrus pouch very long; ventral surface of body with
      rows of glands (Figure 68) .... (B, M) Notocotylidae

9(6) Ventral sucker at posterior extremity of body; fleshy, somewhat cone-shaped flukes (Figure
      69) .... (F, A, R, B, M) Paramphistomatidae
      Ventral sucker not at posterior extremity of body 10

10(9) Genital pore at posterior extremity of body; adhesive body posterior to ventral sucker . 11
       Genital pore not at posterior extremity of body; no adhesive body 13

11(10) Ovary between testes; body oval in outline, not divided into anterior and posterior parts; cirrus
       pouch large (Figure 70) .... (B) Cyathocotylidae
       Ovary anterior to testes; cirrus pouch absent; body divided into anterior and posterior parts..
       12

12(11) Anterior part of body cup-shaped, posterior part ovoid or cylindrical (Figure 71)....
       .... (R, B, M) Strigeidae
       Anterior part of body flattened, posterior region cone-shaped (Figure 72).................
       .... (R, B, M) Diplostomatidae

13(10) With gonotyl (genital sucker) near ventral sucker, latter often modified as a genital sinus;
       body scaly (Figure 73) .... (B, M) Heterophyidae
       Without gonotyl or modification of ventral sucker 14

14(13) Ovary behind testes[2] 15
       Ovary not behind testes 16

15(14) Elongate, cylindrical body; gonads near posterior end of body; uterus pregonadal; vitellaria
       with few follicles at posterior end of body; suckers widely separated (Figure 74).........
       .... (A) Halipegidae

---

1. *Nanophyetus salmincola* and *Sellacotyle mustelae* are minute members of this family which occur in the small intestine of piscivorous mammals.
2. See couplet 34 where some genera of Lecithodendriidae have the testes anterior to ovary. In this case, they are in front of the ventral sucker, whereas in couplet 14 both testes and ovary are behind the ventral sucker.

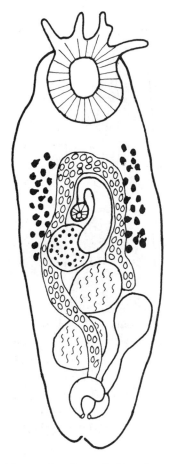

Figure 61. *Bucephalus elegans.*

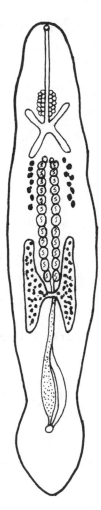

Figure 62. *Sanguini-cola occidentalis.*

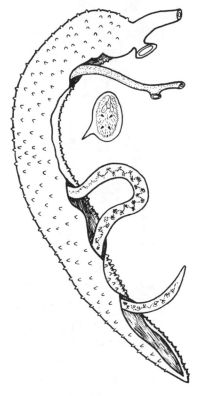

Figure 63. *Schistosoma mansoni.*

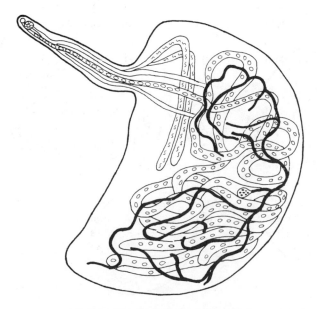

Figure 64. *Didymocystis coatesi.*

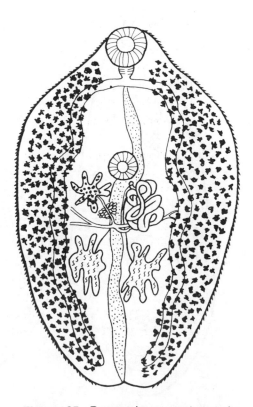

Figure 65. *Paragonimus westermani.*

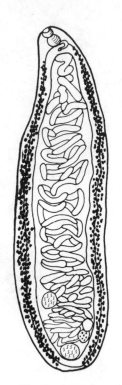

Figure 66. *Cyclocoelium leidyi.*

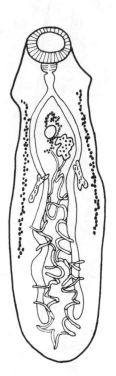

Figure 67. *Eucotyle hassalli.*

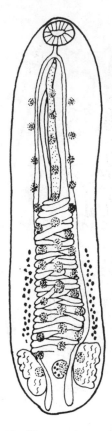

Figure 68. *Notocotylus urbanensis.*

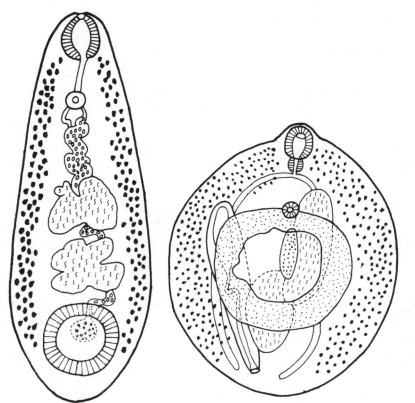

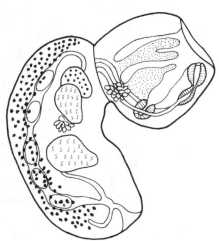

Figure 69. *Paramphistomum cervi.* **Figure 70.** *Cyathocotyle prussica.*   **Figure 71.** *Cotylurus flabelliformis.*

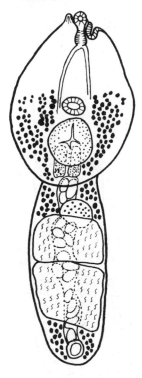

Figure 72. *Diplostomum huronensis.*

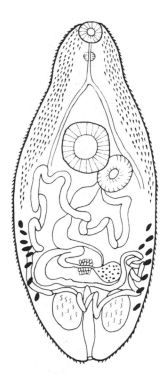

Figure 73. *Heterophyes heterophyes.*

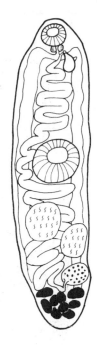

Figure 74. *Halipegus amherstensis.*

Elongated, flattened body; gonads in anterior half of body; uterus almost entirely postgonadal; vitellaria with many follicles near middle third of body; suckers close (Figure 75) . . . . (A, R, B, M) Dicrocoeliidae

16(14) Ovary between testes; genital pore behind ventral sucker 17

Ovary anterior to testes; genital pore not behind ventral sucker 18

17(16) Uterus between anterior testis and fork of intestine; vitellaria lateral, extending back to anterior testis, not intercecal posteriorly; gonads in posterior third of body (Figure 76) . . . . (B, M) Brachylaemidae

Uterus between testes and ventral sucker; vitellaria lateral, extending caudad from gonads, intercecal posteriorly; gonads near middle of body (Figure 77) . . (R, B, M) Clinostomatidae

18(16) One testis 19

Two or more testes 20

19(18) Pharynx small; vitellaria follicular but feebly developed (Figure 78) . . . . (F) Monorchiidae

Pharynx large; vitellaria compact, small (Figure 79) . . . . (F) Haploporidae

20(18) More than 2 testes 21

Two testes 22

21(20) Testes numerous; vitellaria follicular, extending in extracecal spaces behind ventral sucker; uterus pregonadal (Figure 80) . . . . (B) Orchipedidae

Testes (actually 2, divided into several irregular bodies) arranged in 2 longitudinal rows of 4-6 each; vitellaria compact, small, lateral; uterus descends to posterior end of body *(Gorgodera)* (Figure 81) . . . . (A) Gorgoderidae

22(20) Uterus with ascending limb only 23

Uterus with both ascending and descending limbs 30

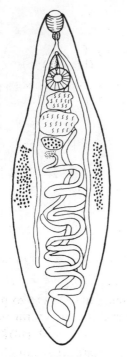

Figure 75. *Dicrocoelium dendriticum.*

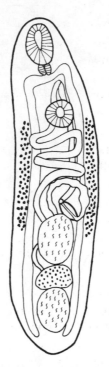

Figure 76. *Brachylaema
virginianus.*

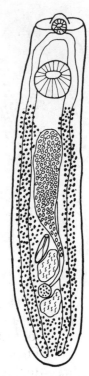

Figure 77. *Clinostomum
complanatum.*

Figure 78. *Monorchis monorchis.*

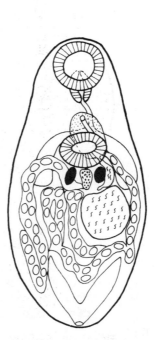

Figure 79. *Haploporus benedeni.*

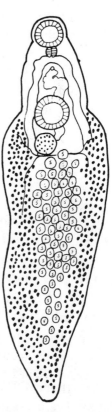

Figure 80. *Orchipedium trachei-
cola.*

23(22) Ovary and/or testes highly branched 24

Neither ovary nor testes branched 25

24(23) Ovary and testes profusely branched; large spiny flukes (Figure 82) . . . . . .(M) Fasciolidae

Only testes branched; medium-sized flukes with cuticle spined ventrally (Figure 83). . . . . .
. . . . (B) Cathaemasiidae

25(23) Head collar with single or double crown of stout spines around margin; suckers close (Figure
84) . . . . (B, M) Echinostomatidae

Head collar absent 26

26(25) Prepharynx long; vitelline follicles small, abundant, lateral and extend to posterior extremity
of body, intermingling medially; excretory bladder I-shaped (Figure 85) . . . .
. . . . (F) Allocreadiidae

Prepharynx short or absent; vitelline follicles fewer, do not reach posterior extremity of body
. . . . 27

27(26) Cirrus pouch absent; seminal vesicle coiled; excretory bladder Y-shaped with long stem and
short arms which do not extend anteriorly to ovary (Figure 86). . . . .
. . . . (R, B, M) Opisthorchiidae

Cirrus pouch present 28

28(27) Body wall thick, with wrinkled cuticle; excretory bladder with stem reaching almost to posterior testis and arms to anterior end of body; cirrus pouch small and anterior to ventral sucker
(Figure 87) . . . . (F) Azygiidae

Body wall thin, not wrinkled; cirrus pouch large, extending far behind anterior margin of ven-
tral sucker 29

29(28) Vitellaria numerous large follicles, filling lateral space behind ventral sucker; small flukes
(Figure 88) . . . . (B) Psilostomatidae

Vitellaria tubular with 6-7 follicles near middle of body; muscular medium-sized flukes (Figure 89) . . . . (B) Philophthalmidae

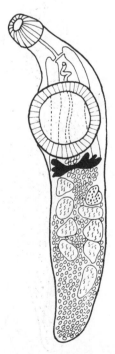

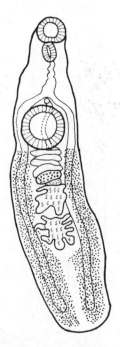

Figure 81. *Gorgodera amplicava.*   **Figure 82.** *Fasciola hepatica.*   **Figure 83.** *Cathaemasia nycti-
coracis.*

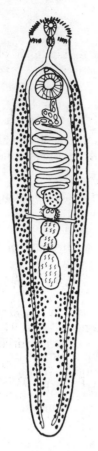

Figure 84. *Echinostoma revolutum.*

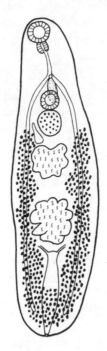

Figure 85. *Allocreadium lobatum.*

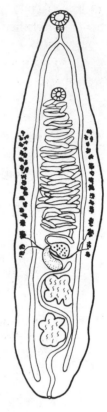

Figure 86. *Opisthorchis felineus.*

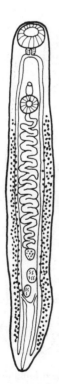

Figure 87. *Azygia acuminata.*

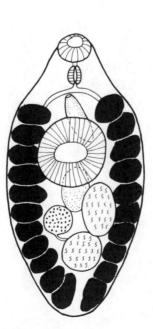

Figure 88. *Psilostoma marilae.*

Figure 89. *Philophthalmus lucipetus.*

30(22) Oral sucker with 6 anterior muscular processes; ventral sucker as large or larger than oral, slightly anterior to middle of body; vitellaria lateral, extend from pharynx to posterior end of body (Figure 90) .... (F) Bunoderidae

Oral sucker without such muscular processes 31

31(30) Vitellaria 2 small compact masses near middle of body; cirrus pouch absent (Figure 91)..... .... (F, A, R) Gorgoderidae

Vitellaria follicular; cirrus pouch present 32

32(31) Genital pore at level of oral sucker; long cirrus pouch extends back to ventral sucker (Figure 92) .... (A, R, B) Cephalogonimidae

Genital pore much posterior to oral sucker 33

33(32) Ceca very short or barely reaching beyond ventral sucker 34

Ceca long, extending well beyond ventral sucker or to posterior extremity of body 35

34(33) Minute, usually pyriform flukes; vitellaria behind ovary and ends of ceca; seminal receptacle absent; testes postacetabular (Figure 93) .... (B, M) Microphallidae

Small flukes, neither very compact nor extended; vitellaria at or near level of ventral sucker or prececal; seminal receptacle present; testes postacetabular to level of esophagus (Figure 94).. .... (A, R, B, M) Lecithodendriidae

35(33) Testes and ovary near ventral sucker; cirrus pouch horizontal, genital pore on left lateral margin of body between suckers; uterus surrounds ventral sucker; ceca extend to posterior extremity of body (Figure 95) .... (B) Stomylotrematidae

Testes and ovary all postacetabular; cirrus pouch vertical, genital pore median; descending and ascending limbs of uterus pass between testes (Figure 96) ... (A, R, B, M) Plagiorchiidae

Figure 90. *Bunodera leucopercae.*

Figure 91. *Gorgoderina attenuata.*

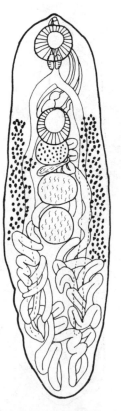

Figure 92. *Cephalogonimus amphiumae.*

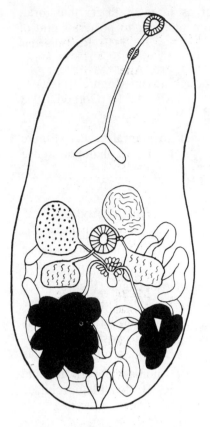

Figure 93. *Microphallus opaca.*

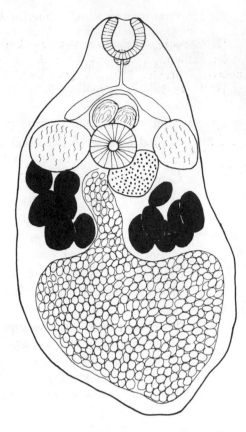

Figure 94. *Lecithodendrium breckenridgei.*

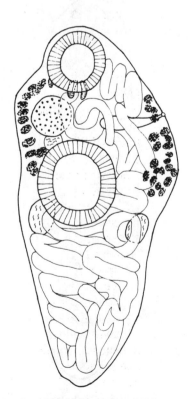

Figure 95. *Laterotrema americana.*

Figure 96. *Plagiorchis muris.*

CLASSIFICATION

The classification of the subclass Digenea is that proposed by LaRue (1957). It is based on life cycles and anatomy, especially the development of the excretory bladder and associated structures. The subclass is divided into 2 superorders: the Epitheliocystidia (orders 1, 2) and the Anepitheliocystidia (orders 3, 4, 5).

Subclass	Order	Suborder	Superfamily	Family
MONOGENEA	Monopisthocotylea		Acanthocotyloidea	Acanthocotylidae
			Tetraoncoidea	Tetraoncidae
			Gyrodactyloidea	Gyrodactylidae
			Dactylogyroidea	Dactylogyridae
				Calceostomatidae
			Udonelloidea	Udonellidae
			Capsaloidea	Capsalidae
				Loimoidae
				Microbothriidae
				Monocotylidae
	Polyopisthocotylea		Megaloncoidea	Megaloncidae
			Avielloidea	Aviellidae
			Polystomatoidea	Polystomatidae
				Hexabothriidae
				Sphyranuridae
			Chimaericoloidea	Chimaericolidae
			Microcotyloidea	Microcotylidae
				Axinidae
			Diplozooidea	Diplozooidae
			Diclidophoroidea	Diclidophoridae
				Dactylocotylidae
				Mazocraeidae
				Discocotylidae
ASPIDOGASTREA				Aspidogasteridae
DIGENEA	1. Plagiorchiida	Plagiorchiata	Plagiorchioidea	Plagiorchiidae
				Dicrocoeliidae
				Prosthogonimidae
				Microphallidae
				Eucotylidae
				Lecithodendriidae
			Allocreadioidea	Allocreadiidae
				Gorgoderidae
				Troglotrematidae
	2. Opisthorchiida	Opisthorchiata	Opisthorchioidea	Opisthorchiidae
				Heterophyidae
		Hemiurata	Hemiuroidea	Hemiuridae
				Halipegidae
	3. Strigeatida	Strigeata	Strigeoidea	Strigeidae
				Diplostomatidae
				Cyathocotylidae
			Clinostomatoidea	Clinostomatidae
			Schistosomatoidea	Schistosomatidae
				Sanguinicolidae
				Spirorchiidae
		Azygiata	Azygioidea	Azygiidae

Subclass	Order	Suborder	Superfamily	Family
		Cyclocoelata	Cyclocoeloidea	Cyclocoelidae
		Brachylaemata	Brachylaemoidea	Brachylaemidae
			Fellodistomatoidea	Fellodistomatidae
			Bucephaloidea	Bucephalidae
	4. Echinostomida	Echinostomata	Echinostomatoidea	Echinostomatidae
				Cathaemasiidae
				Philophthalmidae
				Psilostomatidae
				Fasciolidae
		Paramphistomata	Paramphistomatoidea	Paramphistomatidae
			Notocotyloidea	Notocotylidae
	5. Renicolida	Renicolata	Renicoloidea	Renicolidae

REFERENCES

Dawes, B. 1946. The Trematoda. University Press, Cambridge, 644 p.

Schell, S.C. 1970. How to Know the Trematodes. Wm. C. Brown Company Publishers, Dubuque, Iowa, 355 p.

Smyth, J.D. 1966. The Physiology of Trematodes. Oliver & Boyd, Edinburgh, 256 p.

Stunkard, H.W. 1963. Systematics, taxonomy, and nomenclature of the Trematoda. Quart. Rev. Biol. 38: 221-233.

Monogenea and Aspidogastrea

Bychowsky, B.E. 1957. Monogenetic Trematodes. Their Systematics and Phylogeny. Akad. Nauk SSSR. (W.J. Hargis, Jr., ed. 1961. Amer. Inst. Biol. Sci., Washington, 627 p.).

Kearn, G.C. 1971. The Physiology and Behaviour of the Monogenean Skin Parasite *Entobdella soleae* in Relation to its Host (*Solea solea*). *In* Ecology and Physiology of Parasites, ed. A.M. Fallis. University of Toronto Press, Toronto, pp. 161-187.

Llewellyn, J. 1963 & 1968. Larvae and Larval Development of Monogeneans. *In* Advances in Parasitology, ed. B. Dawes. Academic Press, New York, vol. 1, pp. 287-326; vol. 6, pp. 373-383.

_____. 1970. Monogenea. J. Parasitol. 56 (4, sect. 2, pt. 3): 493-504.

Rohde, K. 1972. The Aspidogastrea, Especially *Multicotyle purvisi* Dawes, 1941. *In* Advances in Parasitology, ed. B. Dawes. Academic Press, New York, vol. 10, pp. 77-151.

Yamaguti, S. 1963. Systema Helminthum. Monogenea and Aspidocotylea, vol. 4. Interscience Publishers Inc., New York, 699 p.

Capsalidae

Jahn, T.L., and L.R. Kuhn. 1932. The life history of *Epibdella melleni* MacCallum, 1927, a monogenetic trematode parasitic on marine fishes. Biol. Bull. 62: 89-111.

Polystomatidae

Paul, A.A. 1938. Life history studies on North American freshwater polystomes. J. Parasitol. 24: 489-510.

Sphyranuridae

Alvey, C.H. 1936. The morphology and development of the monogenetic trematode *Sphyranura oligorchis* (Alvey, 1933) and the description of *Sphyranura polyorchis* n. sp. Parasitology 28: 229-253.

Aspidogasteridae

Williams, C.O. 1942. Observations on the life history and taxonomic relationships of the trematode *Aspidogaster conchicola*. J. Parasitol. 28: 467-475.

Digenea

LaRue, G.R. 1957. The classification of digenetic Trematoda: A review and a new system. Exptl. Parasitol. 6: 306-349.

Pearson, J.C. 1972. A Phylogeny of Life-cycle Patterns of the Digenea. *In* Advances in Parasitology, ed. B. Dawes. Academic Press, New York, vol. 10, pp. 153-189.

Skrjabin, K.I., et al. 1947-1962. Keys to the Trematodes of Animals and Man. Akad. Nauk SSSR. (H.P. Arai, ed.; 1964. Univ. Illinois Press, Urbana, 351 p.).

_____. 1960. Trematodes of Animals and Man. Essentials of Trematodology. Adad. Nauk SSSR. (Transl. Z.S. Cole, ed.; 1964. U.S. Dept. Comm., Springfield, Va., vol. 17, 444 p.; 1965. Ibid. vol. 18, 532 p.).

Opisthorchiidae

Cameron, T.W.M. 1944. The morphology, taxonomy, and life history of *Metorchis conjunctus* (Cobbold, 1860). Canad. J. Res. Sect. D, 22: 6-16.

Komiya, Y. 1966. *Clonorchis* and Clonorchiasis. *In* Advances in Parasitology, ed. B. Dawes. Academic Press, New York, vol. 4, pp. 53-106.

Sun, T., S.T. Chou, and J.B. Gibson. 1968. Route of the entry of *Clonorchis sinensis* to the mamalian liver. Exptl. Parasitol. 22: 346-351.

Prosthogonimidae

Macy, R.W. 1934. Studies on the taxonomy, morphology, and biology of *Prosthogonimus macrorchis* Macy, a common oviduct fluke of domestic fowls in North America. Univ. Minnesota Agr. Exp. Sta. Tech. Bull. 98, 71 p.

Dicrocoeliidae

Denton, J.F. 1944. Studies on the life history of *Eurytrema procyonis*. J. Parasitol. 30:277-286.

_____. 1945. Studies on the life history of *Brachylecithum americanum* n. sp., a liver fluke of passerine birds. J. Parasitol. 31:131-141.

Kingston, N. 1965. On the life cycle of *Brachylecithum orfi* Kingston and Freeman, 1959 (Trematoda: Dicrocoeliidae), from the liver of the ruffed grouse, *Bonasa umbellus* L. Infections in the vertebrate and molluscan hosts. Canad. J. Zool. 43:745-764.

_____. 1965. On the morphology and life cycle of *Tanaisia zarudnyi* (Skrjabin, 1924) Byrd and Denton, 1950, from the ruffed grouse, *Bonasa umbellus* L. Canad. J. Zool. 43:953-969.

Krull, W.H., and C.R. Mapes. 1951-1953. Studies on the biology of *Dicrocoelium dendriticum* (Rudolphi, 1819) Looss, 1899 (Trematoda: Dicrocoeliidae), including its relation to the intermediate host, *Cionella lubrica* (Müller). I-IX. Cornell Vet. 41: 382-432, 433-444, 42: 253-276, 277-285, 339-351, 464-489, 603-604; 43: 199-202, 389-410.

Notocotylidae

Herber, E.C. 1942. Life history studies on two trematodes of the subfamily Notocotylinae. J. Parasitol. 28: 179-196.

_____. 1955. Life history studies on *Notocotylus urbanensis* (Trematoda: Notocotylinae). Proc. Penn. Acad. Sci. 29: 267-275.

Plagiorchiidae

Krull, W.H. 1931. Life history studies on two frog lung flukes, *Pneumonoeces medioplexus* and *Pneumobites parviplexus*. Trans. Amer. Micro. Soc. 50: 215-277.

Echinostomatidae

Beaver, P.C. 1937. Experimental studies on *Echinostoma revolutum* (Froelich), a fluke from birds and mammals. Illinois Biol. Monog. 15: 1-96.

Cort, W.W., D.J. Ameel, and A. Van der Woude. 1948. Studies on germinal development in rediae of the trematode order Fasciolatoidea Szidat, 1936. J. Parasitol. 34: 428-451.

Schistosomatidae

Jordan, P., and G. Webbe. 1970. Human Schistosomiasis. Charles C Thomas, Publishers, Springfield, Ill., 212 p.

Price, Helen. 1931. Life history of *Schistosomatium douthitti* (Cort). Amer. J. Hyg. 13: 685-727.

Short, R.B. 1952. Sex studies on *Schistosomatium douthitti* (Cort, 1914) Price, 1931 (Trematoda: Schistosomatidae). Amer. Midl. Nat. 47: 1-54.

Thomas, J.D. 1973. Schistosomiasis and the Control of Molluscan Hosts of Human Schistosomes with Particular Reference to Possible Self-regulatory Mechanisms. *In* Advances in Parasitology, ed. B. Dawes, Academic Press, New York, vol. 11, pp. 307-394.

Spirorchiidae

Wall, L.D. 1941. *Spirorchis parvus* (Stunkard), its life history and the development of its excretory system (Trematoda: Spirorchiidae). Trans. Amer. Micr. Soc. 60: 221-260.

Fasciolidae

Dawes, B., and D.L. Hughes. 1964 & 1970. Fascioliasis: the Invasive Stages of *Fasciola hepatica* in Mammalian Hosts. *In* Advances in Parasitology, ed. B. Dawes. Academic Press, New York, vol. 2, pp. 97-168; vol. 8, pp. 259-274.

Kendall, S.B. 1965 & 1970. Relationships Between the Species of *Fasciola* and their Molluscan Hosts. *In* Advances in Parasitology, ed. B. Dawes. Academic Press, New York, vol. 3, pp. 59-98; vol. 8, pp. 251-258.

Pantelouris, E.M. 1965. The Common Liver Fluke *Fasciola hepatica* Pergamon Press, New York, 259 p.

Taylor, E.L. 1964. Fascioliasis and the Liver Fluke. F.A.O. Agric. Studies, Rome, No. 64, 234 p.

Clinostomatidae

Cort, W.W., D.J. Ameel, and A. Van der Woude. 1950. Germinal material in the rediae of *Clinostomum marginatum* (Rudolphi). J. Parasitol. 36: 157-163.

Hunter, G.W., III, and W.S. Hunter. 1935. Further studies on fish and bird parasites. *In* Biological Survey of the Mohawk-Hudson Watershed, Suppl. 24th Ann. Rept., 1934, New York State Conserv. Dept., Albany, pp. 267-283.

Paramphistomatidae

Krull, W.H., and H.F. Price. 1932. Studies on the life history of *Diplodiscus temperatus* Stafford from the frog. Occ. Papers Mus. Zool, Univ. Michigan, No. 237, 41 p.

Van der Woude, Anne. 1954. Germ cell cycle of *Megalodiscus temperatus* (Stafford, 1905) Harwood, 1932 (Paramphistomidae: Trematoda). Amer. Midl. Nat. 51: 172-202.

Halipegidae

Macy, R.W., W.A. Cook, and W.R. DeMott. 1960. Studies on the life cycle of *Halipegus occidualis* Stafford, 1905 (Trematoda: Hemiuridae). Northwest Sci. 34: 1-17.

Troglotrematidae

Ameel, D.J. 1934. *Paragonimus*, its life history and distribution in North America and its taxonomy (Trematoda: Troglotrematidae). Amer. J. Hyg. 19: 279-317.

Bennington, E., and I. Pratt. 1960. The life history of the salmon-poisoning fluke, *Nanophyetus salmincola* (Chapin). J. Parasitol. 46: 91-100.

Millemann, R.E., and S.E. Knapp. 1970. Biology of *Nanophyetus salmincola* and "Salmon Poisoning" Disease. *In* Advances in Parasitology, ed. B. Dawes. Academic Press, New York, vol. 8, pp. 1-41.

Philip, C.B. 1955. There's always something new under the "parasitological" sun (the unique story of helminth-borne salmon poisoning disease). J. Parasitol. 41: 125-148.

Yokogawa, M. 1965 & 1969. *Paragonimus* and Paragonimiasis. *In* Advances in Parasitology, ed. B. Dawes. Academic Press, New York, vol. 3, pp. 99-158; vol. 7, pp. 375-387.

Allocreadiidae

Awachie, J.B.E. 1968. On the bionomics of *Crepidostomum metoecus* (Braun, 1900) and *Crepidostomum farionis* (Müller, 1784) (Trematoda: Allocreadiidae). Parasitology 58: 307-324.

Choquette, L.P.E. 1954. A note on the intermediate hosts of the trematode, *Crepidostomum cooperi* Hopkins, 1931, parasitic in speckled trout, *Salvelinus fontinalis* (Mitchill), in some lakes and rivers of the Quebec Laurentide Park. Canad. J. Zool. 32: 375-377.

Heterophyidae

Rothschild, Miriam. 1939. A note on the life cycle of *Cryptocotyle lingua* (Creplin) 1825 (Trematoda). Nov. Zool. 41: 178-180.

Stunkard, H.W. 1930. The life history of *Cryptocotyle lingua* (Creplin), with notes on the physiology of the metacercariae. J. Morph. Physiol. 50: 143-191.

Willey, C.H., and P.R. Gross. 1957. Pigmentation in the foot of *Littorina littorea* as a means of recognition of infection with trematode larvae. J. Parasitol. 43: 324-327.

Strigeidae

Campbell, R.A. 1973. Studies on the biology of the life cycle of *Cotylurus flabelliformis* (Trematoda: Strigeidae). Trans. Amer. Micr. Soc. 92: 629-640.

Hoffman, G.L. 1960. Synopsis of Strigeoidea (Trematoda) of fishes and their life cycles. Bur. Sport Fish. Wildlife, Fishery Bull. 175, 60: 437-469.

Ulmer, M.J. 1957. Notes on the development of *Cotylurus flabelliformis* tetracotyles in the second intermediate host (Trematoda: Strigeidae). Trans. Amer. Micr. Soc. 76: 321-327.

Van Haitsma, J.P. 1931. Studies on the trematode family Strigeidae (Holostomidae) No. XXII: *Cotylurus flabelliformis* (Faust) and its life-history. Papers Mich. Acad. Sci., Arts Letters 13: 447-482.

Diplostomatidae

Hoffman, G.L. 1956. The life cycle of *Crassiphiala bulboglossa* (Trematoda: Strigeida). Development of the metacercaria and cyst, and effect on the fish hosts. J. Parasitol. 42: 435-444.

_____, and J.B. Hundley. 1957. The life-cycle of *Diplostomum baeri eucaliae* n. subsp. (Trematoda: Strigeida). J. Parasitol. 43: 613-627.

Hunter, G.W., III. 1933. The strigeid trematode, *Crassiphiala ambloplitis* (Hughes, 1927). Parasitology 25: 510-517.

_____, and W.S. Hunter. 1935. Further studies on fish and bird parasites. *In* Biological Survey of the Mohawk-Hudson Watershed, Suppl. 24th Ann. Rept., 1934, New York State Conserv. Dept., Albany, pp. 267-283.

Rees, Gwendolen. 1957. *Cercaria diplostomi phoxini* (Faust), a furcocercaria which develops into *Diplostomulum phoxini* in the brain of the minnow. Parasitology 47: 126-137.

Class Cestoidea

The class includes the tapeworms, occurring as adults in the alimentary tract and bile ducts, occasionally in the celom, of vertebrates; a few species mature in freshwater oligochaetes. The metacestode stages, the phases of development between oncosphere and sexuality, occur in a variety of invertebrate and vertebrate hosts. Eggs or gravid proglottids are passed in the feces of the final host. The eggs of some species reach water and continue development there; others may be deposited on land, giving the intermediate host access to the embryonated eggs. A few species do not need an intermediate host but are able to complete the life cycle in a single host. A digestive tract, or any trace of it, is lacking.

Typically, an adult tapeworm consists of an attachment organ, the scolex; a zone of proliferation called the neck; and a strobila, with the formation of proglottids developing asexually by budding from the neck. The proglottids just behind the neck are at first indistinct and have no differentiation of internal organs. As they are pushed backward, organogenesis occurs, so that a single strobila shows a complete developmental series of immature, mature, and gravid proglottids.

In a few primitive taxa, the body consists of but a single unit and resembles a trematode in appearance. Members of these taxa are monozoic, since they do not form chains of proglottids, in contrast to the more typical polyzoic species.

With few exceptions, the cestodes are monoecious and somewhat protandric. The gonads are usually situated in the medullary zone of the proglottid, i.e., between the sheets of muscle fibers. While there are many exceptions, the testes tend to be dorsal and anterior; the ovary and Mehlis' gland, ventral and posterior. The vitellaria may be follicular and widely dispersed, or compact and located behind the bilobed ovary. Genital pores are either lateral or ventral in position, generally the former.

The Cestoidea are arranged in 2 subclasses: Cestodaria, with 2 orders and a single family in each; and Cestoda, with 4 to 11 orders, depending upon which authority one follows. The common species of Cestoda are, however, distributed among 4 orders: the Caryophyllidea, from fish and occasionally aquatic oligochaetes; Pseudophyllidea, from all classes of vertebrates; Proteocephalidea, chiefly from fish; and Cyclophyllidea, mainly birds and mammals. Only the Pseudophyllidea and Cyclophyllidea contain species which attack man and domesticated animals. The other orders, except for the Tetraphyllidea, Lecanicephalidea and Trypanorhyncha, which parasitize elasmobranchs, contain only one or a few families, are of uncertain taxonomic status, and are unlikely to be encountered.

Subclass Cestodaria

Adult cestodarians resemble trematodes, except for the lack of digestive organs. There is no scolex or chain of proglottids, and there is only one set of reproductive organs, the cirrus, vagina, and uterine pore opening independently (Figure 97). These monozoic tapeworms differ from the Cestoda in that their oncospheres have 10 rather than 6 hooks and, accordingly, are known as decacanths. They are parasitic in the celom and intestine of lower fishes. Some examples with their hosts include: *Gyrocotyle urna* from ratfish *(Hydrolagus colliei)* and *Amphilina bipunctata* from sturgeon.

Subclass Cestoda

These are the polyzoic tapeworms which have a scolex followed by a succession of proglottids (except in the order Caryophyllidea, which some authors consider a family of the Pseudophyllidea), each containing a set of male and female reproductive organs. An oncosphere with 3 pairs of hooks (hexacanths) hatches from the egg. A wide variety of both invertebrates and vertebrates act as intermediate hosts.

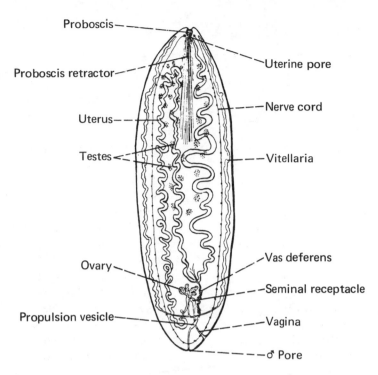

Proboscis

Proboscis retractor

Uterus

Testes

Ovary

Propulsion vesicle

Uterine pore

Nerve cord

Vitellaria

Vas deferens

Seminal receptacle

Vagina

♂ Pore

Figure 97. *Amphilina foliacea,* showing internal anatomy.

Order Cyclophyllidea

Scolex with 4 deeply cupped muscular suckers and usually a rostellum, which may be armed or unarmed; proglottids in all stages of development, gravid ones only near posterior end of strobila; uterus varies in form, but does not open through a special pore; genital pores usually lateral; vitellaria a single or bilobed postovarian mass. The life cycle as a rule involves only 1 intermediate host, usually an arthropod or vertebrate. Adults parasitic in amphibians (though rarely), reptiles, birds, and mammals.

Family Taeniidae

The family includes the majority of the more important adult tapeworms of man and other mammals. This combined with the occurrence of the metacestode stages also in mammals, including man, makes them unrivaled in medical and economic importance. The metacestode of *Taenia* is a cysticercus, of *Hydatigera* a strobilocercus, of *Multiceps* a coenurus, and of *Echinococcus* a hydatid (Figure 98). Eggs of the various species of Taeniidae with their thick, striated shells and hexacanths are characteristic of the family and indistinguishable.

Authorities are not in agreement concerning the number of genera in the family. Some authors place primary importance on the type of metacestodes as the basis for establishing the genera. These workers generally recognize the 4 genera listed above. Other workers maintain that the type of metacestodes is of secondary importance and give primary weight to the similar anatomy of the adult worms and consider *Taenia* as the only valid genus. But they are split in opinion among those who place primary importance on the scolex and hooks as the significant anatomical characters for classification and those who would use only the proglottids. The establishment of a single genus *Taenia* for the group where the adults and their eggs are similar has merit, but so has a system based on the 4 different kinds of metacestode forms, plus other anatomical differences. We shall follow the system that includes the genera *Taenia, Hydatigera, Multiceps,* and *Echinococcus,* each with its separate type of metacestode.

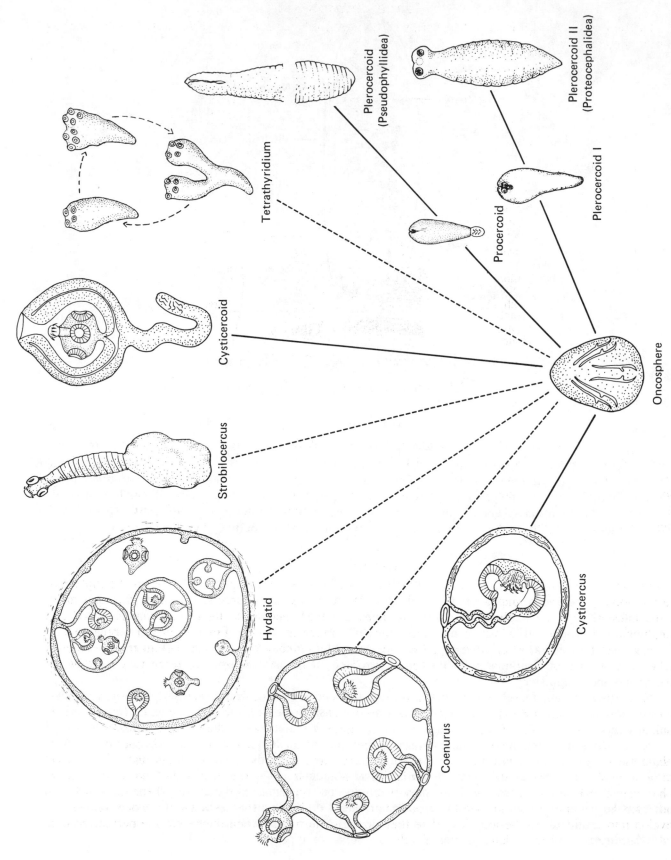

Figure 98. *Infective tapeworm metacestodes:* The basic types are indicated by solid lines.

Plerocercoid
(Pseudophyllidea)

Plerocercoid II
(Proteocephalidea)

Plerocercoid I

Procercoid

Tetrathyridium

Cysticercoid

Oncosphere

Strobilocercus

Cysticercus

Hydatid

Coenurus

108

Taenia pisiformis

This is one of the commonest tapeworms of dogs and related wild carnivores, but it seldom occurs in cats (Figures 99, 100).

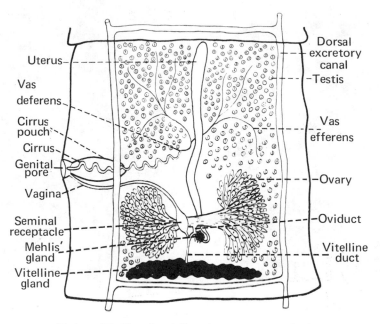

Figure 99. *Taenia pisiformis,* mature proglottid.

Description

Mature worms up to 100 cm long and consisting of about 400 proglottids. Rostellum large and powerful, and armed with a double crown of 34 to 48 hooks of 2 sizes. The longer hooks are 200 to 269 μ (mean 240 μ) and the smaller ones are 114 to 172 μ (mean 140μ) long. Vitellaria often triangular with broad apex extending into interovarian field; do not contact ovary. The vagina bends broadly around distal end of cirrus sac before looping tortuously and entering genital atrium. The gravid uterus has 11 to 15 lateral branches on each side.

Life Cycle

The gravid proglottids become detached and pass out with the feces of the host. With the rupture of the proglottids, the embryonated eggs are released. When ingested by rabbits and hares, the eggs hatch in the intestine. The oncospheres penetrate the intestinal wall, and make their way, via the blood and lymph channels, to the liver, where they grow rapidly, and in about a month the cysticerci break out of the liver into the celom. After remaining free for a short time, they become attached to the mesenteries and enclosed in an adventitious cyst produced by the host. Dogs become infected when they swallow viable cysticerci.

Man serves as final host for the beef tapeworm, *Taenia saginata* (*Taeniarhynchus saginatus* of some authors), and the pork tapeworm, *Taenia solium* (Figure 115). In addition to occurring in different intermediate hosts, specific identification can be made on either the scoleces or gravid proglottids. The scolex of *T. saginata* is without hooks, and the gravid uterus has 15 to 20 lateral branches on each side; in *T. solium* the scolex is armed with 22 to 32 rostellar hooks, and has 7 to 13 lateral branches off the gravid uterus (Figure 101). The cysticerci of *T. solium,* but not those of *T. saginata,* may also occur in man. When man accidentally ingests the eggs, they hatch and the oncospheres bore through the intestinal wall and enter the general circulation via the hepatic portal vein. The oncospheres leave the blood vessels and develop into cysticerci in the muscles. Here they do little harm; but when they enter the nervous system and sense organs, serious damage may result.

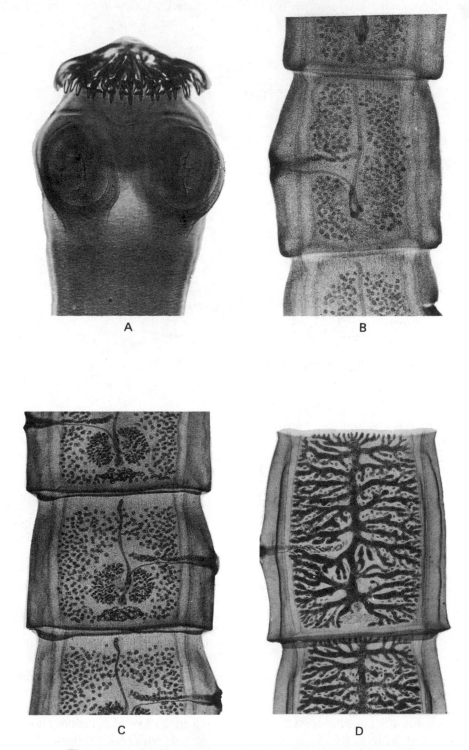

Figure 100. *Taenia pisiformis:* (A) Scolex and neck region; (B) Immature region with male organs developed, female organs beginning to appear; (C) Mature region showing both sets of sex organs; (D) Gravid region, little more than a sac containing the much enlarged egg-filled uterus.

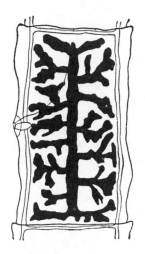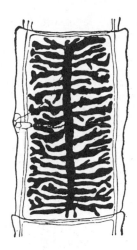

Figure 101. Gravid proglottids of human taenias, showing lateral arms of central uterine stem. *Taenia solium*, left; *T. saginata*, right.

Hydatigera taeniaeformis

This species is of interest because of its metacestode, a strobilocercus (Figure 98).

Description

Mature worms up to 60 cm long. The rostellum is short and armed with a double crown of 26 to 52 hooks of 2 sizes. The larger hooks are 294-429 μ (mean 380 μ) and the smaller ones 215-287 μ (mean 245 μ) long. The neck is very short. Vitelline gland is slightly elongate in transverse plane; does not contact ovary. The vagina is straight, without any loops or bends. The gravid uterus has 5 to 11 lateral branches on each side.

Life Cycle

Adults occur in the intestine of cats and related carnivores. Eggs pass out and when ingested by rodent intermediate hosts, chiefly rats and mice, the oncospheres escape from the embryophores and make their way to the liver, where they develop into a strobilocerci. In the liver, strobilocerci evaginate prematurely, resulting in a bladderworm with a single scolex followed by a series of immature proglottids, and terminated by a small, vesicular bladder. When a liver with the viable strobilocerci is ingested by a cat or other suitable host, the bladder and some of the pseudostrobila are digested off and new proglottids are formed from the neck.

Echinococcus granulosus

This is the smallest of adult Taeniidae tapeworms; consisting of a scolex and neck followed by usually 4 but occasionally 3 or 5 proglottids. The small intestine of the final host, usually domestic dogs, may be "furred" with thousands of worms. The danger to man from an infection is not with the adult of this tapeworm, but with the larval hydatid cyst which is usually in the liver. This is one of the most important human tapeworms throughout the world, certainly the most dangerous. A few human cases of unilocular hydatid disease are diagnosed each year from the United States and Canada.

Description

Worms measure about 5 mm. There may be 2 immature proglottids, a mature and a gravid proglottid; or there may be only 1 immature proglottid and a gravid proglottid, and 2 proglottids in different stages of maturity. The rostellum is armed with a double crown of 30 to 40 hooks of 2 sizes. The larger hooks are 22 to 30 μ long and the smaller ones are 18 to 22 μ. There are about 40 to 60 testes to the proglottid.

Life Cycle

The adult worms occur in the intestine of dogs and related carnivores. The hydatid larval stage occurs in man, sheep, deer, moose, and other large herbivorous mammals (Figures 98, 102). The eggs passed by dogs hatch when they are ingested by a suitable intermediate host. As in the other Taeniidae, the oncospheres enter the hepatic portal vein, whereupon they are carried in the blood stream until becoming lodged in some capillary filter. The first filter is usually the liver, where the greatest number of hydatids occurs; the next filter is the lungs, where a smaller number is found, while still fewer reach more distant loci.

Fibrous tissue develops around the hydatid and grades off into normal tissue cells, which may already be undergoing pressure atrophy, due to the steady increase in the size of the cyst. The mother hydatid produces numerous scoleces and single-layered brood capsules by budding of the germinal layer lining the inside. These are attached at first by a stalk but many break loose and are free in the enclosed hydatid fluid. Brood capsules in turn produce scoleces in a manner similar to that which takes place in the mother hydatid. Thus many thousands of scoleces are formed.

Anatomical and biological data indicate that *Echinococcus granulosus* is a polytypic species with 4 subspecies: *E. g. granulosus*, the adults of which occur in dogs and the hydatid cysts primarily in sheep but occasionally other domesticated animals; *E. g. borealis*, with adults in wolves *(Canis lupus)*, coyotes *(C. latrans)* and dogs, and the hydatids in Cervidae, particularly moose *(Alces alces); E. g. canadensis*, in dogs with hydatids in reindeer *(Rangifer tarandus);* and *E. g. equinus*, in dogs (experimentally) with hydatids in horses.

Echinococcus multilocularis, which in the past has often been confused with *E. granulosus*, has a multilocular or alveolar type of hydatid cyst and utilizes a wider range of final hosts (dogs, foxes, and cats) than does the other species. While ungulates generally serve as intermediate hosts for *E. granulosus*, field mice and related rodents serve for *E. multilocularis*. The latter species, widely distributed in Europe and Eurasia, is known in North America from Canada and Alaska, and several north central states of the contiguous United States, where it is maintained sylvatically by foxes and wild mice (see Leiby et al., 1970).

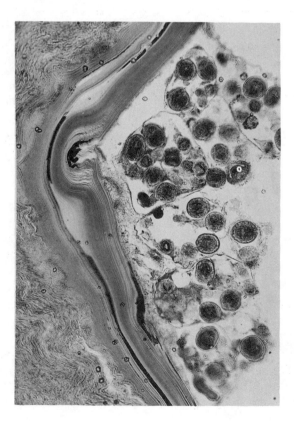

Figure 102. A photomicrograph of a section of unilocular hydatid cyst of *Echinococcus granulosus,* showing: normal to adventitious tissue cells, connective tissue capsule formed by host reaction, external cuticular membrane, internal germinal layer within which are several brood capsules each with scoleces.

Multiceps serialis

Multiceps, like *Echinococcus*, is characterized by polyembryony; this feature may account for the presence of several closely related taxa commonly designated as species, but possibly deserving no more than designation of varieties.

Description

Mature worms up to 70 cm long. Rostellum is weakly developed, and armed with a double crown of 26 to 32 hooks of 2 sizes. The larger hooks are 113-157 μ (mean 139 μ) and the smaller ones are 67-112 μ (mean 91 μ) long. Vitelline gland elliptical; anterior border concave and usually in contact with the ovary. The vagina with or without a reflexed loop in the region of the longitudinal excretory canals. The gravid uterus has 13 to 18 lateral branches on each side.

Life Cycle

Eggs produced by the adult tapeworm in the intestine of a dog or other suitable Canidae pass out and are ingested by a lagomorph (species of *Lepus* and *Sylvilagus*) with contaminated food. Once the oncosphere escapes, it bores into the wall of the intestine and reaches its place of lodgement through the blood vessels or lymph spaces. In the connective tissue under the skin, the intermuscular fascia of the legs, and elsewhere, it comes to rest and develops into a coenurus, with numerous scoleces attached and also with the production of internal and external daughter bladders which in turn develop numerous scoleces (Figure 98). Dogs become infected through eating portions of lagomorph hosts containing the viable coenuri.

Multiceps multiceps occurs in Canidae, especially dogs and foxes, and has a wide distribution. The coenuri develop in the brain and spinal cord of ungulates, especially sheep, and often results in a disease known as "gid" or "staggers." More than 20 human cases of coenurus infection have been reported. On ingestion of the coenurus by a dog, some or all of the scoleces develop into adults. This species is virtually unknown in the United States now.

Taeniidae Life Cycle Exercise

Life cycle exercises may be started with eggs or metacestodes. Adults of *Taenia pisiformis* and *Multiceps serialis* from dogs, and *Hydatigera taeniaeformis* from cats are readily available from veterinary hospitals where the pets are treated for the removal of the tapeworms. Eggs dissected from the gravid proglottids should be fed to the appropriate intermediate host, i.e., eggs of *T. pisiformis* and *M. serialis* to rabbits, and those of *Hydatigera taeniaeformis* to rats or mice. Examination of the experimental intermediate hosts over spaced intervals will show the migration through the circulation and muscles and development of the metacestodes. About a week after ingesting eggs, the metacestodes appear as pearly bodies in the liver and other parts of the body, depending on the species; they are infective to the final host about 60 days postinfection. Dogs and cats may be infected by feeding them mature metacestodes.

As a likely source of cysticerci the following generally available hosts are suggested: rabbits (*Sylvilagus* spp.) and hares (*Lepus* spp.) for cysticerci , the metacestode stage of a dog tapeworm *Taenia pisiformis*; and rats (*Rattus* spp.) for *C. fasciolaris* (as a strobilocercus), the metacestode stage of a cat tapeworm *Hydatigera taeniaeformis*.

Whether starting with eggs or metacestodes, care should be taken to see that the experimental hosts are clean. If eggs or proglottids are in the feces, the intended final hosts should be treated with an anthelmintic before feeding them metacestodes. Only laboratory reared and unexposed hosts should be used in egg feeding experiments.

Family Dilepididae

Dipylidium caninum

The best known member of this family is the widely distributed *D. caninum*, the double-pored tapeworm of dogs, cats, and occasionally man, especially children. Children probably become infected when their mouths are licked by a dog just after the dog has "nipped" a flea. Infections in dogs can be recognized macroscopically by the occurrence on the feces of gravid proglottids, resembling cucumber seeds, which when freshly voided are active. Diagnosis is usually made on the evacuated proglottids, since free eggs are not ordinarily found in the feces of infected animals (Figure 103).

Description

Adults attain a length of 50 cm and a width of 3 mm. The scolex has 4 suckers and a retractable rostellum, armed with 4 to 7 rows of hooks. There are 2 complete sets of reproductive organs, each opening in a common genital pore on the lateral margins of the proglottid. The numerous testes fill the space between the excretory canals. The bilobate ovaries and vitellaria, posterior to the ovary, are in separate clusters. The uterus breaks up into egg-capsules, each containing 15 to 20 eggs, which remain intact after the proglottids disintegrate. The eggs are described under Helminth Eggs, p. 210.

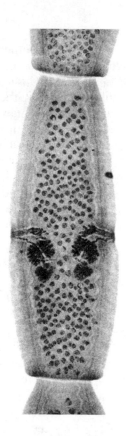

Figure 103. *Dipylidium caninum.*

Life Cycle

Gravid proglottids voided with the feces, or leaving the host spontaneously, disseminate the egg-capsules. The intermediate hosts are the dog flea, *Ctenocephalides canis*, cat flea, *C. felis*, and the human flea, *Pulex irritans*, as well as the dog biting louse, *Trichodectes canis*. When the eggs are eaten by flea larvae, they hatch in the intestine and the oncospheres bore through into the hemocel. Adult fleas have elongated mouthparts, adapted for piercing the skin and sucking blood, so they are incapable of taking in the eggs. Oncospheres develop little in larval fleas; considerable growth occurs in the pupal stage, and development of the cysticercoid is completed in adult fleas when they begin taking blood meals. In biting lice, oncospheres develop quickly to the infective cysticercoids (Figure 98). The final host acquires the infection by swallowing infected adult fleas or biting lice. Prepatency is about a month.

Life Cycle Exercise

Flea eggs can be obtained by placing a pan beneath a caged infested dog or cat. Eggs placed in petri dishes with moist sand at room temperature hatch within a few days. Feed the larvae on dried blood or dog biscuits. Larvae intended for infection should be isolated, and not allowed to eat for about a day. Tease gravid proglottids apart in physiological solution to release the egg-capsules. Add sufficient powdered blood to make a paste and feed it to hungry larvae. After feeding, the larvae must be kept on moist sand until they become adults. Larvae, pupae, and adult fleas should be dissected, in a small amount of physiological solution, to observe the development of the cysticercoids. If final hosts are to be fed experimentally infected fleas, they should be given an anthelmintic first to remove any tapeworms that may already be present.

Another commonly available Dilepididae is *Choanotaenia infundibulum* of chickens. Eggs develop into cysticercoids when fed to houseflies *(Musca domestica)*, grasshoppers *(Melanoplus differentialis)*, and a wide variety of beetles. For life cycle details, consult Horsfall and Jones (1937).

Other often available Dilepididae with their hosts include *Amoebotaenia cuneata*, chickens; and *Metroliasthes lucida*, turkeys.

Family Hymenolepididae

Hymenolepis nana

This species, commonly known as the dwarf tapeworm, is cosmopolitan in distribution and is common in rats and mice. It also occurs in man, especially children. Authors are not in agreement on the identity of the species of worms found in man and those found in rodents. Anatomically the rodent form, sometimes called *H. nana* var. *fraterna* or *H. fraterna*, is identical to the form from man. Their life cycles are also identical. The rodent form can infect man and vice versa, but development is best in the species of host from which the parent worms came. Similar differences occur in specimens obtained from wild rats and mice. The strain obtained from the former is equally infective for rats and mice, while the mouse strain is distinctly more infective for mice: it develops faster and to a larger size and produces a greater incidence and heavier worm burden in mice than in rats. It seems reasonable, therefore, to assume that the forms from mice, rats, and man are the same species, *H. nana*, which has developed different physiological strains.

Description

Adults measure up to 10 cm in length and less than 1 mm in width. The globular scolex has a retractile rostellum with a crown of 20 to 30 hooks. The strobila consists of about 200 proglottids which are considerably broader than long. The 3 testes are arranged in a transverse line and separated by the ovary so that 1 testis is poral and 2 aporal. The uterus develops as a sac which practically fills the proglottids between the excretory canals. The eggs are described under Helminth Eggs, p. 210.

Life Cycle

The dwarf tapeworm, *H. nana*, is peculiar among cestodes in that no intermediate host is required, and the eggs are infective as they pass out with the feces. When the eggs are ingested by a final host, the oncospheres become free in the intestine and burrow into the interior of the villi where they develop into cysticercoids. In about 4 days postinfection, these escape into the lumen of the intestine, attach to the mucosa, and mature in about 15 days.

This species may utilize an intermediate host. When the eggs are ingested by certain insects, such as flour beetles (species of *Tenebrio*) and flea larvae *(Ctenocephalides canis)*, they hatch in the intestine and the oncospheres migrate to the hemocel where they develop into cysticercoids. When the infected insects are eaten by susceptible hosts, the cysticercoids are released, evaginate, attach to the intestinal wall, and develop to maturity.

Life Cycle Exercise

Tapeworms for experimental purposes may be obtained from wild mice and rats. Feed gravid proglottids to flour beetles, from which food has been withheld for a day or so. At room temperature cysticercoids develop in 2 to 3 weeks. Examine the intestinal contents of beetles 4 to 5 hours postinfection for the presence of hatched oncospheres and again 1 or 2 days thereafter to observe the development of the cysticercoids in the hemocel. For further details consult Voge and Heyneman (1957).

Infect mice with cysticercoids from worms taken from mice, and likewise rats with those from rats. Also try cross-infection of rats with larvae originating from mice and vice versa to determine whether there are differences of susceptibility with the use of different physiological stains. Determine the prepatent period.

Also try the direct life cycle, by feeding gravid proglottids directly to mice. Examine the intestinal contents of 1 mouse 4 to 5 hours postinfection for the presence of hatched oncospheres. At daily intervals examine the intestinal mucosa to observe: (1) the penetration of the hexacanths into the villi and the development of the cysticercoids (Figure 98), (2) when the cysticercoids escape into the lumen of the intestine and attach to the mucosa, (3) the rate of development to adults. Determine prepatency. Serial sections are required for the first 2 steps.

Hymenolepis carioca

This, the so-called thread tapeworm, is among the commonest cestodes of chickens in North America, and cosmopolitan in distribution. When the eggs are eaten by several species of beetles, they hatch and develop into cysticercoids (Figure 98). The dung beetles (Scarabaeidae) belonging to *Aphodius, Choeridium, Geotrupes,* and *Onthophagus* are most frequently involved. Infection of the final hosts occurs when they eat beetles containing fully developed cysticercoids.

Other often available Hymenolepididae with their hosts include *Fimbriaria fasciolaris,* a fairly common and cosmopolitan parasite of domestic and wild anseriform birds; *H. diminuta,* rats and mice; and *H. evaginata,* muskrats (Figure 104).

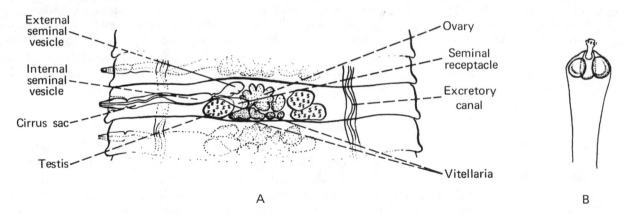

Figure 104. *Hymenolepis evaginata.* (A) Mature proglottid, (B) Scolex.

Family Anoplocephalidae

Moniezia expansa

One of the largest and best known members of the family is the double-pored ruminant tapeworm, *M. expansa,* which occurs commonly in sheep, goats, and cattle in most parts of the world. Lambs may become infected very early in life and pass gravid proglottids when they are 6 weeks old (Figure 114).

Description

Adults attain a length of 600 cm and a width of 1.5 cm. The scolex has 4 prominent suckers, but lacks a rostellum and hooks. Proglottids are wider than long and each contains 2 sets of reproductive organs with the genital pores situated on the lateral margins. The ovaries and the vitellaria, posterior to the ovary, are in separate clusters. The numerous testes are scattered throughout the proglottid or they may be concentrated toward the lateral fields. Situated along the posterior border of the proglottid is a row of interproglottidal glands, arranged around small pits. The eggs are subtriangular in shape and have a pyriform apparatus which contains the oncosphere.

Life Cycle

Segments of gravid proglottids voided with the feces disseminate the embryonated eggs. When the eggs are ingested by soil mites (in North America species of *Galumna, Oribatula, Scheloribates* et al.), they hatch in the intestine. The oncospheres bore through the intestine into the hemocel where they develop into cysticercoids within a short time. When infected mites are taken along with the grass by grazing herbivorous animals, the cysticercoids (Figure 98) are liberated in the intestine where they attach and mature. After becoming gravid, the worms apparently are lost after about 3 months in the host.

Life Cycle Exercise

Adult worms are available from abattoirs where sheep and lambs are slaughtered. Eggs teased from gravid proglottids in physiological solution should be transferred by means of a bulbed pipette onto filter paper and the latter allowed to dry. Place the dried paper with the eggs in a small glass container, add the appropriate mites and allow them to feed for a day or 2. Dissection of the mites in a very small amount of physiological solution under a dissecting microscope and examination of the intestinal contents shortly after infection will show freed oncospheres and empty egg shells. Examination of the hemocel of mites a day or 2 postinfection will reveal the developing cysticercoids.

The appropriate mites can be obtained from sod placed in a modified Berlese funnel with a 60-watt electric bulb over it. In response to the heat from the light, the mites move downward through the soil, pass through the funnel, and fall into the container below.

Since Stunkard (1938) first showed that mites were the intermediate hosts for *M. expansa*, other workers have found that they are also involved in the life cycle of other Anoplocephalidae. Some of the commonly available ones, whose life cycles can be examined in a similar manner, include *Cittotaenia perplexa* from wild rabbits and hares, and *Monoecocestus americanus*, and *M. variabilis* from porcupines.

Other often available Anoplocephalidae with their hosts include *Anoplocephala magna*, equines; *Andrya primordialis*, squirrels; *Mosgovoyia perplexa*, rabbits; and *Thysanosoma actinioides*, western sheep and goats. The life cycle of the last species is unknown, and its identity as an Anoplocephalidae uncertain.

Family Davaineidae

Davainea proglottina

This small, almost microscopic tapeworm occurs in the duodenal area of chickens, and occasionally other gallinaceous birds, throughout the world. It is seen best when the opened duodenum is floated in physiological solution, with the gravid proglottids appearing as raised bumps on the mucosa.

Description

Adults measure up to 4 mm and usually consist of 6 or 7 proglottids, exceptionally 9. The rostellum is armed with 86 to 94 small hooks, in 2 rows, and each of the suckers is armed with 4 to 5 rows of minute hooks along the margins. The genital pores alternate regularly and are situated far anterior on the lateral margin of the proglottids. Large spindle-shaped cirrus pouch extends over 2/3 the width of proglottid. The cirrus is heavily armed with long spines. Uterus ephemeral, with eggs soon breaking out into the parenchyma.

Life Cycle

When the eggs, released from the gravid proglottids and voided in the host feces, are eaten by certain slugs and land snails, they hatch. Intermediate hosts are species of the slugs *Limax, Deroceras,* and *Arion* and of the land snails *Polygyra, Zonitoides,* and *Vallonia*. The oncospheres bore through the intestine to tissues where they develop into infective cysticercoids (Figure 98) within a month or less during the summer. Fowls become infected when they eat infected molluscs. Prepatency takes about 14 days.

Life Cycle Exercise

Worms for experimental purposes can be obtained from farm-reared chickens. The eggs, teased from the gravid proglottids, should be placed in a petri dish with the appropriate molluscan intermediate hosts. To insure against using naturally infected intermediate hosts, they should be collected from localities where chickens have not been kept recently or, better still, reared in the laboratory. Dissection of the intermediary gastropods under a dissecting microscope and examination of the intestinal contents a few hours postinfection will show oncospheres and empty egg shells. Later examinations over spaced intervals will show the passage of the oncospheres through the gut wall and the developing cysticercoids. After about 14 days during the summer, the cysticercoids are infective to baby chicks.

Other often available Davaineidae with their hosts include *Davainea meleagridis*, turkeys; *Raillietina (Raillietina) echinobothrida*, chickens; and *Raillietina (Skrjabinia) cesticillus*, chickens (Figure 118).

Order Pseudophyllidea

Scolex typically with a dorsal and ventral slit or bothrium; majority of proglottids in similar stage of development, shedding eggs from a uterine pore; genital pores on the midventral surface. The testes and vitelline follicles are numerous, small bodies scattered throughout the proglottid except for the region of the ovary and uterus. The life cycle normally involves 2 metacestode stages and 2 intermediate hosts. Eggs are undeveloped when laid, usually operculate, and give rise to oncospheres known as coracidia. The 2 metacestode stages are a procercoid in a copepod crustacean and a plerocercoid (Figure 98) in all classes of vertebrates except birds. Adults are mainly in fish, but also in amphibians, reptiles, birds, and mammals.

Family Diphyllobothriidae

Diphyllobothrium latum
(*Dibothriocephalus latus* of some authors)

The fish tapeworm, *D. latum*, occurs in man, bears, cats, dogs, and possibly other fish-eating carnivores. In North America, it is known from the Great Lakes region (Michigan, Minnesota, and parts of Canada). There are other North American records, but the evidence is not conclusive. *D. latum* is a monster among tapeworms, reaching a length of more than 30 ft, with a width of nearly an inch, and consisting of more than 4,000 proglottids. Östling (1961) reported recovering 16 scolices and 1,006 ft of worm from an adult man in Finland. If the total strobilar length were associated with the 16 scolices, each worm averaged nearly 63 ft (Figure 126).

Description

The almond-shaped scolex has a pair of suctorial grooves, one dorsal and the other ventral, known as bothria. A short neck region separates the scolex from the proglottids. Anterior proglottids are broader than long, but posteriorly they are approximately square. The female system consists of (1) a bilobed ovary, situated posteriorly in the proglottid, (2) a vagina extending in a straight line forward from the ootype and opening in the genital atrium posterior to the cirrus pouch on the midventral surface of the proglottid, (3) follicular vitelline glands situated in layers, dorsal and ventral to the testes, and (4) the rosette-shaped uterus winding forward, in the form of lateral coils, from the ootype and opening through the uterine pore a short distance behind, and often slightly lateral to, the genital pore. The male system consists of (1) many testes and their ducts disposed in a single layer, largely obscured by the vitelline glands and (2) the prominent cirrus pouch, with its cirrus, which opens in the genital pore anterior to the vagina. The eggs are described under Helminth Eggs, p. 210.

Life Cycle

Eggs develop in water and coracidia are formed, each of which consists of a ciliated embryophore enclosing an oncosphere. When the free-swimming coracidium is eaten by copepods (species of *Diaptomus* serve best), the ciliated embryophore is shed and the oncosphere migrates to the hemocel where it develops into a procercoid. The procercoid is liberated when the copepod is eaten by certain freshwater fish and it, in turn, bores through the intestine to reach the body wall or viscera of the second intermediate host, where it develops into a plerocercoid. In North America, pike and walleyes (*Esox* and *Stizostedion*) are the most important hosts, but elsewhere other fish are involved. The plerocercoid is liberated when the fish host is eaten, and develops into the adult in the intestine of the final host. This is the 3 host cycle, determined experimentally, and undoubtedly occurring in nature. But in the case of the larger fish, which are not apt to feed directly on copepods, it is more likely that in nature a smaller forage fish serves as a means of transfer between the copepod and the large carnivorous pike and walleye. Plerocercoids are able to reestablish themselves in fish after fish until they are ingested by a suitable final host in which maturity can be attained.

Life Cycle Exercise

Although *D. latum* is not ordinarily available, diphyllobothriums from gulls, or species of other pseudophyllideans which mature in fish, amphibians, birds, or mammals can occasionally be obtained without much difficulty.

Eggs can be recovered by rupturing or teasing gravid proglottids. Wash eggs and allow them to develop at room temperature in biological water for about 14 days, when hatching begins. If contamination of the culture develops, add a few drops of iodine. Transfer fully embryonated eggs to a microscope slide, apply a cover glass, and observe them under a compound microscope. Note the active coracidia within the eggs, the opening of the operculum and hatching of the eggs, and finally the liberated, free-swimming coracidia. Once the eggs begin to hatch, the harvest can be increased by such stimuli as change of water, change in temperature, and strong sunlight. Place the culture dish under a dissecting microscope, add copepods to dish with the eggs, and observe the copepods feeding on the coracidia. Since feeding is a matter of random contacts, concentration of the coracidia involved in a small amount of water increases the frequency of infection. After exposure of a few hours, remove the copepods to bowls of fresh biological water. Copepods should be examined at spaced intervals to observe the migration of the oncosphere from the gut into the hemocel and development of the procercoid. Examination of infected copepods is accomplished by concentrating them in a small amount of water, transferring a few with a drawn pipette to a slide and adding a rectangular cover glass. Within a few minutes, they will have quieted sufficiently for examination under a compound microscope, after which the sample is washed back into the culture with a wash bottle. After 2 weeks or more, surviving copepods can be fed to threespine stickleback *(Gasterosterus aculeatus)* or fingerling salmonids in an effort to obtain plerocercoids.

If in a *Diphyllobothrium* area, the appropriate host fish should be examined for plerocercoids. The larvae can be removed from the body wall and the surface of viscera with dissecting needles, or by using the HCl-peptic digest method (p. 280). Appropriate final hosts can be infected by feeding them plerocercoids.

Spirometra and Sparganosis

Sparganosis is an infectious disease of man caused by migrating plerocercoids or spargana larvae of tapeworms belonging to the genus *Spirometra*. While there is some confusion about the species of *Spirometra*, in the Orient it is usually regarded as *S. mansoni*, and in the United States, where it is distributed over a broad area of the Atlantic and Gulf States, it is normally recognized as *S. mansonoides*.

When the procercoid-infected copepods (species of *Cyclops* serve best), are eaten by certain amphibians, reptiles, and mammals, the plerocercoid stage develops in their tissues. Infection of the final host, cats and related mammals, comes from eating a second intermediate host containing the viable plerocercoids. As a result of their resistance to both procercoids and plerocercoids, fish are unsuitable intermediate hosts.

When the plerocercoids of *Spirometra*, for which the name *Sparganum* was used before their adult stage was known, get into an unsuitable vertebrate host (frogs, reptiles, and some mammals), they migrate through the intestinal wall and reestablish themselves in the subcutaneous tissues and muscles where they grow in size, until a host is reached in which maturity can be attained in the intestine. This is of much more than academic interest, because man is included among the unsuitable vertebrate hosts.

Thus, in addition to the general susceptibility of *Cyclops* to the coracidia, *Spirometra* differs from *Diphyllobothrium* in 2 important respects: first, the fish, which are obligatory hosts for the plerocercoids of *Diphyllobothrium,* are replaced by amphibians, reptiles, and some mammals (including man) in *Spirometra;* second, man is the preferred final host of *D. latum,* while that of *Spirometra* is the cat.

Sparganosis as a zoonosis results from man: (1) drinking water containing procercoid-infected *Cyclops;* (2) eating amphibians, reptiles, and mammals containing the viable plerocercoids; and (3) applying plerocercoid-infected flesh of frogs, snakes, and possibly mammals to wounds or inflamed eyes as a poultice.

While the use of meat-poultices is a common practice in the Orient, where the spargana of *S. mansoni* is involved, they can be ruled out for America. Until recently it was presumed that zoonotic sparganosis in the United States came from drinking water containing the procercoid-infected copepods. But the finding of spargana in swine in Florida, and the discovery of natural infections of *S. mansonoides* in 2 amphibian, 8 reptilian, and 3 mammalian species in Louisiana by Corkum (1966), who showed experimentally that swine are susceptible to spargana, suggest that human infection may occasionally result from consuming wild animals or pork containing the viable spargana.

Order Caryophyllidea

Scolex is not distinct from the body or may show faint grooves or depressions, but true suckers or bothria are lacking. These are small monozoic forms whose genital pores open on the ventral surface of the body, the uterine pore situated between the openings of the male and female organs. The operculate eggs are unembryonated when laid. The life cycle may be direct or involve an intermediate host; the sexual stages are often considered progenetic procercoids. Parasitic in the intestine of Catostomidae and Cyprinidae fishes. In *Archigetes,* the sexual stages occur in oligochaete annelids as well as in fish.

Family Caryophyllaeidae

Glaridacris catostomi

In North America, *G. catostomi* is a common parasite of the white sucker *(Catostomus commersoni)* and related fish (Figure 106).

Description

Adults measure up to 25 mm long, 1 mm in width, and are somewhat cylindrical in cross section. The scolex is short, with 3 suckerlike depressions on each side. Testes, 150 to 160, are arranged longitudinally and median. The ovary is dumbbell-shaped, with the wings short, compact, and rounded. The preovarian vitellaria are lateral and extend slightly into the area of the testes, as well as anterior to the testes along the side; posterior to the ovary the vitellaria are numerous, forming a V-shaped cluster. The uterus, which opens slightly posterior to the cirrus pouch, is both post- and preovarian. Eggs operculate, with an abopercular boss, 37-48 x 20-31 μ.

Life Cycle

It is generally accepted that the fish becomes infected when it ingests aquatic oligochaetes containing procercoids, but this is based upon circumstantial evidence. The only caryophyllideans for which the life cycles have been fully elucidated, i.e., going from 1 egg stage to that of another, according to Mackiewicz (1972), have been with certain species of *Archigetes.*

Archigetes iowensis

Gravid adults of *A. iowensis* occur seasonally in the intestine of the Cyprinidae fish, *Cyprinus carpio* and occasionally from the seminal vesicles of the oligochaete, *Limnodrilus hoffmeisteri.*

Description

Adults average 1.6 mm in length (without cercomer) by 0.5 mm in width. The scolex is well developed, with 3 pairs of suckerlike depressions. There are about 65 testes anterior to the cirrus pouch and median to the laterally situated vitellaria. Postovarian vitellaria are also present. The uterus, which opens slightly posterior to the cirrus pouch, extends posterior and anterior to the dumbbell-shaped ovary. Eggs operculate, 43 x 32 μ.

Life Cycle

Eggs are produced within the fish host, but procercoids occasionally develop progenetically within the annelid host, *Limnodrilus hoffmeisteri*. The oncosphere stage is reached in water, but further development does not occur until the eggs are ingested by the intermediate host, *L. hoffmeisteri*. The eggs hatch in the posterior region of the intestine of the oligochaete host, after which the oncospheres penetrate the intestinal wall, migrate forward to the region of the seminal vesicles which they enter (through the natural openings). Procercoid development requires about 70 days. The carp apparently first become infected in early spring when they commence feeding, presumably through ingesting infected oligochaetes. While infective procercoids may develop from the eggs produced by progenetic worms in the intermediate host, it is believed that the life cycle in nature is dependent upon the fish hosts. This is in striking contrast with that of *A. sieboldi*, also present in North America, which is primarily a parasite of oligochaetes. Only occasionally are gravid stages found in fish.

Life Cycle Exercise

Eggs obtained by dissecting gravid worms in a dish of biological water kept at room temperature will develop to oncospheres. If the proper uninfected oligochaetes (species of *Limnodrilus* and *Tubifex*) are added to the dishes with the embryonated eggs or the eggs and worms mixed in ooze, the worms will ingest the eggs and become infected. Unless laboratory-reared oligochaetes are used, worms collected in nature should be kept in the laboratory and examined over several weeks to allow any natural infections to mature and be observed. Previously infected worms should be discarded. Hatching of the eggs in the gut, migration, and growth of the larvae may be followed by dissecting infected annelids at appropriate postinfection intervals.

Fingerling suckers and carp may be infected by allowing them to eat the experimentally infected oligochaetes. Unless laboratory-reared fish are used, fish collected in nature should be kept in the laboratory and their feces examined over several weeks to allow any natural infections to be observed. Growth and development of the worms can be followed by dissecting the infected fish at appropriate postinfection intervals.

Other Caryophyllaeidae such as species of *Glaradacris*, *Archigetes*, and *Caryophyllaeus* may be found in Catostomidae and Cyprinidae fishes.

Order Proteocephalidea

Scolex with 4 muscular cuplike suckers, and sometimes a fifth terminal one; genital pores lateral; vitellaria in lateral bands (Figure 109); uterus with numerous lateral outgrowths and 1 or more median ventral openings, made by breaks or clefts in the body wall, through which eggs escape. The life cycle involves a copepod, in which the oncospheres develop into infective plerocercoids (Figure 98). When the infected copepod is eaten by the suitable final host, fish, amphibian, or reptile, the adult worm develops. But these usually final hosts may serve as paratenic hosts when they fall prey to larger fish, amphibians, or reptiles. A second intermediate host is apparently not obligatory.

Family Proteocephalidae

Proteocephalus ambloplitis

This species, known as the bass tapeworm, is widespread throughout much of North America. The extensive range is largely due to stocking infected fish, primarily bass, in uninfected waters. In addition to bass harboring adult worms in the intestine, plerocercoids occur in the celom and viscera. The parenteral plerocercoids in bass are much more damaging than are adults in the intestine. The plerocercoids cause extensive adhesions of the viscera and may injure the gonads, especially in the females, preventing spawning.

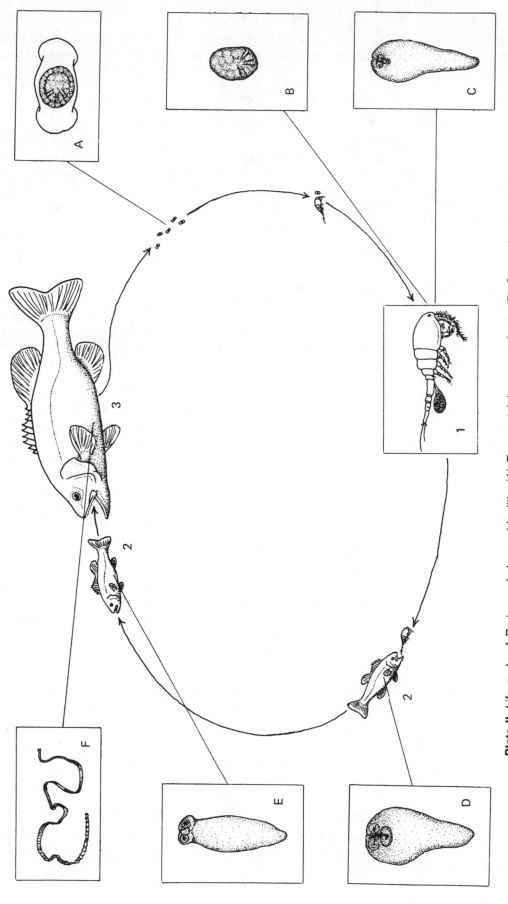

Plate II. Life cycle of *Proteocephalus ambloplitis*: (A) Egg containing oncosphere; (B) Oncosphere in gut of copepod; (C) Plerocercoid I in hemocel of copepod; (D) Plerocercoid II in parenteral cavity of fish; (E) Later stage
of same; (F) Adult in enteral cavity of fish. Numbers 1 and 2 indicate first and second intermediate hosts and 3
final host. After Hunter and Hunter (1929) from Meyer (1954).

Description

The scolex is globose, with 4 simple suckers and a vestigial apical sucker. The vitellaria are arranged in lateral bands along the length of the proglottid. The numerous testes are bounded laterally by the vitellaria and posteriorly by the ovary. The ovary is in the form of a transverse band in the rear of the proglottid. Genital pore is lateral, alternating irregularly; the vagina opens in the genital atrium anterior to the cirrus pouch. Eggs ellipsoidal, with 3 membranes, 43 x 36 μ.

Life Cycle

The life cycle typically involves 3 hosts; a copepod, a forage fish, and the final fish host. The final host, usually a bass *(Micropterus dolomieui, M. salmoides)*, harbors the mature worm in the intestine. Gravid proglottids voided with the feces rupture and release eggs on contact with the water. When eggs are ingested by *Cyclops viridis* and *Eucylops agilis,* according to Hunter and Hunter (1929), they hatch and the oncospheres migrate to the hemocel where they develop into plerocercoids I. When yellow perch *(Perca flavescens)*, pumpkinseed *(Lepomis gibbosus)*, brown bullhead *(Ictalurus nebulosus)*, among other fish, consume infected copepods, the plerocercoids penetrate the gut wall and establish themselves in the celom and viscera. Here they develop into plerocercoid II.

Fischer and Freeman (1969, 1973), using feeding experiments and field data, have shown that: (1) most, if not all, parenteral plerocercoids from fish other than bass, when ingested by bass, leave the gut and reestablish themselves parenterally; (2) after sufficient development in bass, the plerocercoids leave their parenteral sites, return to the gut lumen of the same bass, and mature.

Life Cycle Exercise

Adult *P. ambloplitis* are widely distributed and are usually present where the host fish and copepods occur simultaneously. Place gravid proglottids in biological water, and observe the rupture of the body wall, and escape of the eggs. Remove spent proglottids, change the water, add copepods, and watch them eating the eggs. Examine the intestine of the copepods for oncospheres shortly after they have consumed the eggs, and the body cavity at hourly intervals postinfection for development of plerocercoids. Examination of the copepods is accomplished by concentrating them in a small amount of water, transferring a few in a drawn pipette to a slide, and adding a rectangular cover glass. Within a few minutes, they have quieted sufficiently for examination under the compound microscope, after which the sample is washed back into the culture with a wash bottle. After about 3 weeks postinfection, allow the appropriate fingerling host fish to feed on the copepods. Examine the viscera of the experimentally infected fish for plerocercoids 5, 10, and 15 days postinfection.

Since plerocercoids may occur in the same species of host fish harboring the mature tapeworms, the plerocercoids can be collected and used to reinfect the appropriate host.

Often other available Proteocephalidae with their hosts include *Proteocephalus parallacticus,* salmonids; *P. pearsei,* yellow perch; *P. pinguis,* chain pickerel; *Corallobothrium fimbriatum,* and *C. parvum,* bullheads.

Key to Common Families of Cestoidea

Capital letters in parentheses refer to hosts: F-Fish, A-Amphibia, R-Reptiles, B-Birds, M-Mammals.

1 Large monozoic, linguiform cestodes with proboscis at anterior end but no scolex; testes scattered or in bands; ovary and genital pores at posterior end of body; uterus N-shaped, opening near proboscis; vitellaria lateral, extend full length of body; eggs nonoperculate, embryos 10-hooked; in body cavity (Subclass CESTODARIA) (Figure 105) . . . .
. . . . (F, rarely R) Amphilinidae

Small monozoic or polyzoic cestodes with scolex; embryos with 6 hooks (Subclass CESTODA) . . . 2

2(1) Monozoic; scolex with grooves or depressions; adults with or without tail; testes between lateral bands of vitelline follicles and anterior to ovary and uterus; male and female genital pores and uterine pore midventral at anterior end of uterus (Order CARYOPHYLLIDEA) (Figure 106)
. . . . (F) Caryophyllaeidae

Polyzoic, with segmentation usually distinct externally but occasionally visible only internally 3

3(2) Scolex with single well-developed apical sucker; body cylindrical; genital pores lateral (Order NIPPOTAENIIDEA) One family (Figure 107) .... (F) Nippotaeniidae

Scolex large with enormous glandular rostellum and 4 cup-shaped suckers, or small scolex without suckers but with a rostellum bearing 10 hooks; external segmentation lacking; genital ducts without external openings (Order APORIDEA) One family (Figure 108) ....
.... (B) Nematoparataeniidae

Scolex with 4 cup-shaped muscular suckers, sometimes with a fifth apical sucker or remnant of one; vitellaria in lateral bands; genital pore marginal (Order PROTEOCEPHALIDEA) One family (Figure 109) .... (F, A, R) Proteocephalidae

Scolex with 4 cuplike muscular suckers, with or without apical rostellum bearing 1 or more crowns of hooks; vitelline gland compact, usually behind ovary; genital pore lateral except in Mesocestoididae, when on midventral surface (Order CYCLOPHYLLIDEA) 4

Scolex typically with a dorsal and ventral bothrium (groove); permanent uterine pore on flat surface of proglottid; vitellaria scattered in proglottid (Order PSEUDOPHYLLIDEA) 16

4(3) Rostellum absent 5

Rostellum usually present[1] 9

5(4) Genital pores on midventral surface (Figure 110) .... (B, M) Mesocestoididae

Genital pores marginal, unilateral or alternating 6

6(5) Vitelline gland preovarian, bilobed; suckers commonly with outgrowth from margins; genital pores unilateral (Figure 111) .... (B, M) Tetrabothridae

Vitellinae gland postovarian 7

7(6) Segmentation apparent only posteriorly, body cylindrical; usually 2 testes; encapsulated eggs in several paruterine bodies per segment (Figure 112) .... (A, R) Nematotaeniidae

Segmentation distinct throughout; body usually flat 8

8(7) Gravid proglottids distinctly longer than wide; 1 set of reproductive organs per proglottid; ovary and vitelline gland branched, in anterior third of proglottid; testes mainly or entirely postovarian; gravid uterus with long median stem and irregularly shaped lateral lobes (Figure 113) ......
.... (M, rodents) Catenotaeniidae

Gravid proglottids usually much wider than long; vitelline gland usually unbranched; gravid uterus a transverse tube, reticulate, or unstable; eggs with pyriform apparatus; single or double sets of reproductive organs (Figure 114) .... (R, B, M) Anoplocephalidae

9(4) Rostellum permanently extended, usually armed[2] with 2 circles of hooks; ovary bilobed; vitellaria postovarian, gravid uterus a median stem with lateral branches; eggs with thick, striated shell (Figure 115) .... (B, M) Taeniidae

Rostellum retractile within scolex; eggs without striated shell, gravid uterus not with median stem and lateral branches 10

10(9) Individual strobila with either male or female sex organs but not both, or partially separated, with genital organs of male in anterior parts and of female in posterior parts of same strobila (Figure 116) .... (B) Dioecocestidae

Monoecious forms 11

11(10) Testes large, 1 to 4, usually 3; genital pores unilateral; without or with single crown of rostellar hooks;[3] uterus transverse (*Diplogynia* has double set of reproductive systems) (Figure 117)..
.... (B, M) Hymenolepididae

Testes small, usually more than 4 12

---

1. A few nonrostellate genera are included in families whose members are normally rostellate, e.g., *Anoplotaenia* and *Taenia* in Taeniidae; some species of *Hymenolepis* in Hymenolepididae; *Anonchotaenia, Metroliasthes, Rhabdometra*, etc., in Dilepididae, and *Shipleya* in Dioecocestidae.
2. See couplet 4.
3. Ibid.

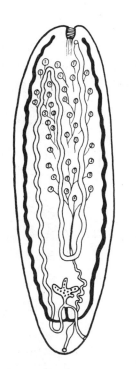

Figure 105. *Amphilina foliacea.*

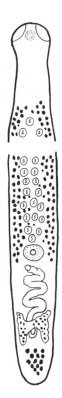

Figure 106. *Glaridacris catostomi.*

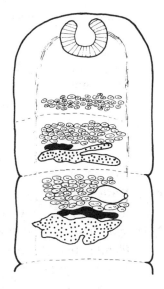

Figure 107. *Nippotaenia chaenogobii.*

Figure 108. *Nematoparataenia southwelli.*

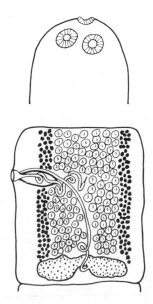

Figure 109. *Proteocephalus pearsei.*

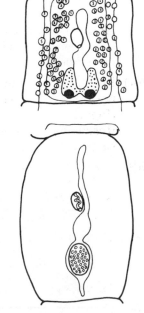

Figure 110. *Mesocestoides lineatus.*

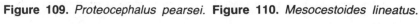

125

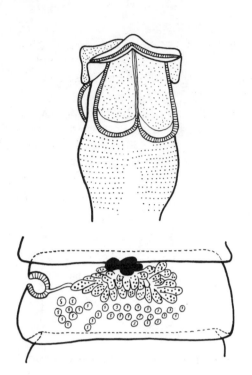

Figure 111. *Tetrabothrius cylindricus.*

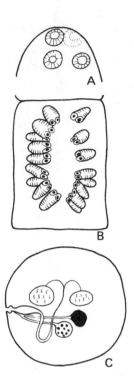

Figure 112. (A) (B) *Baerietta baeri;* (C) *Nematotaenai dispar.*

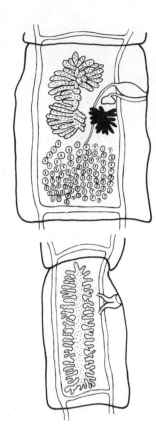

Figure 113. *Catenotaenia peromysci.*

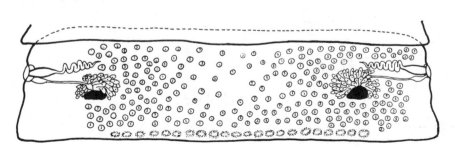

Figure 114. *Moniezia expansa.*

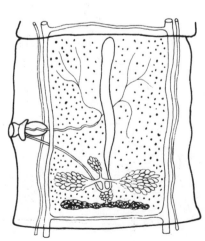

Figure 115. *Taenia solium.*

12(11) Rostellum with 1 or more circles of numerous small T-shaped hooks; margins of suckers commonly spinose; uterus saclike or replaced by egg-capsules or paruterine organ; genitalia single or double (Figure 118) . . . . (B, M) Davaineidae

Rostellar hooks not T-shaped; suckers usually not spinose 13

13(12) Male genital system with 2 separate cirrus pouches; testes few and in 1 or 2 groups; female genital system single, vaginal aperture lacking (Figure 119) . . . . (B) Diploposthidae

Male and female genital systems single (except *Diplopylidium* in Dilepididae and *Diplogynia* in Hymenolepididae where both male and female genital systems are double) 14

14(13) Vaginal apertures present and functional; rostellum usually present and armed with 1 to several circles of hooks of various forms, including rose-thorn shape; suckers aspinose (except *Cotylohipis* with large hooks on margins, nonrostellate); genitalia commonly single but may be double; testes commonly numerous; gravid uterus a transverse, lobed sac, or paruterine organ, or capsules with 1 or more eggs (Figure 120) . . . . (B, M) Dilepididae

Vaginal apertures absent 15

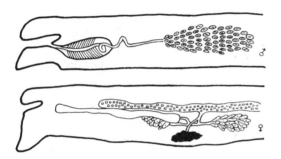

Figure 116. *Dioecocestus fuhrmanni.*

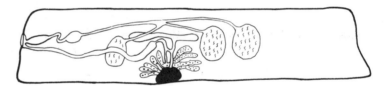

Figure 117. *Hymenolepis diminuta.*

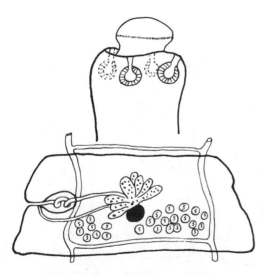

Figure 118. *Raillietina cesticillus.*

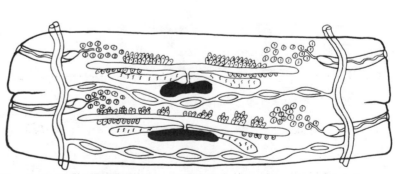

Figure 119. *Diplophallus taglei.*

15(14) Vagina serving as seminal receptacle, no accessory pore developing later to replace vagina (Figure 121) . . . . (B) Acoleidae

Vagina absent, replaced in function later by an accessory canal opening to outside (Figure 122) . . . . (B) Amabiliidae

16(3) Genital pores marginal; uterine pore ventral (rarely dorsal) 17

Genital pores and uterine pore on same surface of proglottid or on opposite surfaces 18

17(16) Scolex subcuboidal or subpyramidal, bothria shallow, apex with or without 4 tridental hooks; uterine pore midventral, anterior to level of marginal genital pore; gravid uterus a tube with many coils; eggs operculate (Figure 123) . . . . (F) Triaenophoridae

Scolex subspherical, bothria well developed, apex without hooks; gravid uterus saclike; uterine pore median; eggs nonoperculate (Figure 124) . . . . (F) Amphicotylidae

18(16) Genital pores and uterine pore on same surface of proglottid 19

Genital pores and uterine pore open on opposite surfaces of proglottid 20

19(18) Small worms; scolex of primary segmented adult worm club-shaped with 4 protrusible tentacles, replaced in secondary segmented adult by rectangular pseudoscolex that resembles first formed segment; cirrus spined; uterus with proximal coils and distal sac (Figure 125) . . . . (F) Haplobothriidae

Large worms, proglottids wider than long; scolex variable in shape, commonly compressed and with bothria weakly developed, without tentacles; uterus tubular throughout length, arranged in median loops rosette-like; genital system usually single (double in *Diplogonoporus* and *Cordiocephalus*); cirrus unspined; eggs operculate, unembryonated when laid (Figure 126) . . . . (F rarely, R, B, M) Diphyllobothriidae

20(18) Scolex with rounded apex, bothria deep and narrow or wide and shallow, eggs nonoperculate, embryonated when laid (Figure 127) . . . . (F) Ptychobothriidae

Scolex with a 4-lobed, fleshy apex and long, oval, shallow bothria; eggs operculate, unembryonated when laid (Figure 128) . . . . (F) Bothricephalidae

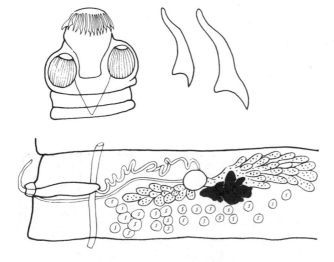

Figure 120. *Dilepis undula.*

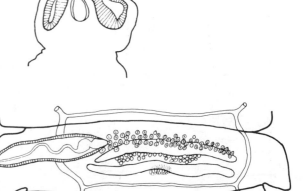

Figure 121. *Acoelus vaginatus.*

Figure 122. *Tatria duodecacantha.*

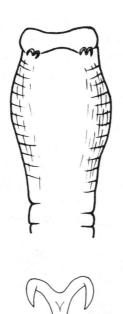

Figure 123. *Triaenophorus nodulosus.*

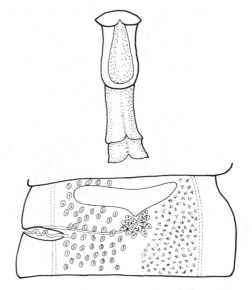

Figure 124. *Abothrium crassum.*

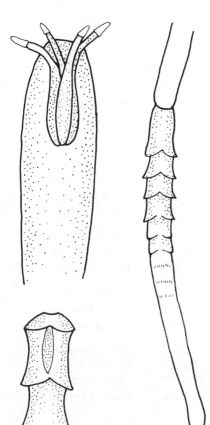

Figure 125. *Haplobothrium globuliforme.*

Figure 126. *Diphyllobothrium latum.*

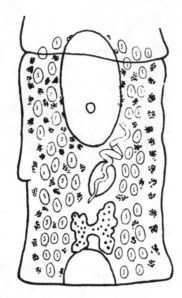

Figure 127. *Clestobothrium crassiceps.*

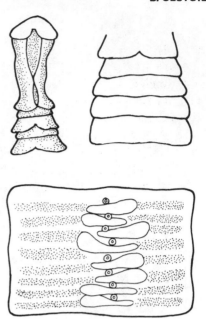

Figure 128. *Bothriocephalus manubriformis.*

CLASSIFICATION

The classification of the Cestoidea is in a much more satisfactory state than that of the Trematoda. There are 2 subclasses: the Cestodaria and Cestoda. Different authors recognize 4 to 11 orders of Cestoda. Following are the principal orders and their common families:

Subclass	Order	Family
CESTODARIA	Amphilinidea	Amphilinidae
	Gyrocotylidea	Gyrocotylidae
CESTODA	Cyclophyllidea	Taeniidae
		Dilepididae
		Hymenolepididae
		Anoplocephalidae
		Davaineidae
	Pseudophyllidea	Diphyllobothriidae
		Amphicotylidae
	Caryophyllidea	Caryophyllaeidae
	Proteocephalidea	Proteocephalidae
	Tetraphyllidea	Phyllobothriidae
		Oncobothriidae
	Trypanorhyncha	Tentaculariidae
		Gymnorhynchidae
		Otobothriidae
		Dasyrhynchidae
	Lecanicephalidea	Lecanicephalidae
		Disculicipitidae

REFERENCES

Freeman, R.S. 1973. Ontogeny of Cestodes and its Bearing on their Phylogeny and Systematics. *In* Advances in Parasitology, ed., B. Dawes. Academic Press, New York, vol. 11, pp. 481-557.

Rees, Gwendolen. 1967. Pathogenesis of adult cestodes. Helminthol. Abstr. 36: 1-23.

Schmidt, G.D. 1970. How to Know the Tapeworms. Wm. C. Brown Company Publishers, Dubuque, Iowa, 266 p.

Slais, J. 1973. Functional Morphology of Cestode Larvae. *In* Advances in Parasitology, ed., B. Dawes. Academic Press, New York, vol. 11, pp. 395-480.

Smyth, J.D. 1963. The Biology of Cestode Life-Cycles. Tech. Comm. No. 34, Commonwealth Bur. Helminthology, St. Albans, Herts, England, 38 p.

———. 1969. The Physiology of Cestodes. Oliver & Boyd, Edinburgh, 279 p.

———, and D.D. Heath. 1970. Pathogenesis of larval cestodes in mammals. Helminthol. Abstr., Series A, 39: 1-23.

Stunkard, H.W. 1962. The organization, ontogeny, and orientation of the Cestoda. Quart. Rev. Biol. 37: 23-34.

Voge, Marietta. 1967 & 1973. The post-embryonic Developmental Stages of Cestodes. *In* Advances in Parasitology, ed., B. Dawes. Academic Press, New York, vol. 5, pp. 247-297; vol. 11, pp. 707-730.

Wardle, R.A., and J.A. McLeod. 1952. The Zoology of Tapeworms. The University of Minnesota Press, Minneapolis, 780 p.

Yamaguti, S. 1959. Systema Helminthum. The Cestodes of Vertebrates, vol. 2, Interscience Publishers, Inc., New York, 860 p.

Taeniidae

Leiby, P.D., W.P. Carney, and C.E. Woods. 1970. Studies on sylvatic echinococcosis. J. Parasitol. 56: 1141-1150.

———, and D.C. Kristsky. 1972. *Echinococcus multilocularis:* A possible domestic life cycle in central North America and its public health implications. J. Parasitol. 58: 1213-1215.

Pawlowski, Z., and M.G. Schultz. 1972. Taeniasis and Cysticercosis *(Taenia saginata). In* Advances in Parasitology, ed., B. Dawes. Academic Press, New York, vol. 10, pp. 269-343.

Rausch, R.L. 1967. On the ecology and distribution of *Echinococcus* spp. (Cestoda: Taeniidae), and characteristics of their development in the intermediate host. Ann. Parasitol. Hum. Comp. 42: 19-63.

Smyth, J.D. 1964 & 1969. The Biology of the Hydatid Organisms. *In* Advances in Parasitology, ed., B. Dawes. Academic Press, New York, vol. 2, pp. 169-219; vol. 7, pp. 327-347.

Dilepididae

Gleason, Neva N. 1962. Records of human infections with *Dipylidium caninum,* the double-pored tapeworm. J. Parasitol. 48: 812.

Horsfall, M.W., and M.F. Jones. 1937. The life history of *Choanotaenia infundibulum,* a cestode parasitic in chickens. J. Parasitol. 23: 435-450.

Venard, C.E. 1938. Morphology, bionomics, and taxonomy of the cestode *Dipylidium caninum.* Ann. New York Acad. Sci. 37: 273-328.

Hymenolepididae

Heyneman, D. 1958. Effect of temperature on the rate of development and viability of the cestode *Hymenolepis nana* in its intermediate host. Exptl. Parasitol. 7: 374-382.

Voge, M., and D. Heyneman. 1957. Development of *Hymenolepis nana* and *Hymenolepis diminuta* (Cestoda: Hymenolepididae) in the intermediate host *Tribolium confusum.* Univ. California Publ. Zool. 59: 549-580.

Anoplocephalidae

Freeman, R.S. 1952. The biology and life history of *Monoecocestus* Beddard, 1914 (Cestoda: Anoplocephalidae) from the porcupine. J. Parasitol. 38: 111-129.

Kates, K.C., and C.E. Runkel. 1948. Observations on oribatid mite vectors of *Moniezia expansa* on pastures, with a report of several new vectors from the United States. Proc. Helminthol. Soc. Washington 15: 10, 19-33.

Stunkard, H.W. 1938. The development of *Moniezia expansa* in the intermediate host. Parasitology 30: 491-501.

Davaineidae

Abdou, A.H. 1958. Studies on the development of *Davainea proglottina* in the intermediate host. J. Parasitol. 44: 484-488.

Enigk, K., E. Sticinsky, and H. Ergün. 1958. Die Zwischenwirte von *Davainea proglottina*, (Cestoidea). Zeit. Parasitenk. 18: 230-236.

Wetzel, R. 1932. Zur Kenntnis des weniggliedrigen Huhnerbandwurmes *Davainea proglottina*. Arch. Wissensch. Prakt. Tierheilk. 65: 595-625.

Diphyllobothriidae

Corkum, K.C. 1966. Sparganosis in some vertebrates of Louisiana and observations on a human infection. J. Parasitol. 52: 444-448.

Meyer, M.C. 1970. Cestoda zoonoses of aquatic animals. J. Wildlife Dis. 6: 249-254.

_____. 1972. The pattern of circulation of *Diphyllobothrium sebago* (Cestoda: Pseudophyllidea) in an enzootic area. J. Wildlife Dis. 8: 215-220.

Mueller, J.F. 1974. The biology of *Spirometra*. J. Parasitol. 60: 1-14.

Östling, G. 1961. Treatment of tapeworm infection with desaspidin, a new phloroglucinol derivative isolated from Finnish fern. Amer. J. Trop. Med. Hyg. 10: 855-858.

Vik, R. 1964. The genus *Diphyllobothrium*. An example of the interdependence of systematics and experimental biology. Exptl. Parasitol. 15: 361-380.

Wardle, R.A. 1935. Fish-Tapeworm. Bull. Biol. Bd. Canada, No. 45, 25 p.

Caryophyllaeidae

Mackiewicz, J.S. 1972. Caryophyllidea (Cestoidea): A review. Exptl. Parasitol. 31: 417-512.

Proteocephalidae

Fischer, H., and R.S. Freeman. 1969. Penetration of parenteral plerocercoids of *Proteocephalus ambloplitis* (Leidy) into the gut of smallmouth bass. J. Parasitol. 55: 766-774.

_____. 1973. The role of plerocercoids in the biology of *Proteocephalus ambloplitis* (Cestoda) maturing in smallmouth bass. Canad. J. Zool. 51: 133-141.

Hunter, G.W., III, and W.S. Hunter. 1929. Further experimental studies on the bass tapeworm, *Proteocephalus ambloplitis* (Leidy). *In* Biological Survey of the Erie-Niagara System, Suppl. 18th Ann. Rept., 1928, New York State Conserv. Dept., Albany, pp. 198-207.

Phylum Acanthocephala

The phylum consists of a small group of unsegmented, cylindrical, dioecious, endoparasitic worms lacking a digestive tract throughout their development. Anteriorly the body terminates in a retractable proboscis, usually armed with recurved hooks, that becomes embedded in the intestinal wall of the final host.

The body is divided into an anterior region, the presoma, and posterior region, the trunk, which lies free in the intestinal lumen of the host. Externally the presoma consists of the proboscis, followed by the glabrous neck; internally the neck continues as the proboscis receptacle into which the proboscis and neck in most species is retractable. Associated structures of the presoma include the small cephalic ganglion in the receptacle, a pair of elongated lemnisci originating from the neck and hanging free in the body cavity, and a pair of retractor muscles. The proboscis is everted by means of hydrostatic fluids under pressure by body contraction.

A pair of relatively broad muscles, the proboscis invertors, extends back from the tip of the proboscis and inverts it upon contracting. Each of these muscles extends through the proboscis and the neck, and continues to the posterior extremity of the proboscis receptacle. Here they penetrate the receptacle, after which they are known as receptacle retractors, continue in the body cavity and eventually attach to the trunk of the body wall.

The remainder of the body comprises the trunk, containing the excretory and reproductive systems, nerve fibers arising from the cephalic ganglion, and the body fluid. The body wall consists of an outer cuticle, a thick middle layer composed of 3 outer layers of fibrils and an inner syncytial layer containing a few nuclei, and an innermost muscular layer. A lacunar system of dorsal and ventral, or lateral, canals extends the length of the body and is connected by a network or transverse system of small canals. The body cavity lacks a definite epithelial lining and is regarded as a pseudocelom.

Extending from the base of the proboscis receptacle to the posterior end of the trunk is a strand of single or double connective tissue, the ligament, which encloses the reproductive organs.

The male reproductive system consists of a pair of testes in tandem and genital ducts, cirrus, and accessory organs (cement gland complex and copulatory apparatus). A single syncytial cement gland or several unicellular glands, a cement reservoir in some species, and their ducts comprise the cement gland complex. The ducts of the cement glands, individually or collectively, enter the common genital duct. The copulatory apparatus consists of a Saefftigen's pouch, a muscular fluid reservoir, and an eversible copulatory bursa, with a cirrus, at the posterior end of the trunk. During copulation the everted bursa fits over the posterior end of the female, the cirrus enters the vagina, discharging sperm into the uterus, after which the cement gland secretion plugs up the female following mating, thus preventing loss of the sperm. Eversion of the bursa results from the hydrostatic pressure of fluid from the Saefftigen's pouch; inversion is by muscular action (Figure 129, A).

The female reproductive system consists of an ovary, which develops from the ligament, uterine bell, uterus, and vagina. Early in development, the ovary breaks up into ovarian balls, or floating ovaries. The uterine bell is a bell-shaped structure with its large opening directed anteriorly and a small opening posteriorly, through which the mature eggs are passed into the uterus, whence they move into a short vagina and out through the gonopore. Immature eggs are returned to the body cavity or dorsal ligament (when present), through lateral pores near the base of the uterine bell (Figure 129, B). The mechanism by which the uterine bell "distinguishes" between mature and immature eggs, at least in *Polymorphus minutus*, is apparently based on their length (Whitefield, 1970).

Eggs have 3 or 4 enveloping membranes and are embryonated when laid. They are normally very resistant and do not hatch until ingested by the proper crustacean or insect host. The fusiform larva, the acanthor, is covered with small spines and anteriorly with hooks; inside are many small nuclei (Figure 129, C). After hatching the acanthor penetrates the host's gut wall and reaches the body cavity. Here the hooks of the previous stage are lost and a progressive development ensues; the proboscis and its sac, the

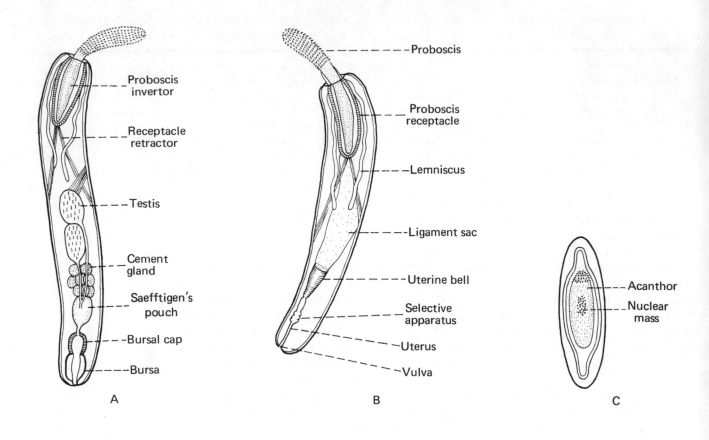

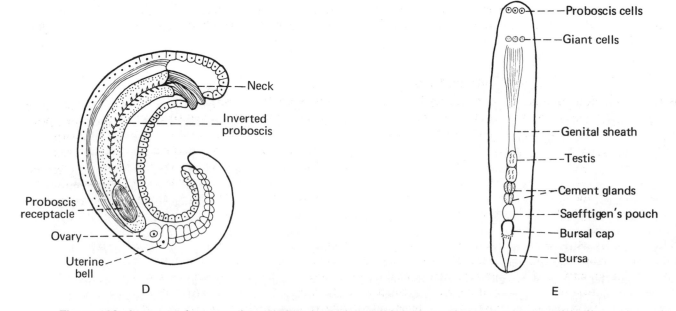

Figure 129. Anatomy of an acanthocephalan, *Leptorhynchoides thecatus*. (A) Adult male; (B) Adult female; (C) Embryonated egg showing spiny acanthor with internal nuclear mass; (D) Female acanthella removed from cyst; and (E) Young excysted male acanthella.

lemnisci, and the anlagen of the reproductive system appear. This preinfective stage is known as the acanthella. After undergoing further development and encystment, it is known as cystacanth. The sex of the now infective larva is apparent; the proboscis is formed but inverted, and it is enclosed within a hyaline membrane (Figure 129, D, E). Acanthocephalan larvae are very versatile in re-encysting, without undergoing further development, in various animals not appropriate as final hosts. Thus, a number of paratenic hosts may be interposed in the completion of the life cycle.

Order Palaeacanthocephala

Body usually small; proboscis with few or many hooks, lacunar canals lateral; nuclei of body wall usually fragmented or ramified; proboscis receptacle double-layered; ligament sacs rupture in females; males normally with 6 or fewer cement glands. Adults mostly in fish and aquatic birds and mammals; cystacanths in Crustacea.

Family Echinorhynchidae

Leptorhynchoides thecatus

This species parasitizes a variety of fish throughout the United States, but it occurs primarily in Centrarchidae: small mouth bass (Micropterus dolomieui), largemouth bass (M. salmoides), and rock bass (Ambloplites rupestris).

Description

When fully extended, the proboscis, armed with 12 longitudinal rows of 12 to 13 hooks each, is perpendicular to the long axis of the body; in extreme extension it may form an acute angle with the body. The hooks are surrounded throughout much of their length by an ensheathing cuticular collar. The shape and angle of the extended proboscis and the cuticular collar of the hooks are specifically diagnostic. The lemnisci are tubular to filiform, and considerably longer than the proboscis receptacle. The male has 8 cement glands, closely compacted at the posterior border of the hind testis. Eggs 80-110 x 24-30 μ.

Life Cycle

The eggs, fully developed when passed in the host feces, hatch within 45 minutes after being ingested by the amphipod Hyalella azteca. The acanthors soon make their way through the intestinal wall and come to rest beneath the serosal covering, where development proceeds to the acanthella stage. After about 14 days, during which there has been much increase in size and internal organization, the acanthellas become detached and drop into the hemocel. The cystacanths are infective about a month postinfection.

The cycle is completed when amphipods harboring cystacanths are eaten by appropriate fish hosts. The larvae are freed in the stomach of the fish, after which they enter the pyloric ceca, where they mature. Prepatency takes about 8 weeks, but the males are sexually mature in about a month (Plate III).

Life Cycle Exercise

The eggs, obtained by dissecting gravid females in a dish of biological water, should be placed in a small jar and the amphipods added. After exposure for about an hour, wash the amphipods to remove any eggs that may be adhering and place them in a clean container. If left much longer, overinfection results, which kills many of the amphipods or causes abnormal and retarded development of the parasites. In order to insure that clean Hyalella are used, they should be collected from a habitat lacking fish. Amphipods may be kept in battery jars or similar containers with a few willow rootlets and duck weed. The addition of a bit of yeast to the culture each week, and aeration are recommended.

Dissection of the intermediary crustaceans under a dissecting microscope and examination of the intestinal contents a few hours after exposure will show acanthors and empty egg shells. Later examination over spaced intervals will show the passage of the acanthors through the gut wall and subsequent larval development.

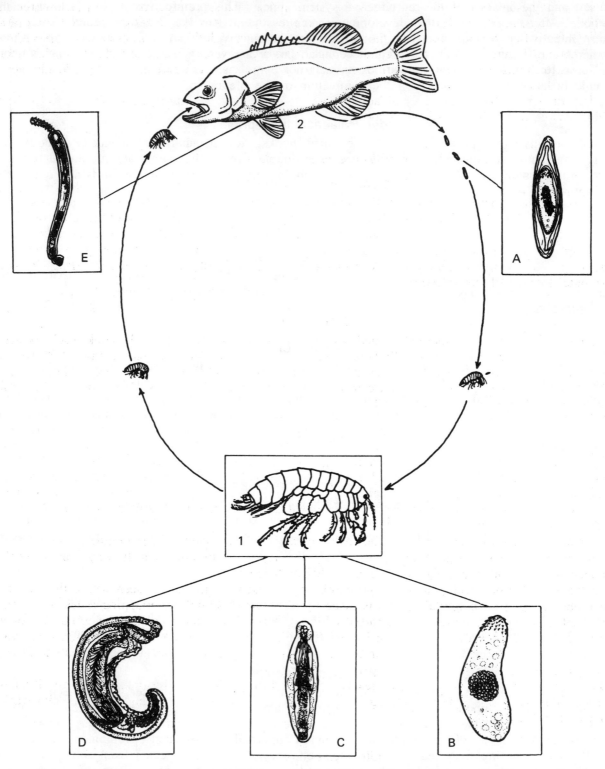

Plate III. Life cycle of *Leptorhynchoides thecatus:* (A) Egg containing acanthor; (B) Acanthor in gut of amphipod; (C) Acanthella in hemocel of amphipod; (D) Cystacanth in hemocel of amphipod; (E) Adult in entral cavity of fish. Numbers 1 and 2 indicate intermediate and final hosts respectively. After DeGiusti (1949) from Meyer (1954).

When the cystacanths mature, they should be fed to small rock bass or sunfish. Fish can be infected by feeding them infected amphipods or by dissecting out the cystacanths and feeding them by means of a bulbed pipette. After infecting the fish, they should be placed in individual containers and observed to check whether the cystacanths are regurgitated. To insure having clean fish, they should be held for about 2 months during which the feces are examined for parasite eggs. For further life cycle details, consult De Giusti (1949).

Other often available Echinorhynchidae with their hosts include *Echinorhynchus leidyi*, freshwater fish; and *Acanthocephalus jacksoni*, freshwater fish.

Family Pomphorhynchidae

Pomphorhynchus bulbocolli

This species is a common parasite of the white sucker *(Catostomus commersoni)* and is occasionally found in closely related fish. The long cylindrical neck, between the trunk and the proboscis, distinguishes it from other species in fish. At the distal end of this thin, cylindrical neck there is a large, globular bulb, which is an accessory attachment organ, since the tissues of the host intestine grow around it (Figure 134). Macroscopically, the fibrous capsules surrounding the proboscis and bulb are clearly visible on the outer surface of the intestine.

Description

Proboscis long, almost cylindrical, armed with 12 longitudinal rows of 12 to 14 hooks each; posterior hooks more slender than anterior. Proboscis receptacle double-walled, inserted at base of proboscis and extending throughout the length of the neck. Lemnisci short, claviform. Testes oval, tandem, near middle of trunk. Cement glands 6, rounded to oval. Eggs fusiform, 53-83 x 8-13 μ, with a prominent prolongation at each pole of middle shell.

Life Cycle

The eggs, fully developed when passed in the host feces, hatch when they are ingested by the amphipod *Hyalella azteca*. A recently released acanthor has 55 to 60 small accessory spines at the anterior end, and a central syncytial mass of nuclei in a surrounding syncytial area containing 18 giant nuclei. Within about 5 hours, the acanthors make their way through the intestinal wall of the host into the hemocel.

After about 14 days of development, the central embryonic mass shows the primordia of the proboscis, proboscis sheath, anterior ganglion, gonads, and other reproductive structures. It is now an acanthella. Sixteen days postinfection, the sexes can be differentiated.

On the 19th day of development, the primordia of proboscis hooks can be seen within the still inverted proboscis. The male genital system is made up of a pair of slightly overlapping testes, 6 cement glands, Saefftigen's pouch, cirrus, and a partially formed bursa. The female genital system consists of 4 to 6 embryonic ovarian masses within the ligament, uterine bell, and beginnings of the uterus and vagina. The basic structures of the adult worm are attained by the 33rd day. After 35 to 37 days, the larva, now a cystacanth, is infective to the proper fish host.

Based upon sperm production, males reach sexual maturity in 3 weeks postinfection, while it takes 10 weeks for females to produce viable eggs. At 10 weeks, females reach a length of 14.5 mm while males measure 10.6 mm.

Life Cycle Exercise

Since this species utilizes the same amphipod host as does *Leptorhynchoides thecatus*, obtain the worms from suckers and follow experimental procedure given for intermediate host of *L. thecatus*. If suckers are available, they should be fed the amphipods harboring the infective cystacanths. To insure having clean fish, they should be held for about 2 months during which time the feces are examined repeatedly for parasite eggs.

Order Archiacanthocephala

Body medium to large; proboscis with few or many hooks; main lacunar canals median; nuclei in body wall few and large; proboscis receptacle single- or double-layered; ligament sacs persistent in females; males with 8 uninucleate cement glands. Adults in terrestrial vertebrates; cystacanths in insects (grubs, cockroaches, etc.).

Family Oligacanthorhynchidae

Macracanthorhynchus hirudinaceus

Description

This species, the giant thorny-headed worm of swine, is the most widely known member of the phylum. This is attributable to the facts that it has attained practically cosmopolitan distribution through human agencies and that the species has direct economic importance. The body is circular in cross section when preserved, tapering from a rather broad anterior region to a slender posterior end. Females may reach 48 cm, males 11 cm in length. The proboscis is globular, with 6 spiral rows of 6 hooks each. Lemnisci flat, ribbonlike, with a few giant nuclei anteriorly. The testes are elongate and located in the midbody or slightly anterior to it. The 8 cement glands are elliptical, often arranged in pairs as an elongate series (Figure 140). The eggs are described under Helminth Eggs, p. 210.

Life Cycle

The embryonated eggs are passed in the feces of swine and are very resistant to adverse conditions, being able to survive for several years. When the eggs have been ingested by the larvae of a number of species of Scarabaeidae beetles of the genera *Cotinus* and *Phyllophaga*, hatching occurs within an hour. The released acanthors migrate through the intestinal wall into the body cavity. They usually remain attached to the outer surface of the gut for 5 to 20 days, after which the larvae become free in the body cavity. In the meantime, the larval hooks have been lost and the primordia of the various organs of the adult have appeared, the most conspicuous of which is the proboscis with its hooks. At 24 C, cystacanths develop in experimentally infected grubs in 60 to 90 days, but the minimum developmental period in nature is nearer the upper limit. The infective cystacanths are essentially juvenile parasites, possessing the structures and organ systems of adult worms. After entering the final host, further development is merely an elaboration of the organ systems already formed.

Swine become infected by eating either the grubs or the adult beetles which harbor the infective cystacanths. Prepatency in swine is believed to be from 2 to 3 months, and patency may continue for 10 months.

Life Cycle Exercise

Adult worms are available from abattoirs where swine, which have had access to pasture, are slaughtered. The eggs, teased from gravid females in physiological solution, should be mixed with slightly moist soil. Add the larval beetles and allow them to feed for several days. Grubs for experimental purposes should be collected from soil to which swine have not had access for several years. To observe the hatching of the eggs, the passage of the acanthors through the intestinal wall, and subsequent development of the larval stages in the body cavity, examine the experimentally infected grubs, kept at room temperature, at spaced intervals varying from a few hours to about 90 days postinfection.

Other often available Oligacanthorhynchidae with their hosts include *Pachysentis canicola*, rodents; and *Oncicola canis*, carnivores.

Family Moniliformidae

Moniliformis moniliformis

This species (*M. dubius* of some authors) is a common parasite of rats throughout the world, especially in regions where the cockroach intermediate hosts are present.

Description

The body shows prominent sexual dimorphism; females up to 32 cm, males 15 cm in length. Both sexes show conspicuous pseudosegmentation, except at the extremities. The proboscis is cylindrical and club-shaped, with 12 to 14 longitudinal rows of 10 or 11 hooks, occasionally 9 or 12 hooks, in each longitudinal series. The lemnisci are long and narrow, each with a few giant nuclei. Male organs are confined to the posterior nonannulated region of the body. The testes are very long, elliptical, in contact in young males, but may be widely separated in old males. Eight cement glands, pyriform in shape, commonly arranged in an elongate series but rarely in pairs (Figure 144). Eggs 90-125 x 50-62 µ.

Life Cycle

When the eggs, fully developed when passed in the host feces, are ingested by cockroaches *(Periplaneta americana)*, they hatch in the midintestine within 24 to 48 hours. The released acanthors, bearing 6 bladelike hooks, penetrate the intestinal wall and appear as minute iridescent specks on the outer surface in 10 to 12 days, drop into the body cavity, and there continue their development. The infective cystacanths appear about 7 to 8 weeks after infection.

The cycle is completed when cockroaches harboring cystacanths are eaten by rats. The released cystacanths attach to the wall of the intestine, and mature in 5 to 6 weeks.

Life Cycle Exercise

Adults can be obtained from rats in habitats where cockroaches are also present. The eggs, teased from gravid females, in physiological solution, should be mixed with a bit of pabulum in a small glass jar. Introduce clean cockroaches, from which food has been withheld for 24 hours, to the jar with the egg-contaminated pabulum. After being fed, the roaches should be removed to battery jars or similar containers with moist sand over which are placed wet paper towels, to maintain the moisture and provide shelter for the roaches. Cover jars to retain the moisture. Roaches thrive on pabulum and are fond of apples.

Infect a sufficient number of roaches so that dissections can be made over spaced intervals, say from 2 days through 2 months, to observe the hatching of the eggs, the passage of the acanthors through the intestinal wall, and subsequent development of the larval stages in the body cavity. To insure against using naturally infected roaches, they should be collected from rat-free localities or reared in the laboratory. Dissection of an adequate sample of a wild population will determine if they are suitable for experimental purposes.

White rats can be used as experimental final hosts. The cystacanths can be fed orally by means of a bulbed pipette. The experimental rats should be kept in a roach-proof enclosure to prevent further infections. Beginning about the fifth week postinfection, the prepatency period can be determined by examining the rat's feces.

Moniliformis clarki, chiefly from squirrels, may also be available.

Order Eoacanthocephala

Body usually small; proboscis with few or many hooks; main lacunar canals median; nuclei of body wall few, usually large; proboscis receptacle single-layered; ligament sacs persistent in females; in males cement glands syncytial, rarely divided into 2 lobes. Adults mostly in fish, rarely in turtles; cystacanths in Crustacea.

Family Neoechinorhynchidae

Neoechinorhynchus emydis

This species, one of the few thorny-headed worms occurring as common parasites of turtles in the United States, is of special interest because 2 intermediate hosts are involved in the life cycle.

Description

The proboscis, globular in shape, measures about 0.18 mm in length by 0.22 mm in width. It bears 3 circles of 6 hooks each. There are 6 subcuticular nuclei, of which 5 are in the middorsal, and 1 in the midventral line near the anterior end. The longer of the lemnisci contains 2 nuclei and the shorter, a single nucleus.

Gravid females are 10.2 to 22.2 mm long by 0.8 to 1.3 mm wide. The genital pore is subterminal, opening on the ventral surface of the body. Adult males measure 6.3 to 14.5 mm in length and 0.5 to 1.0 mm in width. The reproductive organs occupy about the posterior half of the body. The testes are contiguous, in tandem, and immediately followed by the large, syncytial cement gland typically with 8 large nuclei. Posterior to the cement gland is a large globular cement receptacle, with a thin-walled cement duct passing posteriorly to the bursa. Eggs ovoidal, 18-22 μ.

Life Cycle

Adult females in the intestine of turtles (*Emys blandingi* and species of *Graptemys*) lay fully embryonated eggs which are voided with the feces. The ostracod, *Cypria maculata*, ingests the eggs which hatch within 8 to 12 hours. Upon release from the egg, the acanthors promptly penetrate the intestinal wall, arriving in the body cavity within 24 hours. After about 19 days, the larvae have reached the definitive shape, the proboscis with its fully formed hooks is inverted in the proboscis receptacle, and the reproductive apparatus is developed to the extent that the sex can be readily determined. Completely developed unencysted larvae are present in the ostracod 21 days postinfection. The proboscis is capable of full eversion and inversion, the reproductive structures are well developed, and the 6 giant subcuticular nuclei occupy the same position as in the adult worm.

Snails, *Campeloma rufum*, serve as the second intermediate host. They become infected by eating the ostracods harboring the fully developed acanthellas. In the intestine of the snail, ostracods are digested and the larvae freed. Those which have developed to the infective stage penetrate the intestinal wall and burrow into the tissues, especially the foot, where the cystacanths appear as opaque spots. Some growth of the larvae occurs within the snail, which serves as the principal source of infection of the turtle, whose food is largely mollusks.

Life Cycle Exercise

Adult worms can be obtained from the appropriate turtle hosts. Eggs, teased from gravid females, should be placed in a small dish of water to which the ostracods are added. In order to insure that clean *Cypria* are used, they should be collected from a habitat lacking turtles.

Dissection of the ostracods under a dissecting microscope and examination of the intestinal contents a few hours after exposure will show the acanthors and empty eggs shells. Later examinations over spaced intervals will show the passage of the acanthors through the gut wall and further larval development.

Other often available Neoechinorhynchidae with their hosts include *Neoechinorhynchus cylindratum* and *N. rutili*, freshwater fish; *N. chrysemydis*, turtles; *Octospinifer macilentus*, freshwater fish.

Key to Common Families of Acanthocephala

In this key and those like it which follow, numbers in parentheses indicate the alternative condition. For example, if the proboscis of worm in 1(2) has hooks instead of fine spines, proceed to 2(1).

Capital letters in parentheses refer to hosts: F-Fish, A-Amphibia, R-Reptiles, B-Birds, M-Mammals.

1(2) Proboscis with fine spines, not invaginable; without proboscis receptacle. Order SPHENACAN-THOCEPHALA (Figure 130) .... (B) Apororhynchidae

2(1) Proboscis with hooks, invaginable; with proboscis receptacle 3

3(6) Proboscis usually with few hooks; proboscis receptacle single-layered; cement gland syncytial, usually single; embryo unarmed; mostly in fish, rarely in turtles ....
.... Order EOACANTHOCEPHALA 4

4(5) Trunk glabrous (Figure 131) .... (F) Neoechinorhynchidae

5(4) Trunk spinose (Figure 132) .... (F) Quadrigyridae

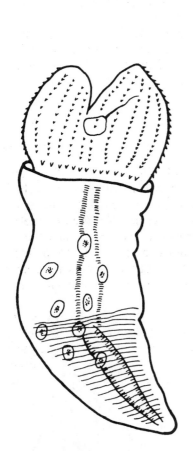

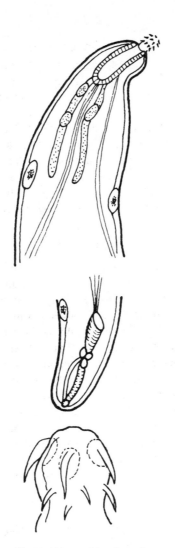

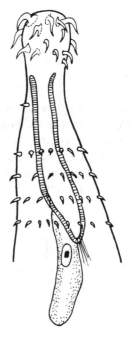

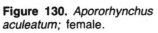

Figure 130. *Apororhynchus aculeatum;* female.

Figure 131. *Neoechinorhynchus rutili.*

Figure 132. *Quadrigyrus torquatus.*

6(3) Proboscis with few or many hooks; proboscis receptacle single- or double-layered; cement gland not syncytial, not single; not limited to fish 7

7(20) Proboscis usually with many hooks; proboscis receptacle double-layered; cement gland divided into 2 or more lobes; egg fusiform, usually with polar prolongations of middle shell; embryo with hooks at one end only; mostly in fish and aquatic birds and mammals. . . . .
. . . . .Order PALAEACANTHOCEPHALA 8

8(9) Trunk glabrous 10

9(8) Trunk spinose 16

10(11) Cement glands 4, occupying greater length of trunk; testes cylindrical (Figure 133) . . . .
. . . .(F) Fessisentidae

11(10) Cement glands 3 to 8, not occupying greater length of trunk; testes not cylindrical 12

12(13) Neck very long, cylindrical or spirally twisted, with or without bulbous swelling (Figure 134)
. . . .(F) Pomphorhynchidae

13(12) Neck short, if present, never cylindrical or spirally twisted, without bulbous swelling 14

14(15) Cement glands usually compact or pyriform (Figure 135) . . . .(F) Echinorhynchidae

15(14) Cement glands usually tubular, long (Figure 136) . . . .(B, occ. R) Plagiorhynchidae

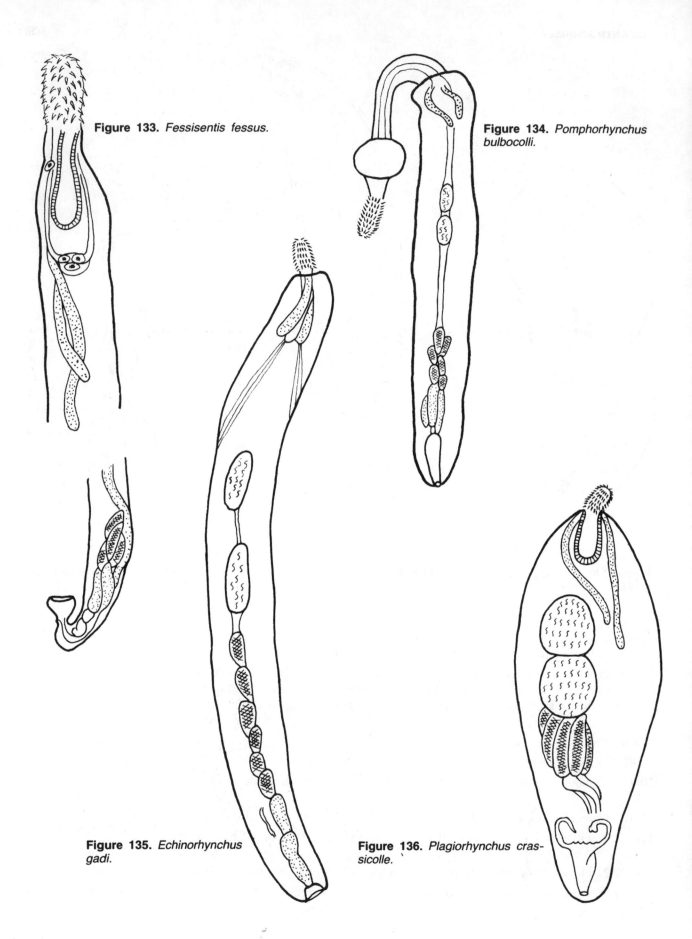

Figure 133. *Fessisentis fessus.*

Figure 134. *Pomphorhynchus bulbocolli.*

Figure 135. *Echinorhynchus gadi.*

Figure 136. *Plagiorhynchus crassicolle.*

16(17) Trunk spines extensive or limited; proboscis hooks very numerous; cement glands 2 to 8, usually elongate (Figure 137) .... (F) Rhadinorhynchidae

17(16) Trunk spines limited to anterior portion; proboscis hooks usually not numerous; cement glands 2 to 6, exceptionally 8, usually tubular 18

18(19) Eggs with polar prolongation of middle shell (Figure 138) .... (B, M) Polymorphidae

19(18) Eggs without polar prolongation of middle shell (Figure 139) .... (B) Filicollidae

20(7) Cement gland divided into 3 or more lobes; egg oval to elliptical, without polar prolongations of middle shell; embryo with hooks at both ends and numerous minute spines elsewhere; in terrestrial vertebrates .... Order ARCHIACANTHOCEPHALA 21

21(22) Proboscis with few hooks, in spiral rows; protonephridial organ present (Figure 140) .... (B, M) Oligacanthorhynchidae

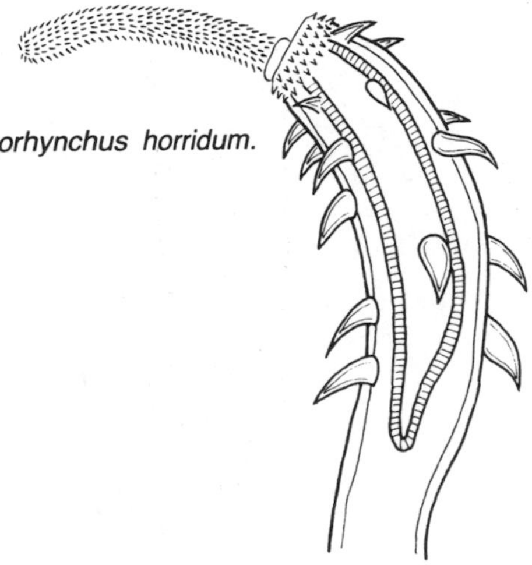

Figure 137. *Rhadinorhynchus horridum.*

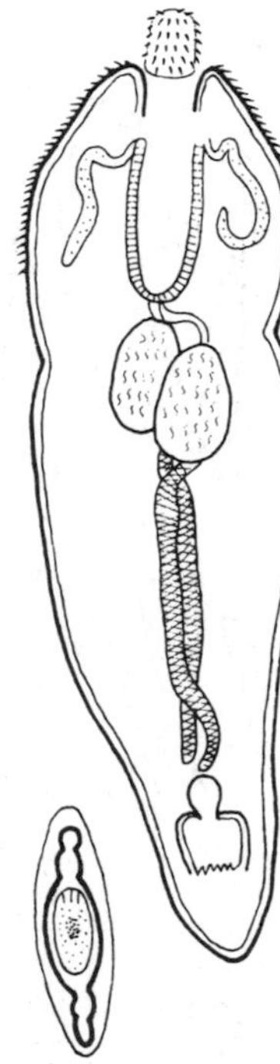

Figure 138. *Polymorphus minutus.*

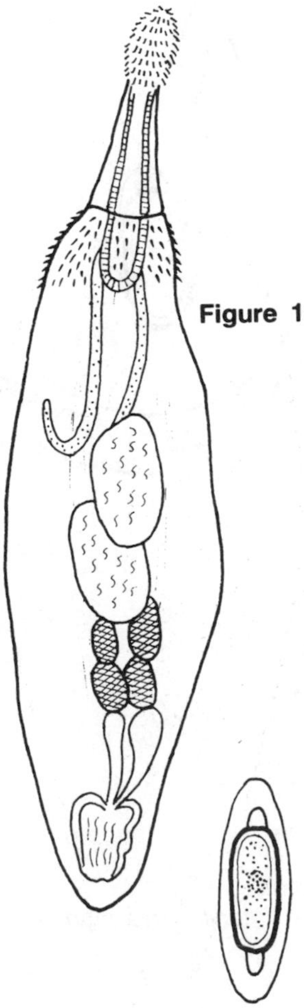

Figure 139. *Filicollis anatis.*

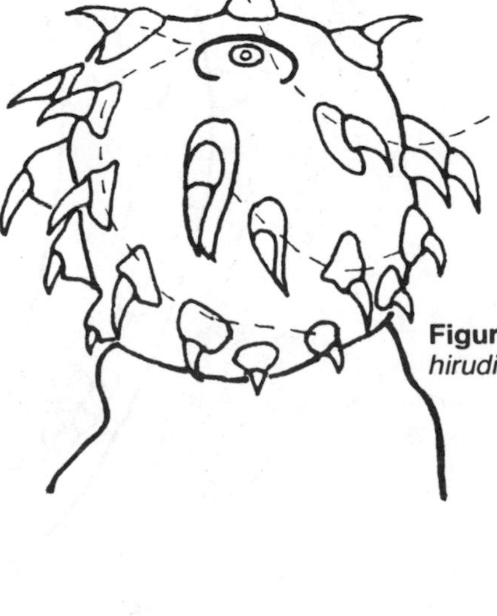

Figure 140. *Macracanthorhynchus hirudinaceus.*

22(21) Proboscis with numerous hooks, mostly in longitudinal rows; protonephridial organ absent 23

23(24) Proboscis receptacle inserted anterior to base of proboscis and dividing the latter into 2 regions (Figure 141) . . . . (B, M) Centrorhynchidae

24(23) Proboscis receptacle inserted at base of proboscis 25

25(26) Cement glands compact 27

26(25) Cement glands long, tubular; proboscis hooks more or less uniform (Figure 142) . . . . (B, M) Prosthorhynchidae

27(28) Proboscis hooks divided into 2 groups of different size and shape (Figure 143) . . . . (B, M) Gigantorhynchidae

28(27) Proboscis hooks otherwise (Figure 144) . . . . (M, occ. B) Moniliformidae

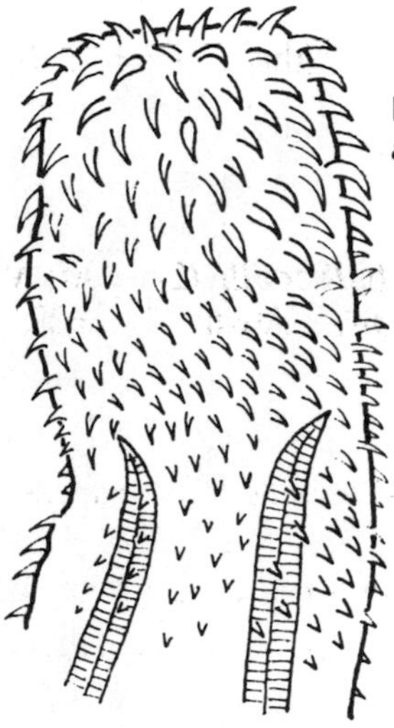

Figure 141. *Centrorhynchus aluconis.*

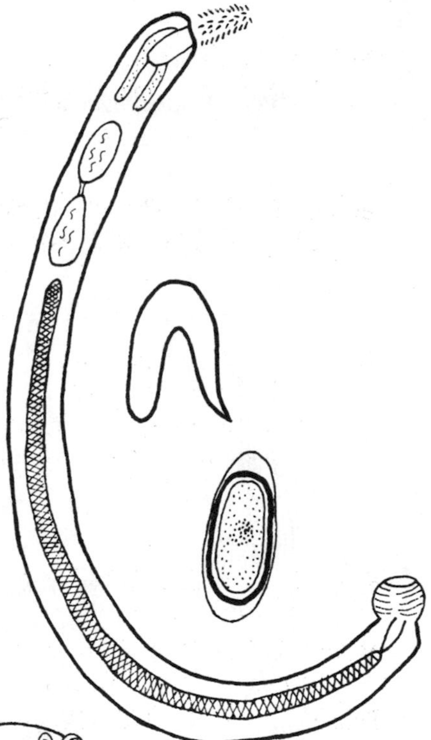

Figure 142. *Prosthorhynchus scolopacidis.*

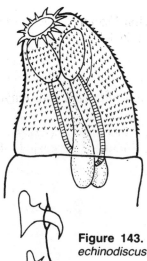

Figure 143. *Gigantorhynchus echinodiscus.*

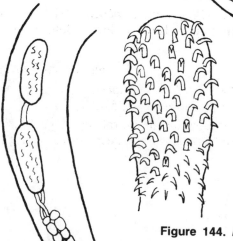

Figure 144. *Moniliformis moniliformis.*

REFERENCES

Bullock, W.L. 1969. Morphological Features as Tools and Pitfalls in Acanthocephalan Systematics. *In* Problems in Systematics of Parasites, ed. G.D. Schmidt. University Park Press, Baltimore, pp. 9-43.

Nicholas, W.L. 1967 & 1973. The Biology of Acanthocephala. *In* Advances in Parasitology, ed. B. Dawes. Academic Press, New York, vol. 5, pp. 205-246; vol. 11, pp. 671-706.

Van Cleave, H.J. 1948. Expanding horizons in the recognition of a phylum. J. Parasitol. 34: 1-20.

Whitefield, P.J. 1970. The egg sorting function of the uterine bell of *Polymorphus minutus* (Acanthocephala). Parasitology 61: 111-126.

Yamaguti, S. 1963. Systema Helminthum. Acanthocephala, vol. 5, Interscience Publishers, Inc., New York, 423 p.

Echinorhynchidae

DeGiusti, D.L. 1949. The life cycle of *Leptorhynchoides thecatus* (Linton), an acanthocephalan of fish. J. Parasitol. 35: 437-460.

Pomphorhynchidae

Jensen, T. 1953. The life cycle of the fish acanthocephalan, *Pomphorhynchus bulbocolli* (Linkins) Van Cleave, 1919, with some observations on larval development *in vitro*. Diss. Abstr. 12: 607.

Oligacanthorhynchidae

Kates, K.C. 1943. Development of the swine thorn-headed worm, *Macracanthorhynchus hirudinaceus*, in its intermediate host. Amer. J. Vet. Res. 4: 173-181.

_____. 1944. Some observations on experimental infections of pigs with the thorn-headed worm, *Macracanthorhynchus hirudinaceus*. Amer. J. Vet. Res. 5: 166-172.

Moniliformidae

King, D., and E.S. Robinson. 1967. Aspects of the development of *Moniliformis dubius*. J. Parasitol. 53: 142-149.

Moore, D.V. 1946. Studies of the life history and development of *Moniliformis dubius* Meyer, 1933. J. Parasitol. 32: 257-271.

Schaefer, P.W. 1970. *Periplaneta americana* (L) as intermediate host of *Moniliformis moniliformis* (Bremser) in Honolulu, Hawaii. Proc. Helminthol. Soc. Washington 37: 204-207.

Neoechinorhynchidae

Fisher, F.M., Jr. 1960. On Acanthocephala of turtles, with the description of *Neoechinorhynchus emyditoides* n. sp. J. Parasitol. 46: 257-266.

Hopp, W.B. 1954. Studies on the morphology and life cycle of *Neoechinorhynchus emydis* (Leidy), an acanthocephalan parasite of the map turtle, *Graptemys geographica* (Le Sueur). J. Parasitol. 40: 284-299.

Phylum Nematoda

The body of nematodes is cylindrical except in the case of some females which are spindle-shaped (Figure 190) or twisted when gravid. There are no appendages in the parasitic species. Sexes are separate, with females generally larger than males. A pseudocel, or false body cavity, is present (Figures 145, A; 146). The tail consists of that part of the body posterior to the anus.

The anterior end of the body forms the head. The mouth is terminal and associated with it are various structures such as lips, pseudolabia, odontia, cephalic papillae, amphids, collarettes, and cordons.

The primitive number of lips is 6 of which 2 are dorsal, 2 lateral, and 2 ventral. On each of these triangular lips are 3 sensory papillae, 1 at the tip near the opening of the mouth and 2 at the base, 1 on each side. Papillae at the tips of the lips form the inner circle and those at the base constitute the outer circle (Figure 145, G).

Through fusion, the number of lips is reduced and a new arrangement established. When the 2 dorsal lips fuse and 1 lateral and 1 ventral lip on each side fuse, 3 lips occur (Figure 145, G). This is the situation in the suborders Ascaridina, Oxyurina, and Heterakina, which have a dorsal and 2 ventrolateral lips. On the other hand, 1 dorsal, 1 lateral, and 1 ventral lip on each side may fuse to form 2 large lateral lips, as occurs in the Spirurina (Figure 193). These new structures in the Spirurina are known as pseudolabia.

The lips may be lost entirely and not replaced by any other structures, as occurs in the Camallanina (Figure 181) and some Filariina (Figure 202). Other filarioids have some type of development such as a circumoral elevation (Figure 205) or toothlike projections known as odontia, or there may be lateral projections known as helmets (Figure 194).

While the number of labial papillae may be reduced by fusion or loss, the inner and outer circles are usually recognized. They are tactile in function.

The amphids consist of 2 pores, 1 on each side of the head near the base of the lips (Figure 205). They are chemoreceptors.

The posterior end of males varies in structure. In the Strongylina (Figure 145, H), it is a bilateral, cuticular bursa supported by a system of paired muscular rays and in the Dioctophymatina, it is a fleshy, rayless, cup-shaped sucker (Figure 215). Ascaridina have blunt, ventrally curved tails (Figure 169), and Oxyurina long, attenuated ones (Figure 172). The tails of Spirurina and Filariina are often coiled and somewhat tendrillike. Sensory papillae are present ventrally or laterally, or both, on the posterior end of the body (Figure 145, I), except in the Strongylina.

The body wall consists of 3 layers: the outermost cuticle, middle hypodermis, and innermost somatic muscles (Figure 146). The cuticle is a thin, transparent, elastic acellular layer covering the outside of the body. It extends inwardly at the mouth, anus, and vulva. The cuticle has several types of external markings. Fine transverse striations on the surface are present in most species. Strong annulations presenting the appearance of external segmentation occur in some species (Figure 188) and longitudinal ridges are present in others. Lateral expansions of the cuticle are called alae (Figure 171). They may be present at the anterior end of the body as cervical alae or the posterior end as caudal alae, or the full length of the body as lateral alae. Inflated areas of the cuticule resemble an irregular arrangement of plaques or warts (Figure 185). A preanal circular structure having the appearance of a sucker is present in the males of some species (Figure 176). Expansions of the cuticle outward from the head are termed helmets or hoods (Figure 194).

Several other cuticular structures are common. Cordons and epaulets are ribbon-shaped bands of cephalic cuticle that begin much as cephalic alae and then loop anteriorly on the ventral side where they may unite (Figure 198). Spines occur on the posterior margins of body annulations, forming spined collarettes (Figure 199), or expanded head balloons (Figure 195) or in rows on the body. Plectanes are thickened cuticular supports for genital papillae.

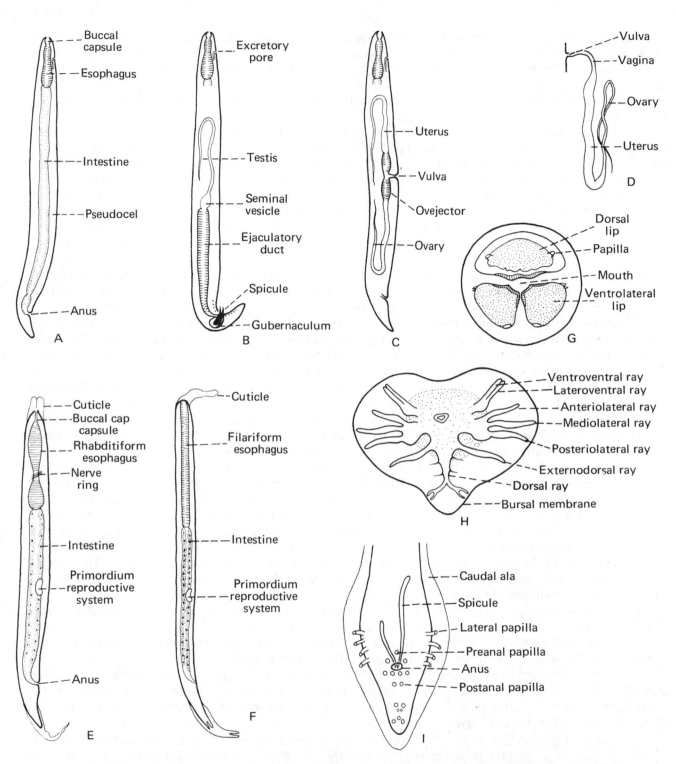

Figure 145. Basic morphology of nematodes. (A) Diagrammatic representation of digestive system; (B) Diagrammatic representation of male reproductive system typical of *Ascaris* type; (C) Diagrammatic amphidelphic reproductive system of a female nematode; (D) Monodelphic reproductive system of a female nematode; (E) Second stage rhabditiform larval nematode, showing loosening cuticle of first stage larva; (F) Ensheathed third stage infective filariform larva enclosed in cuticle of second stage rhabditiform larva; (G) *En face*, or end, view of adult *Ascaris suum*, showing lips and papillae; (H) Copulatory bursa of bursate male strongyle nematode *(Castrostrongylus castoris)* spread to show the basic arrangement of the rays; and (I) Ventral view of a nonbursate nematode *(Physaloptera* sp.), showing caudal alae and basic location of the various caudal papillae.

The outermost cellular layer of the nematode body is not a true epidermis since it is covered by cuticle. It is designated as the hypodermis and forms a thin layer of cells lying between the cuticle and the somatic muscles. On the dorsal, ventral, and each lateral side of the body is an elongate thickening of the hypodermis that extends inward between the muscle cells into the body cavity. These are known as the hypodermal chords and may extend the full length of the body, or be present only in the anterior part (Figure 146).

The somatic muscle cells lie on the hypodermis and form the innermost layer of the body wall. They are separated into 4 major fields by the hypodermal chords. The terms polymyarian, meromyarian, and holomyarian refer to the number of muscle cells between the hypodermal chords. These terms serve in a general way to designate groups of nematodes, i.e., polymyarian forms have many muscle cells between each hypodermal chord (Figure 146) and meromyarian few and large cells; holomyarian forms have no hypodermal chords except in the anterior part of the body so that in a section from elsewhere the chords are absent and the cells appear closely packed, numerous, and of uniform size. The arrangement of the interchordal muscle cells is not consistent with any taxon but is useful in identifying groups of worms appearing in sections of tissue.

The external openings include the mouth, anus, amphids, and excretory pore in both sexes (Figures 145, A, B, C; 205). Females possess a vulva in addition to the other openings (Figure 145, C). It is located on the ventral side of the body anywhere from the anterior to the posterior end. In the Secernentea, the phasmidial cells open through a minute pore on each side of the body in the anal region.

The alimentary canal is a more or less straight tube consisting of a stoma (buccal capsule, vestibule, or mouth cavity), esophagus, intestine, and rectum. In males, the rectum opens into a cloaca. The anus is ventral and usually subterminal (Figure 145, A, E, F).

The stoma is the cavity into which the mouth opens (Figure 145, A, E). It may be well developed with thick walls (Figure 157), rudimentary (Figure 151), or absent (Figure 172). When present, it may be at the extreme anterior end of the esophagus or partially surrounded by esophageal tissue.

The esophagus is an elongate muscular organ of variable shape. It may (1) be rhabditiform, in which there is an enlarged anterior portion connected by an isthmus, or a thin part, to a posterior bulb or enlargement (Figure 145, E); (2) have a posterior bulb with a cuticularized valve inside (Figure 171); (3) be cylindrical and muscular (filariform) (Figure 145, F); (4) cylindrical and composed of a short anterior muscular and long posterior glandular part (Figure 190); (5) a short anterior muscular part and a long posterior part consisting of a single row of cuboidal cells in which is embedded a capillary tube (Figure 214). The lumen of the esophagus is lined with cuticle and is triradiate, except in the capillary tube, with one ray directed ventrally.

The intestine is a more or less straight, thin-walled tube composed of epithelial cells. The type of cells vary in the different groups of nematodes and is helpful in distinguishing them, especially in sections of tissue containing the worms.

The intestine may be composed of 3 parts: a short part next to the esophagus that is separated from the remainder by a constriction in the ventriculus (Figure 167), the intestine proper, and the prerectal part. In some species, an anteriorly or posteriorly directed cecum originates from the ventriculus. The rectum is the terminal part of the intestine and is lined with cuticle.

The male reproductive system normally consists of 3 principal parts: the testis (there may be 2 in some species), the seminal vesicle, which is usually dilated and serves as a sperm storage organ, and the vas deferens. The terminal part of the vas deferens is generally a muscular structure that functions as an ejaculatory duct (Figure 145, B).

Accessory structures are associated with the male reproductive system. They include 1 or 2 cuticularized spicules in a spicular pouch that opens into the dorsal part of the cloaca (Figure 145, B). The spicules are protrusible and function in mating. Sometimes a gubernaculum, a thickening of the dorsal wall of the spicular pouch (Figure 145, B), is present and serves as a guide in directing the spicules during protrusion. A somewhat complex structure formed by a thickening of the lining of the cloaca of lungworms is termed the telamon (Figure 165).

The female reproductive system consists of ovaries, oviduct, uterus, and a cuticular lined vagina opening to the outside of the body through the vulva which is usually located on the ventral side of the body. There may be 1 ovary, which is monodelphic; 2 ovaries, which is amphidelphic; or more than 2, which is polydelphic (Figure 145, C, D). In amphidelphic forms, the ovaries may be arranged opposite each other so that one lies in each end of the body, or they may be parallel and both lie in either the anterior or posterior parts of the body. The terminal part of the uterus may be muscular, forming an ovejector which forces the eggs into the vagina. The uterus of mature worms generally is filled with eggs or larvae.

The overt parts of the excretory system consist of the minute excretory pore opening ventral to the esophagus and its slender duct directed posteriorly (Figure 145, B). The remaining parts of the system consist of ducts in the lateral chords. The arrangement of the ducts may vary considerably in pattern from the basic H-shaped type found in the oxyurids. In some forms, there is a gland that lies posterior to the cross bar of the H into which it empties. In others, parts of the basic H-shaped organ are missing.

The conspicuous nerve ring surrounding the esophagus about midway between the anterior and posterior ends represents the central nervous system (Figure 145, E). From it, fibers extend (1) anteriorly to innervate the cephalic papillae of the lips, amphids, deirids (cervical papillae), and esophagus, and (2) posteriorly to the genital papillae, phasmids, and intestine. Peripheral nerves originate from the fibers and extend to the body muscles. Generally, only the nerve ring is seen.

The eggs of parasitic nematodes vary in shape, size, thickness of shell, and external markings, as well as the stage of development when laid (See p. 212). Each ovum comprises an egg cell that is enclosed in 3 primary layers which consist of (1) the thin vitelline membrane surrounding the egg cell, (2) the middle thick egg shell, and (3) the outer protein layer which may be stained yellow with bile. The stage of development when laid may be a single cell or with all succeeding stages up to and including ensheathed third stage larvae. Some precocious eggs hatch in the intestine or lungs, in which cases larvae appear in the feces. Others hatch in the uterus of the female worms and active larvae are born.

General Biology

The life cycles of nematodes are of 2 basic types: direct or monoxenous (i.e., with only 1 host in the cycle) and indirect or heteroxenous (i.e., with 2 or more hosts in the cycle). The basic pattern of development is similar whether the life cycle is direct or indirect. Larvae hatching from the eggs progress through a series of stages in their development. Beginning with the first stage, each one is separated by a molting of the cuticle. There are 4 larval stages, i.e., first, second, third, and fourth, followed by the adult. The third stage is infective to the final host. The pattern of growth of the larvae (L) and occurrences of the successive molts (M) may be expressed as follows:

$$\text{Egg} \longrightarrow L_1 + M_1 \longrightarrow L_2 + M_2 \longrightarrow \boxed{L_3 + M_3} \longrightarrow L_4 + M_4 \longrightarrow \text{Adult}.$$

In many parasitic nematodes (Orders Rhabditida, Strongylida) with a direct life cycle, the first, second, and third stages are free in the soil. The first 2 stages feed on organic material. The third stage which retains the shed cuticle of the second stage as an enclosing sheath, is unable to ingest food (Figure 145, E, F). In some species, the first (*Ascaris*) and second molts (some hookworms, strongyles, and trichostrongyles) take place inside the egg shell.

Third stage larvae enter the final host when swallowed with food or by penetrating the skin, as in the case of hookworms. The third and fourth molts are completed inside the final host.

Nematodes with an indirect life cycle have the same stages but reach the final host through the means of a vector or an intermediate host, usually an arthropod, but other invertebrates are utilized. Vectors are transmitters of parasites. If the transmitter is essential in the life cycle of the parasite, it is a biological vector; if it is unessential, it is a mechanical vector.

In oviparous forms with the indirect cycle, the eggs are embryonated when laid and in most cases hatch only when ingested by the intermediate host, in whose body the third or infective larval stage is reached through the necessary molts. In some of the lungworms, the eggs hatch on the ground or in the feces of the final host and the larvae penetrate land snails, which serve as the intermediate host. Infection of the final host occurs when the intermediaries containing the infective larvae are swallowed. The third and fourth molts take place in the final host.

In the ovoviviparous nematodes with an indirect cycle, the intermediaries, such as copepods, become infected by ingesting early larval stages free in the water, such as those of dracunculids or mosquitoes and other bloodsucking arthropods feeding on animals containing microfilariae in the blood, which are the larval stages of filariae worms. The larvae develop to but not beyond the third, or infective stage, in the vector and intermediate hosts. Infection of the final hosts occurs when infected copepods are swallowed or when infective larvae are injected into or onto the body when mosquitoes and other hematophagous arthropods are feeding.

Class SECERNENTEA

The esophagus is club-shaped, cylindrical, or with a posterior bulb, and muscular, or it may be muscular in the anterior and glandular in the posterior parts. A bilobed copulatory membranous bursa, when present, is supported by 6 pairs of lateral rays. Phasmids are present, but difficult to see in adult worms.

Order Ascarida

Members of this order are generally stout worms, usually with 3 lips. The esophagus is muscular, with or without a caudal bulb, and may or may not have a ventriculus at the posterior end; the intestine may or may not have a diverticulum at the anterior end. Spicules are small and equal or unequal. Parasitic in the alimentary canal of all classes of vertebrates.

Family Ascaridae

Esophagus is somewhat club-shaped, lacking both a bulb and a ventriculus posteriorly. Male without precloacal sucker. Cervical alae usually absent (present in *Toxascaris*).

Ascaris suum

Ascaris suum and *A. lumbricoides* occur in the small intestine of swine and man, respectively. Anatomically, specimens recovered from the 2 hosts are virtually indistinguishable, but physiologically, they differ in that the eggs from worms in swine do not readily infect man and vice versa.

Description

This is the largest of the swine intestinal nematodes: females up to 35 cm in length and about 5 mm in diameter; males up to 17 cm long by 3 mm in diameter. Males are distinguished by the ventrally curved tail, and 2 copulatory spicules, but no gubernaculum, while the females have a blunt tail. The vulva is situated near the end of the first third of the body. The mouth is guarded by 3 lips (Figure 145, G): a dorsal and 2 ventrolateral lips, each with 2 basal papillae. Each lip bears on its inner margin 2 rows of small denticles. The eggs, identical with those of *A. lumbricoides,* are described under Helminth Eggs, p. 213.

Directions for Study

Secure an entire preserved specimen from the preparation table. Follow the outline, making the dissection and identifying the parts. The females are larger than the males and have a straight, blunt tail, while that of the males is curved ventrally into a hooklike shape. Locate the anus near the posterior end on the ventral side of the body. Sometimes a pair of hyaline, needlelike spicules protrudes from the anus of the male. Examine the surface of the body to see the differentiation into the lateral, dorsal, and ventral surfaces. Note the fairly conspicuous light colored stripes running along both sides of the body. These are the lateral lines. Cut off the anterior end at right angles to the long axis of the body close behind the lips with a razor blade, and mount in Berlese's fluid or glycerine for an *en face* view. Note the 3 roughly triangular lips arranged so there is a broad dorsal one and a pair of smaller ventrolateral ones. Each lip bears 2 sensory papillae located near the outer margin of the base. The inner margin of each lip is finely denticulate. The small triangular buccal opening appears where the apexes of the lips come together.

With fine pointed scissors open the specimen along the middle of the dorsal surface, being careful to avoid injury to the internal organs; then pin the worm out flat in a dissecting pan. The cavity which this slit exposes is the pseudocel. The alimentary canal extends as a straight tube through the length of this cavity. The anterior end consists of a small, muscular esophagus. Near the posterior end of the esophagus is a pair of large yellowish, stellate cells, the celomocytes. The mass of coiled tubules is the reproductive organs, either male or female. The genital opening of the female occurs midventrally near the union of the first and second quarters of the body while in the male the reproductive organs share a common opening with the digestive system in the cloaca near the posterior end of the body.

The female reproductive organ is roughly Y-shaped, its branches becoming finer distally. The two free ends of the Y constitute the ovaries. The ovary, oviduct, and uterus form a continuous and very long tube representing the branches on either side of the Y. Externally, their boundaries are not discernible. The extent of each can be made out only through the study of sections. The stem of the Y, where the two branches meet, is a very short structure called the vagina. It leads to the vulva, which is located anteriorly, as stated above. Remove a portion of the uterus where it joins the vagina, tease apart with needles and examine the eggs on a slide under the low power of a compound microscope. Note the thick shell. Focus carefully on the upper surface of an egg and observe the small irregular ridgelike elevations. Sometimes live and moving larvae are present in the eggs.

For the histology of the female reproductive system, study a cross section of an adult worm. Between the intestine and the body wall 4 (ovary, oviduct, uterus, and intestine) types of structures appear. The ovary is the germinal zone and consists of a mass of protoplasm with an abundance of nuclei, or germinal vesicles, scattered through it. A short distance from the germinal zone, the protoplasm is formed around the vesicle and cells or ova are seen. This portion of the ovary is followed by a developmental zone in which the ova have become elongated and are arranged in wedgelike shapes around a central stalk or rachis. Further down, the rachis disappears as the ovary passes into the oviduct and the ova separate from each other, assuming a more or less oval form. In the uterus, the eggs are large, possess shells with surface ridges, and often show early stages of cleavage, or even larvae. The eggs are fertilized in the seminal receptacle, which is between the uterus and the oviduct. The intestine is somewhat flattened and lined with cells of uniform height.

The male reproductive organs are in the form of a long single tube or thread. The free end of this thread is the testis. It joins an enlarged seminal vesicle, which extends to the posterior end of the body and unites with the alimentary canal to form a cloaca. Frequently, spicules, used in copulation, may protrude from the anus.

Cross sections show the basic structural pattern, characteristic of nematodes generally (Figure 146). The body wall consists of 3 parts: (1) the outer noncellular cuticle composed of 3 layers; (2) cellular hy-

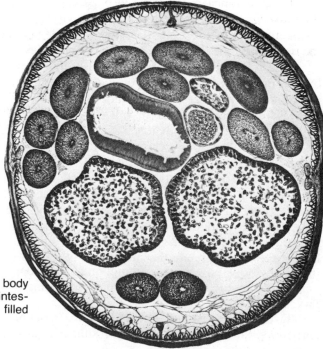

Figure 146. *Ascaris suum,* cross section of female showing body wall, chords in dorsal, ventral, and lateral locations, oval intestine, ovary with rachis, narrow oviduct, and large uterus filled with eggs.

podermis with a thickening on each lateral side and 1 each on the dorsal and ventral sides; these are the longitudinal hypodermal chords; the lateral chords contain the excretory canals and are more prominent than the dorsal and ventral ones; and (3) the innermost layer of body-wall muscle fibers; they are arranged in 4 principal groups between the chords. When the number of muscle fibers between the chords is great, as appears here, the arrangement is said to be polymyarian and when few meromyarian. If there are no chords present and the muscle fibers are small and closely packed, it is holomyarian. Recognition of these arrangements is useful in identifying cross sections of worms in tissues.

Life Cycle

Eggs passed in the feces are undeveloped and must undergo a period of incubation outside the host. Infection results from ingesting eggs containing second stage larvae. Upon hatching in the small intestine, the larvae burrow into the intestinal mucosa, enter the hepatic portal system, and are carried to the liver, then the heart, and eventually the lungs. Here they undergo the second and third molts, break out of the capillaries into the alveoli and make their way through the bronchi, trachea, throat, esophagus, and back to the small intestine, meanwhile having increased about 10 times in length. Here they undergo their final molt and become mature, some 8 weeks after ingestion of eggs.

Life Cycle Exercise

Living gravid females placed in Kronecker's solution ovulate for 6 to 12 days. Eggs may also be obtained quickly by removing them from the uteri of gravid females but many undeveloped ones will be present. Eggs kept in 1 to 2% formalin at 20 to 33 C develop to the second stage larvae in 3 to 4 weeks but do not hatch. Care must be taken in handling embryonated eggs as they are able to hatch and the larvae migrate in man.

First stage larvae are easily liberated mechanically by gently rolling the eggs back and forth under a cover glass, or tapping the cover glass with the handle of a teasing needle. Watch for developing larvae, which appear about the eighth day, and liberate them mechanically from the eggs, beginning when they first appear and continue at frequent intervals until satisfied that you have second stage larvae. They will be the largest ones enclosed in the loose cuticle of the first molt.

Third stage larvae do not appear until eggs containing fully developed second stage larvae have hatched in the gut of the host and the larvae have reached the lungs about 5 days after ingestion.

Young guinea pigs are good experimental hosts, but before infecting them, some eggs embryonated for 4 weeks should be tested for infectivity by giving them to white mice. Examine the contents of the small intestine 6 to 8 hours later for empty egg shells and newly hatched larvae. If there are no larvae in the intestine, the eggs are not sufficiently developed.

Infect 10 guinea pigs weighing 140 to 150 gm each with 40,000 to 50,000 fully embryonated eggs by placing them back of the tongue in a minimum volume of water in a bulbed pipette. Make estimates of the number of eggs by aliquot counts.

Examine each animal in the same manner and look for the following: (1) intestinal contents for eggs, shells, and newly hatched larvae; (2) wall of small intestine for hemorrhages, congestion, and second stage larvae; (3) liver for hemorrhages, white spots, pale color, swelling, congestion, and second stage larvae; (4) heart for swelling; (5) lungs for hemorrhages, congestion, and second, third, and fourth stage larvae; (6) congestion and pale color of kidneys, and (7) note whether labored breathing occurs.

To examine experimentally infected hosts, put finely diced intestine, liver, and lungs in Baermann funnels kept at about 37 C for several hours. Press bits of intestinal wall, liver, and lungs between 2 pieces of glass about 4 inches square and examine each under a microscope for migrating larvae. Wash intestinal contents from gut and clean by repeated sedimentation and decantation in water, using a hand centrifuge to precipitate the material. Put a drop of cleaned sediment on a microscope slide and examine under the low power of a compound microscope for eggs, shells, and larvae.

Examine each of the guinea pigs according to the protocol given below and search for the migrating larvae.

Two hours postinfection: A few macroscopic lesions in the small intestine. Many unhatched eggs and empty shells, and larvae in the stomach and small intestine. A few larvae in the mucosa; none in the liver and lungs.

Four hours: Minute hemorrhages in intestine; newly hatched larvae but none in lymph glands, liver or lungs.

Eighteen hours: Numerous small hemorrhages in the intestine; liver pale, with numerous hemorrhages; few larvae in the liver and lungs.

One day: No unhatched eggs but many shells in the gut; lymph nodes, liver, and lungs with many larvae; liver pale and with white spots.

Three days: Many intestinal hemorrhages, congestion of intestine and liver. Many second stage larvae in liver and lungs. Lungs with many hemorrhages; kidneys pale.

Five days: Many larvae in lungs, few in liver. Liver enlarged and with white spots. Second molt begins in lungs and third stage larvae appear. Many second stage larvae still present.

Eight days: Lungs with excessive damage. Heart and spleen enlarged. Many third stage larvae in lungs and some in stomach and intestine. Larvae in trachea. Labored breathing.

Ten days: Third molt begins in lungs. Intestine with hemorrhages, liver congested, lungs with hemorrhages. Labored breathing.

Eleven days: Liver englarged, congested, and dark. Lungs with many hemorrhages and area of consolidation. Kidneys pale. Small intestine with hemorrhages and congestion. Are larvae still in the intestine?

Sixteen days: Host recovering. Liver normal in size but pale and with white spots. Lungs with small area of hemorrhages. Are larvae in intestine? Development to the adult does not occur in guinea pigs.

Some other common Ascaridae with their hosts include: *Baylisascaris columnaris*, skunks and raccoons; *B. devosi*, fisher and marten, *Parascaris equorum*, horses, *Toxascaris leonina*, carnivores.

Family Toxocaridae

Esophagus with posterior ventriculus. Cervical alae present: prominent in *Toxocara*, inconspicuous in *Neoascaris*.

Toxocara canis

This large roundworm of carnivores is cosmopolitan in distribution. The adult occurs in the small intestine of dogs, foxes, and coyotes. It is often a serious problem in puppies, not only because the infection is so common, but also because the infection is usually very heavy. The worms often enter the stomach and cause vomiting, with the vomitus containing many worms. On account of its general availability and resourcefulness in the various ways the host may acquire a patent infection, it is of special interest.

Description

Adult females measure up to 18 cm long and males up to 10 cm. The posterior end is usually curved ventrally. There are broad cervical alae, which give the anterior end a spearhead shape, but caudal alae are lacking. The tail of the male is abruptly reduced in diameter and bears 5 pairs of lateral papillae. There are about 20 pairs of preanal lateral ones (Figure 168). The vulva is situated in the first quarter of the body. The eggs are described under Helminth Eggs, p. 213.

Life Cycle

The life cycle may be relatively simple, completed in one host, or it may be more complicated, involving 2 or more hosts. When the infective eggs are swallowed by the host, they hatch in the small intestine, a few of the second stage larvae enter the lacteals but most enter the hepatic portal system, and are carried to the lungs via the liver and heart. From this point on, the migratory behavior of the larvae varies depending upon the age and the species of the host.

In puppies, 3 weeks and younger, the larvae in the capillaries of the lungs enter the alveoli, migrate up the trachea, are swallowed, and go back to the small intestine where they mature in 4 to 5 weeks.

In older dogs, 5 weeks and more, however, the larvae mainly return from the lungs to the heart and are dispersed throughout the body where they are encapsulated in the various tissues.

Larvae embedded in the tissues of bitches become reactivated during pregnancy and some reenter the blood stream. Those that enter the placental circulation and make their way into the liver of the fetuses produce prenatal infection. In newborn puppies, third stage larvae are present in the lungs; within another day, the larvae, having migrated up the trachea and been swallowed, are in the stomach; at 3 days after birth fourth stage larvae are already in the intestine. In puppies so infected, patency is reached as early as 4 to 5 weeks after birth. Prenatal infection is apparently the usual mode of infection in dogs, according to Sprent (1958).

As a sequel to the prenatal mode of infection, Sprent (1961) has shown that postparturient infection of bitches may occur immediately after parturtion. Larvae migrate from the lungs of newborn pups into the intestine, but some of them fail to maintain a hold in the intestine and are swept out in the feces. Thus when the third stage larvae, beyond the stage requiring migration through the lungs, are swallowed by the bitch they develop to adults in the alimentary canal. The explanation for human infection with adult *T. canis* is that the children involved ingest the advanced stage larvae from the feces of a puppy or young dog.

Eggs swallowed by mice hatch in the small intestine. The larvae enter the hepatic portal system, are transported to the lungs via the heart, back to the heart, and out into the tissues. Many larvae are lodged in the brain but apparently cause little disturbances of locomotion. This is in marked contrast to infection with larvae of *Baylisascaris columnaris*, in which such disturbances are frequently observed in mice. When infected mice are eaten by dogs, the larvae are freed in the intestine and undergo the hepato-pulmono-tracheal migration to the small intestine where they mature.

Thus there are 4 ways in which a dog may acquire a patent infection: (1) puppies ingesting eggs; (2) prenatally, resulting in a high infection among dogs less than a year old; (3) bitches swallowing third stage larvae in the feces of their puppies; and (4) by ingesting tissues of animals harboring second stage larvae. Since larvae have been observed in the milk of nursing bitches (Stone and Girardeau, 1967), it is possible that the transmammary route may serve as still another way in which patent infections are acquired.

When human beings swallow infective eggs, the larvae behave in much the same manner as in mice, causing a condition known medically and parasitologically as visceral larval migrans. The larvae occur primarily in the internal organs and musculature.

Life Cycle Exercise

The objective of this experiment is to follow the migration of the larvae in white mice, determining the course taken and the time required for them to complete the somatic migration to the various tissues of the body. From the information obtained, it is possible to extrapolate what happens in children when infective eggs are swallowed.

Obtain gravid female *T. canis* from a small animal clinic where young dogs are treated to remove the worms. Dissect eggs from the terminal part of the uteri and incubate them in moist charcoal at room temperature. They will be infective when the second stage larvae appear in them.

Infect each of 16 to 20 white mice of equal size and 5 to 6 weeks old with 2,500 embryonated eggs. Counts of eggs are made from dilution aliquots. Mix the eggs with a small amount of meal and serve in a shallow dish to hungry mice. Kill one mouse each day for 14 consecutive days and at 21 and 28 days and examine it for larvae. From the results, it is possible to construct the route of migration by the larvae, the time required for each phase, final destination of the larvae, and their ultimate fate in the mouse.

Kill the mice by decapitation; remove and discard the skin, feet, and tail. Save the liver, lungs, alimentary tract, kidneys, carcass, and brain in separate containers for individual examination and enumeration of larvae. Grind the parts individually in physiological saline containing pepsin or trypsin in a blender. Use trypsin, pH adjusted to 7.0 with 0.1 N sodium hydroxide, for larvae less than 8 days old and pepsin, pH adjusted to 1.0 with hydrochloric acid, for larvae 8 days or older.

Grind carcasses in a blender for 45 seconds in 50 ml of saline containing 1 gm of pepsin or trypsin and digest at 37 C for 2 hours. Grind liver and lungs for 45 and 30 seconds, respectively in 25 ml of saline, pour into a pointed 50 ml centrifuge tube containing 0.05 gm trypsin in 5 ml saline, and digest for 3 to 4 hours. Stomach, intestine, and cecum are freed of contents, which are saved, and all 3 parts are ground together for 45 seconds in 20 ml of saline and treated similarly to the liver and lungs. Treat the kidneys and brain in a similar manner for counts of larvae in them. Before grinding the brain, note the hemorrhagic spots on the surface caused by the migrating larvae.

Each digested organ is carefully screened to remove tissue debris, sedimented and decanted several times by suction, and the remainder fixed by adding an equal volume of boiling 6% formalin. The volume of material from each organ is reduced after final sedimentation to 10 ml and stored in screw top vials.

For counting the larvae, reduce the completely sedimented contents of each vial to 5 ml by suction. Shake the vial vigorously and quickly remove 1 ml of the contents with a graduated, bulbed pipette to a shallow, lined, glass dish, and count the larvae in one-half of it. If the larvae are too numerous to count, or the mixture too opaque to see them, reduce the volume in the vial to 3 ml and use 0.1 ml on a microscope slide, covering the fluid with a cover glass. To arrive at the approximate number of larvae per organ, multiply the number in 1 ml aliquots by 10 and the average of 3 counts of 0.1 ml by 30.

From the data acquired, prepare a report in which the migratory route, time of appearance and disappearance of larvae in each organ, and their final destination will be made. Begin the account with larvae in the intestine. For fuller details, consult Sprent (1952).

Some other common Toxocaridae with their hosts include: *Toxocara cati*, cats; *Neoascaris vitulorum*, cattle.

Family Oxyuridae

Characterized by a nearly spherical, posterior bulblike enlargement of the esophagus. A single spicule is present but no cervical spines. Parasites of the large intestine of mammals.

Enterobius vermicularis

The human pinworm, *E. vermicularis* (Figure 172), is cosmopolitan in distribution but unlike most helminthic infections is relatively rare in the tropics. The pinworm is probably the most prevalent nematode of man in the United States, occurring in about 50% of the children; but fortunately it is more of a nuisance than a serious health problem.

Description

The mouth is surrounded by 3 lips, as in *Ascaris*, and there is no buccal capsule. The anterior end of the body has alae (cuticular expansions) on the dorsal and ventral sides. The excretory pore, seen from the side, is near the middle of the esophagus. The basal end of the esophagus bears a distinct bulb with a valvular apparatus. The intestine is straight.

Males are up to 5 mm long. The tail is curved ventrally and the lateral caudal alae extend around the end. From the ventral view, the tail is broadly rounded and bears a pair of lateral adanal and postanal papillae and 2 pairs of ventral postanal ones. The single spicule with curved tip is 0.07 mm long. The threadlike testis begins somewhat posterior to the middle of the body, extends anteriorly, reflexes backward forming the thin tubular vas deferens, and is followed by a thicker muscular ejaculatory duct.

Females measure up to 13 mm long. Gravid ones are somewhat spindle-shaped and have a long, slender, pointed tail. The vulva is ventral, slightly anterior to the equator of the body, from which the narrow vagina extends backward to the middle of the body where it divides into an anteriorly and posteriorly directed uterus. Each uterus extends to the end of the body, where it bends back on itself to form the threadlike ovary. The eggs are described under Helminth Eggs, p. 213.

Life Cycle

At night gravid females migrate out through the anus, deposit their eggs in the perianal region, and retreat to the rectum, but some worms are passed in the feces. Since eggs are seldom found in the feces, the standard method for their recovery involves the application of strips of cellophane tape around the perianal region (see p. 277). Upon ingestion of the infective eggs, the larvae hatch and reach maturity in the cecal region.

Other common species of Oxyuridae include *Passalurus nonannulatus* of rabbits, and *Oxyuris equi* of equines.

Family Syphaciidae

Characterized by males with both a spicule and a gubernaculum. Parasites of the cecum and large intestine of rodents.

Syphacia obvelata

This species (Figure 171), common in colonies of white mice, is an ideal model for laboratory study.

Description

Three lips surround the triradiate mouth. There are 4 cephalic papillae in the outer circle.

Males measure up to 1.6 mm long by 0.2 mm in diameter with the caudal end bent ventrally into a spiral. There are 3 mamelons on the ventral surface of the body with the middle one near the equator. The tail is 0.1 mm long and the ratio of the length to the greatest width is 0.9. The ratio of the distance from the anterior tip of the body to the excretory pore to the total length of the body is 1:4.4. The spicule is slightly curved and measures 0.09 mm long and 0.007 mm thick at the base; the ploughshare-shaped gubernaculum is located at the base of the spicule and is 0.04 mm long. There are 2 pairs of preanal and 1 pair of postanal papillae.

Females are up to 5.7 mm long by 0.3 mm in greatest diameter. The posterior end terminates in a long, slender tail. The ratio of the distance from the anterior end of the body to the excretory pore to the total body length is 1:10. The vulva is on a prominence about 0.9 mm from the anterior end of the body. The ovaries are in the middle third of the body; 1 is pointed anteriorly and the other posteriorly. The parallel oviducts extend posterior beyond the hind ovary, and unite to form the long single uterus that extends anteriorly to the vulva located a short distance behind the esophagus. Eggs are flat on 1 side and average 134 x 36 μ. The embryos are not formed when the eggs are laid.

Life Cycle

The life cycle is direct and of short duration. Infection is by mouth. Eggs may be picked up from the surroundings or licked from the perianal region. This life cycle is an ideal model for laboratory study of the Oxyurina nematodes, especially since it parallels that of *Enterobius vermicularis*, the human pinworm.

Life Cycle Exercise

Eggs dissected from females do not develop *in vitro* but those laid by them do. To obtain egg donors, place 10 to 12 parasite-free mice 3 to 4 weeks old on contaminated fecal trays with infected mice for a period of 24 hours. The maximum number of gravid females appear in the colon and cecum on the 12th day postinfection. When the female worms are transferred quickly from the intestine to physiological solution, they promptly lay about 300 eggs each (Chan and Kipilof, 1958). Such worms are in the process of leaving the host to oviposit and terminate their lives. Eggs begin to appear in the perianal region on the 11th day, as shown by the transparent adhesive tape examination method (see p. 277), indicating that the worms are in the colon and rectum.

When kept at room temperature, these eggs contain first stage infective larvae in about 20 hours and at 37 C in about 3 to 5 hours. Infect 10 to 12 young mice 3 to 4 weeks old with 300 to 400 embryonated eggs by means of a stomach tube or by putting eggs on a small piece of bread for individual hungry mice to eat.

Since the eggs hatch in the presence of trypsin rather than pepsin, newly hatched first stage larvae appear in the small intestine. Inasmuch as development can proceed only in the environment provided by the cecum, the larvae quickly enter it, where subsequent development takes place.

Having selected an infected animal, dispatch it with deep anesthesia by ether, chloroform, or natural gas. Separate the parts of the small intestine from the cecum and large intestine and put in individual dishes of physiological solution. For examination, each segment is slit, spread open in a dish and scrutinized for worms. After this examination, each segment of gut is put in a jar of physiological solution, shaken vigorously, allowed to stand sufficiently long for the worms to settle to the bottom, and the supernatant fluid decanted. Clean the samples by repeated sedimentation and decantation until they are clear. Examine the sediment in a petri dish in good light, under a dissecting microscope for worms. Examine mice according to the protocol given below.

Two hours postinfection, empty egg shells and first stage larvae are present. Larvae up to 0.2 mm long are in the intestine and cecum. Where are the greater number of larvae? Why are there more in one place than the other? Do you see any larvae with loose cuticular sheaths, indicating commencement of the first molt?

At 24 hours, the majority of the larvae are in the cecum but they are undifferentiated sexually. Enveloping cuticular sheaths are present. This is probably the first molt and the larvae are second stage. They are up to 0.4 mm long.

At the third day, sexual differentiation is apparent together with early development of the genital system. Males show the cuticular mamelons and females a column of cells leading to the vulva, which is not open. Some larvae enveloped in sheaths are still present. These may be in the process of undergoing the second molt. The males are up to 0.5 mm long; females are up to 0.6 mm long. Note the relative proportion of males to females on this and succeeding days.

By the fourth day, males show distinct mamelons; the spicule, preanal, and postanal papillae are forming. The female reproducive system shows little change over the previous day. These are probably third stage larvae. Males are up to 0.9 mm long.

By the fifth day, about a third of the females have brown plugs on the vulva indicating that fertilization has occurred. The female reproductive system shows differentiation. These are probably fourth stage larvae. Males measure up to 1.1 mm long; females are up to 2.1 mm long.

At the sixth day, the majority of the females are inseminated. The males have attained adult size of up to 1.5 mm long. Females are up to 2.6 mm long.

By the seventh day, the reproductive system of the female shows distinct uteri, oviducts, and ovaries, and the males are sexually mature. Males are up to 1.5 mm long by 150 μ in diameter, with the esophagus 229 μ long and the tail 122 μ. Females measure up to 3.8 mm by 173 μ in diameter with the esophagus 312 μ long.

At the eighth day, eggs appear in the uteri for the first time. Having completed their function, the males begin to disappear. This feature is also observed in *Enterobius vermicularis*. Females are up to 4.0 mm long.

By the ninth day, nearly all of the females are gravid and the eggs are in the early stages of cleavage. They are still in the cecum.

By the tenth day, practically all females are gravid and the eggs show further development. Females are up to 4.7 mm long. They are still in the cecum. Check the perianal region with transparent adhesive tape for eggs. What results did you get? How do you interpret this?

By the eleventh day, larvae enclosed in eggs show further development, some with the gut formed. Where are the females? What results do you get with the transparent tape? What is your interpretation of this?

On the twelfth day, adult female worms migrate from the rectum, as revealed by the presence of eggs on the tape. Sometimes, worms may be seen emerging from the anus. Many of them die in the perianal folds after oviposition. The manner of migration is similar to that of *Enterobius vermicularis*.

By the thirteenth day, larval stages appear in the intestine and cecum. These are the result of the availability of eggs from the experimental infections 2 weeks earlier. The mice are becoming reinfected.

Other common species of Syphaciidae include *Wellcomia evaginata* of porcupines, and *Syphacia muris*, normally of white rats but capable of infecting white mice. In *S. muris*, the anterior mamelon is near the midbody, and the eggs are much smaller (75 x 29 μ). For further descriptions of *S. muris* and *S. obvelata*, consult Hussey (1957).

Order Rhabditida

Females are the only known parasitic forms. The esophagus of the adult is muscular, cylindrical, and long or short.

Family Strongyloididae

The esophagus is long and slender in the parasitic females, there is no buccal capsule, and the vulva is in the posterior part of the body. Adult females are parasitic in the mucosa of the small intestine of reptiles, birds, and mammals.

Strongyloides stercoralis

This human intestinal threadworm is cosmopolitan in distribution, but is primarily found in warm climates (Figure 152).

Description

Parasitic females are small, slender, threadlike nematodes about 2 mm long, with both ends pointed. The vulva is near the junction of the middle and posterior thirds of the body. The anus is near the posterior end, leaving a short, pointed tail. The alimentary canal consists of a small buccal capsule, a long, slender filariform esophagus extending through the anterior third of the body, and a thin intestine. Cuticularized buccal spears appear in the anterior end of the esophagus. The short vagina divides into an anterior and a posterior uterus and ovary; each ovary begins near the middle of the body and extends anteriorly or posteriorly, and loops back upon itself to form the oviduct and uterus. Each branch of the uterus is filled with a short string of thin-shelled, embryonated eggs that hatch within the intestinal epithelium of the host. Parasitic males are unknown.

The free-living adults have a short rhabditiform esophagus. In the female, the vulva is near the middle of the body. The short vagina receives the anterior and posterior uteri, which contain a single string of thin-shelled, embryonated eggs. The smaller male has a long, rather thick testis that opens into an equally wide seminal vesicle which narrows into a terminal muscular ejaculatory duct. A pair of short spicules and a gubernaculum are present. The eggs are described under Helminth Eggs, p. 213.

Life Cycle

Eggs of 2 kinds are produced by the parasitic females. The homogonic line has 3 chromosomes and produces rhabditiform first and second stage and filariform third stage larvae with a notched tail. These filariform larvae are infective and enter the host through the skin and develop to parthenogenic females. The eggs that produce the heterogonic line are of 2 kinds. Those with 1 chromosome develop into free-living males and those with 2 chromosomes into free-living females. All stages of larvae hatching from the 1N and 2N eggs are rhabditiform, as are the adults. The eggs produced by the free-living females are 3N and the larvae develop into infective larvae similar to those of the homogonic line.

There are 2 types of life cycles, each with different forms. They are the homogonic, or parasitic, in which a parthenogenic female produces larvae that enter the host or larvae that develop into heterogonic, or free-living males and females, whose larvae develop into the infective stage and enter the host. Upon contact with skin, the filariform larvae from either direct or indirect mode of development, penetrate the epidermis, enter the small blood vessels, and are carried to the lungs. Here they break out of the capillaries into the alveoli and make their way via the trachea and throat to the intestine, where they molt twice and reach maturity in about 2 weeks.

Strongyloides ratti

This is a common intestinal parasite of colonies of white rats and populations of wild ones.

Description

Parthenogenic females are up to 3.0 mm long. There are 6 lips. The finely pointed tail is about twice as long as the diameter of the body at the anus. The vulva is near the union of the third and last quarters of the body. There is a very shallow buccal capsule followed by a filariform esophagus a fifth to a third the total length of the body. There are 2 ovaries, 1 anterior with the reflexed part near the base of the esophagus and the other posterior and near the anus. The vagina is very short. Each uterus contains 10 to 12 thin-shelled, embryonated eggs measuring 47-52 x 28-30 μ.

Life Cycle

Larvae of the homogonic line are 3N, rhabditiform in the first and second stages, and filariform in the third stage.

Free-living stages are not well known, as studies have been directed mainly toward the parasitic phase. In general, they are short and plump worms. Adult males are around 0.8 mm long and have a ventrally curved, pointed tail. There is a pair of prominent, pointed spicules, and a simple testis. Females have an anterior and posterior ovary.

Larvae of the heterogonic line developing from eggs produced by the parasitic females are all rhabditiform and develop into free-living adult males (1N) and females (2N). Their offspring (3N) are rhabditiform in the first and second stages, and filariform and infective in the third stage.

In view of its ready availability, this is an ideal model for class use in demonstrating the anatomy and biology of this group of parasites. For life cycle details consult Spindler (1957), and Abadie (1963).

Family Rhabditidae

In this group, the buccal capsule is present and the esophagus is short. The vulva is near the middle of the body. Found in the lungs of amphibians and reptiles (Figure 151).

Rhabdias ranae

This species, the lungworm of frogs, is widespread in North America.

Description

Adult parthenogenic females are up to 4.5 mm long. The circular mouth is surrounded by 3 groups of low papillae. There are 2 pairs of lateral postanal papillae. The vulva is up to 2.5 mm from the tip of the tail and the anus is up to 0.3 mm. The buccal capsule is broadly crescent-shaped, with the concave side posterior. A club-shaped esophagus up to 0.6 mm is followed by a dark colored intestine. A pair of long cervical glands is located at the esophago-intestinal junction and 3 rectal glands are near the anus. Each ovary begins opposite the vulva, 1 extending anteriorly almost to the esophagus and the other posteriorly almost to the anus. Each bends back to form a short oviduct, seminal receptacle, and wider uterine portion that is filled with eggs. The vagina is extremely short. Embryonated eggs measure 40 x 75 μ and the larvae are 0.3 mm long at the time of hatching.

Life Cycle

The life cycle is incompletely known. Eggs hatch in the lungs and the upper part of the small intestine into first stage rhabditiform larvae. According to Walton (1929), the first stage larvae grow rapidly in the lower part of the intestine where they attain a length of 1.5 mm. The genital primordium is recognizable but the sexes are not. The large size suggests that these might be second stage larvae and that the first molt was not seen. Walton states that the first molt produces filariform or infective larvae. They are ensheathed, burrow through the gut wall into the celom, and migrate through the mesenteries into the lungs. He claims 4 molts occur, the last in the lungs leaving the young adults ensheathed. Walton believed that the heterogonic line might not occur in this species, as it does in R. bufonis from European amphibians.

Order STRONGYLIDA

Males are characterized by a caudal membranous copulatory bursa supported by 1 dorsal ray, usually with a divided tip and 6 pairs of lateral rays. Some rays may be fused or lost. This order is known collectively as the bursate nematodes. Many species are of great medical and veterinary importance.

Family Ancylostomatidae

The dorsal cutting plates, which surround the inner margin of the mouth, have teeth on their free margin and a head that is bent sharply dorsad, giving a hooked appearance.

Ancylostoma duodenale

This is known as the Old World hookworm of man and is of great public health importance. It occurs commonly in tropical and subtropical countries, bounded in general by the latitudes 36° N and 30° S.

Description

Stocky worms with cuticle having fine transverse striations. A pair of cervical papillae, 1 on each side of the body, is located near the middle of the esophagus. The funnel-shaped buccal capsule is large and has a thick wall. There is a pair of ventral cutting plates, each with 2 large toothlike structures and a small dorsal plate with a median indentation (Figure 147, B). A pair of internal teeth is located in the depth of the capsule. A short club-shaped esophagus opens through paired valves into the long tubular intestine. A pair of long cervical glands, each with a distinct nucleus lies just posterior to the esophagus.

Males are up to 11 mm long. The copulatory bursa consists of a small dorsal lobe and 2 large lateral ones. Beginning with the single dorsal ray whose tip is divided into several small digitations, the paired rays are as follows: external dorsal rays originating from each side of the base of the dorsal ray; 3 lateral rays arising from a single base are the posteriolateral, mediolateral, and anteriolateral (sometimes called the externolateral), and 2 pairs of ventral rays, the lateroventral and the ventroventral, having a single basal stalk separate from that of the lateral rays. The posterio- and mediolateral rays originate from a single stalk (See Figure 145, H).

The typical arrangement of the strongylid bursal rays can be represented roughly by placing both hands, palms down, on one's thigh with the thumbs together their full length and the fingers spread. The thumbs represent the dorsal ray, the index fingers the externodorsal rays; and the middle, ring, and little fingers the lateral complex of rays—posterio-, medio-, and anteriolateral rays, respectively. With fingers lacking to represent the ventral rays, they must be visualized.

The reproductive system consists of the coiled, threadlike testis lying at the posterior ends of the cervical glands, followed by the posteriorly directed long, wide seminal vesicle, which empties into a long, muscular ejaculatory duct surrounded midway by cement glands. The ejaculatory duct narrows posteriorly and empties into the cloaca. A pair of long, cuticularized, brownish, needlelike spicules, associated with a gubernaculum in the dorsal wall of the cloaca, function in transferring seminal fluid from the male to the female.

Females are up to 13 mm long and the tail ends in a small spine. The vulva is slightly behind the middle of the body. The 2 threadlike ovaries, 1 anterior to and the other posterior to the vulva, and their associated parts, the oviducts and uteri, are intricately wound about the intestine. Each uterus terminates in a slender muscular ovejector and these join to form the short vagina. The ovejectors and terminal parts of the uteri are filled with elliptical, hyaline, thin-shelled eggs, typical of the Strongylina, in the early stages of cleavage. The eggs are described under Helminth Eggs, p. 213.

Life Cycle

Eggs, laid in the 4-cell stage, pass out in the feces onto the soil. Under favorable conditions of temperature and moisture, the larvae hatch within 24 hours. After 2 molts, the larvae lose their typical rhabditiform esophagus, and become strongyliform. The free-living larvae of hookworms in all stages are distinguishable from the larvae of Strongyloides by their longer and narrower buccal cavity. The third stage infective larvae normally burrow through the skin, enter the lymphatics or veins, and are then carried by the blood stream to the heart and thence to the lungs. They break out of the capillaries into the alveoli and crawl up the bronchi and trachea to the throat and are finally swallowed. In the intestine, the larvae undergo the third molt within 3 to 5 days. They grow rapidly, up to 5 mm, and then molt for the fourth and last time. Prepatency usually takes about 6 weeks.

Ancylostoma caninum

This is a common hookworm of dogs in temperate climates (Figures 147, C, E, F; 154). It is larger than A. duodenale (males up to 12 mm and females 16 mm long). Each of the 2 dental plates has 3

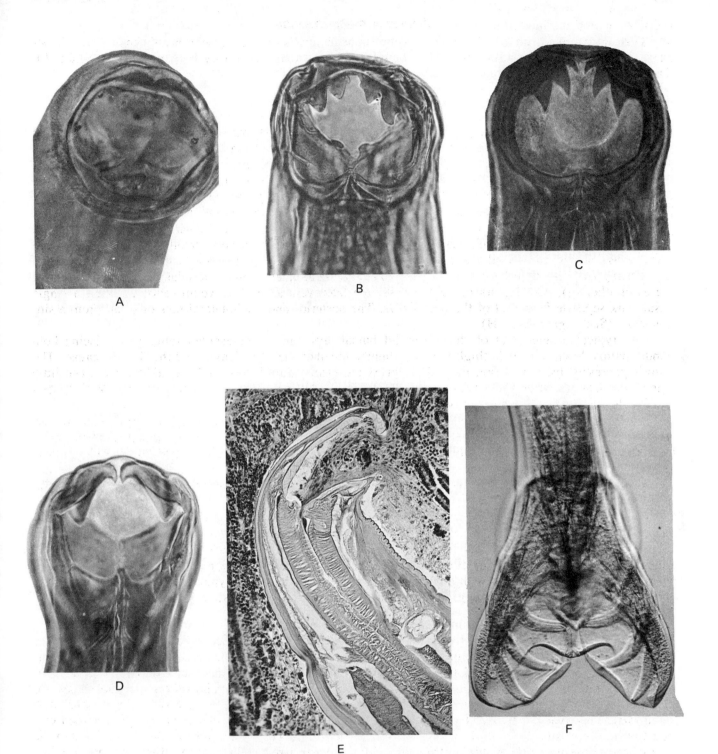

Figure 147. Hookworms: (A) Mouth parts of *Necator americanus;* note the two pairs of characteristic cutting plates; (B) Mouth parts of *Ancylostoma duodenale;* note the two pairs of teeth and the accessory processes on the median teeth; (C) Mouth parts of *A. caninum;* note the three well-developed pairs of teeth; (D) Mouth parts of *A. braziliense;* note the two pairs of teeth, both pairs without accessory processes; (E) *A. caninum,* longitudinal section showing manner of attachment to intestinal mucosa; (F) *A. caninum,* bursa of male.

teeth, of which the innermost is the smallest. The eggs are described under Helminth Eggs, p. 213. The life cycle is similar to that of *A. duodenale* but transmammary infection is common in dogs. The larvae of *A. caninum*, together with those of *A. braziliense* (Figure 147, D), the adults of which occur in cats, cause cutaneous larva migrans or creeping eruption when they penetrate the skin of man.

Life Cycle Exercise

In warm regions where *A. caninum* occurs commonly in dogs, eggs for experimental studies may be obtained from feces of infected animals or gravid female worms expelled by dogs treated in small animal clinics. Fecal material should be mixed with sand or sphagnum moss to provide good aeration and reduce putrefaction, kept moist, and incubated at room temperature. Eggs dissected from gravid females may be incubated in petri dishes in shallow water. The larvae are easily recovered from the culture with the Baermann funnel. Place a single layer of fecal material on the submerged screen of the funnel containing water about 40 to 42 C. The larvae will soon forsake the feces and appear in the rubber tubing on the stem of the funnel. By frequent examinations of the material, find the first and second stage rhabditiform larvae, and the third stage strongyliform, infective larvae. When the buccal capsule of the strongyliform larvae is viewed in optical section, it appears to consist of a pair of spears, extending from the mouth caudad for a short distance. These are actually the walls of the buccal capsule seen in optical section.

For experimental infection of mice, place 4,000 to 5,000 third stage larvae in a small volume of water on a clean, dry, shaved area of skin of each of 3 restrained mice. A desk lamp placed near them will hasten evaporation of the water and facilitate the penetration of the larvae.

The first mouse should be examined 12 hours postinfection, by grinding the skin finely in a food chopper and baermannizing for recovering the larvae. Each of the remaining mice should be examined on consecutive days after exposure for the recovery of migrating larvae, according to the following procedure.

Heart Blood: ligate the hepatic portal vein near the intestine and the pulmonary artery near the lungs. Remove the heart and the 2 tied blood vessels intact, wash out the blood, and examine for larvae after sedimenting and decanting until clear. Allow 30 minutes for sedimentation so larvae will reach the bottom of the container.

Lungs: examine lungs for hemorrhages due to blood diffusing from wounds produced by larvae migrating from blood vessels into the air sacs. Heavy infections will cause pneumonia. Grind the lungs and place them in Baermann funnels to recover migrating larvae.

Trachea and esophagus: slit open the trachea and esophagus, wash, and scrape them in separate dishes. Examine the sediment for migrating larvae.

Stomach and intestine: examine separately the contents and scrapings of the stomach wall for larvae, using the high power of a dissecting microscope.

Body muscles and organs: grind and baermannize separately voluntary muscles and various major organs for larvae.

Central nervous system: examine the brain for larvae by macerating and baermannizing it.

In comparison with the free-living third stage larvae, do the ones recovered from the tissue of mice appear similar or have they developed beyond the third stage? How do you account for larvae that have entered the mice via the skin being in the body muscles and central nervous system?

Another common Ancylostomatidae is *Ancylostoma tubaeforme* of cats.

Family Uncinariidae

Large ventral cutting plates are smooth and semilunar in shape along the free margin.

Necator americanus

This, the so-called New World hookworm of man, is the predominant species in the United States and has a wide distribution in the tropical and subtropical regions of the world.

In general, the anatomy is similar to *Ancylostoma duodenale* except that *N. americanus* is smaller and the free margin of the cutting plates is smooth and semilunar in shape. A pair of single lancets is at the bottom of the buccal capsule (Figure 147, A).

Males are up to 9 mm long. The bursa is much wider than long, with the dorsal lobe having a broad median indentation so that the dorsal ray is divided almost to its base into 2 widely divergent branches; the externodorsal rays are somewhat club-shaped and near the laterals. The posterio- and mediolateral rays originate from a common stalk with the ventral rays arising from that same stalk. The spicules are long and slender. Their distal ends are fused and tipped with a barb.

Females are up to 11 mm long. The vulva is slightly anterior to the middle of the body. There is no spine at the tip of the tail. The eggs are described under Helminth Eggs, p. 213. The life cycle follows closely that of *A. duodenale*.

Other common Uncinariidae are *Uncinaria stenocephala* of dogs (Figure 153), *U. lucasi* of fur seals and Steller's sea lions, and *Bunostomum trigonocephalum* of sheep, goats, and cattle.

Family Syngamidae

Parasites of the trachea (birds) and kidneys (swine). The head extends forward and has a wide mouth without lips or leaf crowns. The buccal capsule is cup-shaped, has a thick hexagon-shaped rim, and several small teeth in the base.

Syngamus trachea

This is the common gapeworm of domestic fowl (Figure 155). Adults occur in the trachea and bronchi of chickens, turkeys, pheasants, and related gallinaceous birds, together with many wild species of birds. Gapeworms are becoming less common in domestic fowl due to better management but still occur frequently in wild birds, especially pheasant chicks reared in captivity.

Description

Living worms are red. The small males are joined in permanent copula with the large females so that the united pair present a Y-shaped appearance. The buccal capsule is shallow and cup-shaped with a wide mouth whose rim is thickened and hexagonal in shape. There are neither lips nor leaf crowns. Six to 10 small teeth occur in the bottom of the buccal capsule. The esophagus is short and club-shaped.

Males are up to 10 mm long and possess a short bursa with stout rays; the posteriolateral ray is separate from the other laterals. Females measure up to 40 mm long. The vulva opens in the anterior sixth of the body of gravid worms. The conical tail is slightly more than twice as long as the diameter of the body at the level of the anus. The reproductive organs fill the body. Ellipsoidal eggs possess an operculum at each end and are in an advanced stage of embryonation when laid under the bursa of the male. They are carried up the trachea by the action of the ciliated epithelium, swallowed, and voided with the feces.

Life Cycle

The life cycle is direct but infection of the final host may take place in any of 3 ways, i.e., by swallowing (1) eggs containing third stage larvae, (2) hatched third stage larvae, and (3) earthworms, and other invertebrates, harboring third stage larvae. For life cycle details, consult Wehr (1937).

Other species of Syngamidae include *Cyathostoma bronchialis* of Canada geese and *Mammonogamus laryngeus* of bovines in tropical regions.

Family Heligmosomatidae

The anterior end of the body has a cuticular expansion and the copulatory bursa has large lateral lobes, which may be asymmetrical, and the spicules are long and slender (Figure 159). Females have a single ovary and uterus with the vulva near the anus. They are primarily parasites of rodents.

Nematospiroides dubius

This species occurs commonly in the intestine of mice in colonies and in populations of wild mice.

Description

Live worms are red and coil in a spiral. The anterior end of the slender body is inflated. The cuticle has about 30 longitudinal striations. A buccal capsule is present but weakly developed.

Males are up to 10 mm long, have elongate prebursal papillae and a large asymmetrical bursa in which a dorsal lobe is lacking. Lateral and ventral rays originate from a large but short common trunk that divides into a large ventral and a smaller lateral branch. The large ventral rays are strongly divergent; the anterio- and mediolateral rays are nearly parallel, but the posteriolateral rays are divergent. Externodorsal rays originate near the base of a slender dorsal ray that is quadrifurcate at the tip. Spicules are slender, filariform, and up to 0.6 mm long. A gubernaculum is lacking.

Females are up to 21 mm long. There is a single reproductive organ with the vulva near the anus. The tail is fairly long, conical, and bears a small spine at the tip. Eggs 75-90 x 43-58 μ.

Life Cycle

Eggs passed in the feces are in the 8 to 16 cell stage, and hatching occurs about 24 hours after leaving the host. The free-living larval stages require about 4 days to become infective. Upon being ingested, the larvae lose their sheaths, and 24 to 48 hours later they have penetrated the intestinal mucosa and are lying near the longitudinal muscle layer of the gut wall. Here the third molt is completed in 6 to 8 days, after which the fourth stage larvae leave the mucosa and return to the intestinal lumen where they undergo the last molt. The prepatent period is about 9 days, and the cycle from egg to egg takes about 15 days.

Life Cycle Exercise

This species is a good illustration of a Trichostrongylina life cycle in which the parasite enters its host passively through the mouth and has a limited migration that extends into the mucosal lining of the intestine for only a short time before returning to the lumen. An additional advantage is its ready availability.

To collect eggs, allow feces from infected mice to accumulate for about 12 hours on moist paper towels placed beneath the cages in trays containing a shallow layer of water. Transfer the pellets to a flask or jar containing decinormal NaOH and leave in the refrigerator overnight to comminute. The next morning, add water, shake vigorously, and remove the coarse material by straining through several layers of cheese cloth. The strained portion contains fine particulate material, coloring matter, and eggs.

Next remove the fine material and coloring matter, and concentrate the eggs. This is done by repeated sedimentation, by gravity or centrifugation, and decantation. Divide the strained material into several tall cylinders, fill each with water and let stand for 30 minutes. At the end of this time, decant the supernatant fluid and repeat until the water is clear. Concentrate the sediment which contains the eggs by a short period of centrifugation, transfer to a vial, and store in a refrigerator until ready to study them.

The centrifugation method of cleaning is more rapid. Fill 2 or 4 centrifuge tubes, preferably those of 50 ml capacity, spin briefly to precipitate the eggs but not the fine material, decant, and repeat until the water is clear. The trick is to spin the tubes only enough to throw the eggs down but leave the maximum amount of particulate and coloring matter in suspension for removal with the supernate. When the water is clear, concentrate the sediment from the various tubes and store in a vial in a refrigerator for subsequent observation on the development of the larvae.

Eggs incubated in clean water in a small dish with cover are good for observing embryonation, hatching, and development of the 3 stages of larvae. Examine eggs and larvae at frequent intervals by placing them on glass slides in water during the first 3 to 4 days. Use a finely tipped, bulbed pipette to transfer them. Sealing the edges of the cover glass with Vaseline will maintain the water mounts, enabling one to observe the development of individual eggs over a long period. Hatching occurs in 20 to 24 hours after the feces are passed and third stage larvae are present in 3 to 4 days.

Large numbers of third stage larvae for infection of experimental hosts can be obtained from eggs collected as described above. It is easier, however, to prepare fecal cultures from infected mice in which the eggs hatch and the larvae develop to the third stage in 3 to 4 days at room temperature.

To obtain larvae, the fecal pellets may be handled in several ways, among them: (1) in the petri dish-filter paper method, feces comminuted in a small amount of water are spread in a thin layer over moist discs of filter paper, placed in a petri dish, covered, and kept at room temperature; (2) in charcoal-fecal cultures, comminuted pellets are mixed with an equal amount of powdered charcoal so as to be moist discs of filter paper, placed in a petri dish, covered, and kept at room temperature; (3) in charcoal-fecal cultures, comminuted pellets are mixed with an equal amount of powdered charcoal so as to be moist and incubated (sand or sphagnum moss may be used in lieu of charcoal); and (4) fecal pellets alone may be incubated in a moist condition in petri dishes. In each case, larvae appear in large numbers.

Upon hatching, the active rhabditiform first stage larvae are up to 0.4 mm long. The intestine is composed of indistinct cells. Ten hours after hatching, the larvae become lethargic for an equally long time in preparation for the first molt, which begins about 48 hours after hatching. It is indicated by the loosened cuticle appearing at the ends of the body.

The third stage larvae are enclosed in the shed cuticle of the second stage larvae (Figure 145, F). They are present between 48 to 56 hours after the beginning of the second molt, and measure up to 0.6 mm long. The buccal capsule is much shorter than in the preceding stages and the eosphagus has lost its rhabditiform character and becomes strongyliform, i.e., long and slender with a slight swelling at the posterior end. The tail is shorter, and blunter. The strongyliform esophagus, short blunt tail, and ensheathing loose cuticle are characteristic of this stage. Further development takes place only in the body of the host.

Third stage ensheathed larvae may be exsheathed chemically by placing them in a fresh solution of 1 part commercial Clorox and 3 parts of water for 10 to 15 minutes. Transfer the sheathed larvae from water into the exsheathing solution on a glass slide, cover, and watch the process under a microscope. Normal molting of third stage larvae takes place in the mucosal lining under chemical conditions very different from those on the microscope slide.

The route of infection is by mouth only. Infect 10 mice 3 to 4 weeks old by placing about 200 third stage larvae, as determined by aliquot counts, in 0.2 ml of water on a small piece of bread which is fed to the hungry animals.

The objective of examining mice after they have swallowed the third stage larvae is to: (1) see third stage larvae free in the lumen of the small intestine; (2) see them in the mucosal lining; (3) observe the third molt and subsequent rapid growth in the mucosa; (4) see them return to the lumen; (5) observe the fourth and final molt; and (6) note the development to sexual maturity.

Remove the small intestine, slit it open, and wash out the contents to obtain the 3 stages of worms at appropriate times. Press sections of it between 2 glass plates to see the worms in the mucosa.

According to the following protocol, examine the mice in search of the developing worms. Examine 1 mouse 24 hours after infection and look for larvae still free in the lumen and those that have entered the intestinal mucosa. On the fourth day, necropsy another mouse and look for larvae in the lumen and mucosa. What stage are they, and is there any indication of molting? Examine a mouse on the sixth day and another on the eighth day. Where are the larvae, and what stage are they? Do any show evidence of the final molt approaching? Note the stage of development of the reproductive organs in both sexes. Are the sexes distinguishable?

Larvae may be removed from the intestinal wall by digesting the gut in a solution of 1% HCl and 1% pepsin in tap water for 8 to 10 hours at 37 C, with periodic shaking. When the digestive process is completed, wash and decant several times, and finally add sufficient formalin to make a solution of 10% (10 ml of commercial formalin in 90 ml of water) for fixing and preserving the worms.

Counting the worms is accomplished best after staining for 30 minutes in bulk in the formalin solution with iodine (30 gm I, 40 gm KI in 100 ml of hot water). Excessive coloring of the debris may be removed by washing in 5% aqueous sodium thiosulphate solution. Count all of the worms in the intestinal wall and those found in the lumen to get the percentage of those swallowed that infected the mouse and those that failed to do so.

After determining the prepatent period, sacrifice the remaining mice to recover the adult worms, which should be prepared for study, using 1 of the standard methods.

Families of Lungworms

Adults of this group occur in the respiratory system of mammals. They are long, filiform worms in which the buccal capsule is absent or rudimentary. The bursa is usually small and the rays atypical, being reduced in either size or number, often by fusion, or both. The vulva is behind the middle of the body and the females are oviparous or ovoviviparous. The basic characteristics of a few common genera are given.

Dictyocaulus spp. (Dictyocaulidae) have a well-developed bursa in which the posterio- and medio-lateral rays and both ventral rays are fused except at the tips. The spicules are large, equal, and sox-shaped (Figure 160). The vulva is near the middle of the body. They are parasites of equines and ruminants. The life cycle is direct.

Metastrongylus spp. (Metastrongylidae) have 2 lateral trilobed lips. The bursa has large lateral lobes. The ventral rays are distinctly separated from each other; the posteriolateral rays are very large, long and separate; the mediolateral rays are short and thick with the very small anteriolateral rays arising from the base; the dorsal ray is double and minute, as are the externodorsal rays (Figure 163). The spicules are long and delicate with transversely striated wings. The body of the female narrows abruptly behind the anus and the vulva opens near the anus. They are parasites of swine. Earthworms serve as intermediate hosts.

Protostrongylus spp. (Protostrongylidae) males have a short bursa. The dorsal ray is very thick and bears 6 papillae on the ventral side; the externodorsal rays are spikelike; the posterio- and mediolateral rays are fused except at the tip; the other rays are separate. The spicules are long with striated wings. A strong cuticularized arch, the telamon, strengthens the posterior end of the male body (Figure 165). The vulva is near the anus. They are parasites of sheep, goats, deer, bighorns, hares, and rabbits. Land snails serve as intermediate hosts.

Order SPIRURIDA

Characterized by a cylindrical esophagus composed of a short anterior muscular and a long posterior glandular part. The spicules usually are unequal and dissimilar. The mouth is bordered by 2 large lateral pseudolabia or no lips at all. There may be caudal alae but no copulatory bursa. The females may be oviparous, in which case the eggs are embryonated when laid, or ovoviviparous. First and second stage larvae have a cylindrical esophagus. Parasites of the alimentary canal, tissues, and circulatory system. The life cycles are indirect with arthropods commonly serving as intermediate hosts.

Family Dracunculidae

These are ovoviviparous worms of the subcutaneous tissues of reptiles, birds, and mammals. There are 8 cephalic papillae in the external circle and 6 in the internal circle. The esophagus has a marked constriction encircled by the nerve ring. The spicules are subequal and the vulva is immediately behind the head or back as far as the posterior third of the body.

Dracunculus medinensis

The dragon worm, Medina worm, or Guinea worm, *D. medinensis*, has been known since ancient times. It is a parasite of man that occurs especially from central India to Arabia, and is locally important in Indonesia, Egypt, and tropical Africa. The adult female lives in the subcutaneous connective tissue, and cutaneous swellings which later become ulcers develop around its anterior extremity (Figure 183).

Description

Males are up to 29 mm long by 0.4 mm wide and the females up to 53 cm long by 1 mm wide. The mouth is small and surrounded by an internal circle of 4 to 6 well-developed papillae and an external circle of 8 (4 double) papillae; the amphids are well developed. The posterior end of the male is coiled ventrally and bears 10 pairs of genital papillae, 6 postanal and 4 preanal. The spicules are subequal, 0.5-0.7 mm long and the gubernaculum 0.2 mm long.

Life Cycle

The females, when gravid, migrate to the parts of the skin which frequently come into contact with water, such as the arms and legs. A small vesicle soon appears and ulceration follows. A small hole may be seen at the base of the ulcer from which the anterior end of the worm may protrude. When the affected parts are wetted with water, the body wall ruptures, allowing a loop of the gravid uterus to prolapse and, on contact with water, burst open, discharging a mass of first stage larvae into the water.

The motile larvae are apparently attractive to copepods, and when eaten by several species of *Cyclops* the larvae penetrate the gut wall into the body cavity and undergo 2 molts. After about a month in the copepod, the larvae are infective for man. The cycle is completed when the copepods containing the infective third stage larvae are swallowed with drinking water. The larvae, which are freed when the copepod is digested, penetrate the intestinal wall and migrate to the loose connective tissues. Maturity in man is slow and it is not until the worm is about a year old that it seeks the surface of the body to discharge the young.

Dracunculus insignis

This guinea worm, originally described from raccoons, occurs commonly in the hind feet, particularly during the winter months when ulcers are produced by the protruding gravid females. It occurs in dogs, cats, foxes, skunks, otters, and mink as well as raccoons, over much of the United States.

Description

The anterior end of the body forms a large, raised, circular, cephalic cap on which are papillae, mouth, amphids, and prominences. Surrounding the small, circular mouth is a transversely located rectangular border on the outside margin of which is located a large, double papilla at each corner, a large amphid at each end, and a large prominence on the dorsal and on the ventral sides. The esophagus consists of a short, narrow, muscular portion and a long, thick, glandular part with a marked constriction toward the anterior end encircled by the nerve ring.

Males are up to 22 mm long. The tail is conoid and bent ventrally. The cervical papillae (deirids) are 0.7 mm and the excretory pore 0.8 mm from the anterior extremity of the body. The genital papillae consist of 5 pairs of preanals arranged in an arc over the sides and anterior margin of the anus, 2 pairs of ventral postanals in a transverse row at the posterior margin of the anus, followed closely by 1 pair of lateroventrals and then 2 pairs of subventrals, and finally nearest to the tip of the tail are the paired pore-like phasmids opening laterally. The spicules are slender, equal, and 460-465 μ long; the gubernaculum is 119 μ long.

Females measure up to 28 cm long with the vulva near the middle of the body. The tail is short, blunt, and curved ventrally. Embryos are large, being up to 0.6 mm long, with the tail consisting of about one-third the total length of the body. A pair of large pocket-shaped phasmids, 1 on each side of the body, is located near the anus.

Life Cycle

Insofar as known, the life cycles of the dracunculids are basically similar in that they are indirect, first stage larvae are born, a copepod serves as the intermediate host, and the adult females protrude through an ulcer on the body of the host to deposit her first stage larvae.

Other Dracunculidae which may be encountered include *Dracunculus ophidensis* of water and garter snakes, *Avioserpens bifidens* of ducks, *A. nana* of great blue herons, and *Chelonidracunculus globicephalus* of snapping turtles.

Family Philometridae

The anterior extremity of the body lacks a cuticular shield and the esophagus has no median constriction. Adults occur in the celom, serous membranes, and connective tissue of fresh- and salt-water fish.

Philometroides nodulosa

Adults occur in the fins and cheeks of suckers (*Catostomus*) and occasionally in other closely related fish (Figure 184).

Description

The body is long, cylindrical, and with blunt extremities. The mouth, surrounded by 3 elevated lips, lies in a depression. The anterior end of the esophagus is slightly bulbous.

Males are up to 2.7 mm long and have a smooth cuticle. The mouth is surrounded by 3 small lips. The posterior end is blunt, rounded, and with 2 ventral and 2 dorsal swellings. The spicules are slender and nearly equal in length. A gubernaculum with a terminal barb is present.

Females are up to 45 mm long and the body is covered by irregularly spaced cuticular bosses. The head bears 3 elevated fleshy lips and an inner ring of 4 and an outer ring of 8 evenly spaced papillae. There is an anterior and a posterior ovary, and a single uterus packed with developing eggs and first stage larvae. The vulva is atrophied. Intrauterine eggs are rounded to ovoid in shape and 44 x 43 μ in size.

Life Cycle

The life cycle resembles that of other Dracunculoidea in the requirement of a copepod intermediary and in its initial development in the tissues of the final host. Females, gravid with fully developed first stage larvae, lie in the subcutaneous tissues of the host, usually between the fin rays. When the gravid females break out of the surrounding tissue, they burst and liberate many thousands of larvae. When the motile larvae are eaten by certain species of *Cyclops*, they undergo molts to the infective stage in about 3 weeks at room temperature. The cycle is completed when the infected copepods are eaten by a susceptible final host fish. The larval worms are liberated in the host's intestine and migrate anteriorly to their eventual location. Fertilization occurs during migration, after which the males die.

Family Tetrameridae

There is great sexual dimorphism in members of this family. Gravid females are globular with the small pointed ends extending beyond the enlarged part of the body *(Tetrameres)* (Figure 190), or the enlarged, long axis of the body is coiled in a complex manner *(Microtetrameres)*. The males are free in the lumen of the proventriculus and the gravid females are in the glands of Lieberkühn of birds.

Microtetrameres centuri

Members of this genus are distinct from most all other Spiruroidea in the nature of the tightly coiled gravid females and the filariform males. Adults occur in the proventricular glands of meadow larks *(Sturnella magna, S. neglecta)* and bronzed grackles *(Quiscalus versicolor)*. Living females are bright red and tightly coiled so that the body presents a globose configuration.

Description

Males are up to 1.4 mm long with the right spicule almost the length of the body and the left about 1/10 that of the other. There is no gubernaculum. There are 2 pairs of pre- and 2 pairs of postanal papillae. The tail is pointed.

Females are up to 1.4 mm long with the pointed ends extending beyond the body coils. Fixed specimens are pinkish and slightly longer than living ones. The buccal capsule is prominent and expanded midway between the anterior and posterior ends. An intestinal diverticulum is present at the esophago-intestinal junction. The vulva is near the anus. Embryonated eggs are thick-shelled, flattened on 1 surface, with a boss on at least 1 end, and measure about 35 x 51 μ in size.

Life Cycle

The eggs contain fully developed first stage larvae when laid. They hatch when eaten by nymphs of certain species of *Melanoplus* grasshoppers. The larvae migrate into the hemocel and undergo the first molt, beginning about the eighth day, which is followed closely by the second molt. By the end of the ninth day, the earliest third stage larvae are present in the hemocel and begin encystment in clear, thin-walled cysts. The worms are infective after 40 days at room temperature. When released from the grasshoppers in the intestine of the host, the larvae migrate forward to the crop, traveling beneath the keratinoid lining of the gizzard en route.

Other Tetrameridae whose life cycles are known to some extent include *Tetrameres americana* in poultry, pigeons, ducks, and geese, with grasshoppers and cockroaches as intermediate hosts; *T. crami* of ducks, with gammarid crustacea as intermediaries; and *Microtetrameres helix* from crows, with grasshoppers and cockroaches as invertebrate hosts.

Family Gnathostomatidae

Characterized by a large cuticular head bulb covered with spines or strong, transverse striations and separated from the body by a constriction. The head bulb consists of 4 intercommunicating compartments called ballonets. The body may or may not be spined in the anterior portion. The species are parasitic in the stomach and intestine of fish, reptiles, and mammals.

Gnathostoma spinigerum

This species (Figure 195), the classical representative of the family, occurs in stomach tumors of carnivores (tigers, domestic cats, bobcats, dogs, leopards) and occasionally in the human stomach as adults or in skin lesions as larvae. Geographically, it is confined primarily to Asia. Because of its presence in man, much attention has been directed to its biology and epidemiology.

The life cycle is unusual in that 2 intermediate hosts are required. Copepods serve as the first intermediary, with fish, frogs, snakes, and possibly other poikilothermous animals as the second.

Gnathostoma procyonis

This species occurs in raccoons throughout the southeastern United States, and may be as widespread as the host. Adults occur in the stomach where they do great mechanical damage.

Description

They are large, stout worms with a head bulb bearing 9 or 10 transverse rows of 90 to 100 spines in each. The anterior half of the body is covered with large trident scales, which fuse toward the posterior end of the body to form encircling ridges or striations. The posterior part of the body is marked by coarse transverse wrinkles and noticeably less spinose or aspinose. The esophagus is club-shaped and about one-fifth the length of the body. There are 4 elongate cervical sacs attached inside the anterior of the body.

Males are up to 19 mm long and 0.8 mm in diameter where the cuticle is not inflated. The head bulb is up to 0.7 mm broad and 0.3 mm long. A bursalike expansion of the caudal end of the body is supported by 4 large, lateral, nipple-shaped papillae. Ventral papillae are absent. The right spicule is up to 3.0 mm long and the left 0.7 mm.

Females measure up to 26 mm in length and 2 mm in diameter. The head bulb is up to 0.8 mm broad and 0.4 mm long. The vulva is up to 8.5 mm from the tip of the tail. The ovejector extends forward, widens into a thin-walled tube filled with eggs that continues forward then bends back and divides into 2 posteriorly directed uteri. Posterior end of body is bluntly rounded. Eggs are brownish in color, with a thin, finely granulated shell, and bear a glassy, pluglike operculum at 1 end. They are unembryonated when passed in the feces and measure 64-76 x 37-41 μ.

Life Cycle

The eggs are passed in the feces of the host. At room temperature, hatching begins at about 12 days and continues for about 2 weeks. The motile larvae are sheathed and die in a few days unless they are eaten by copepods. Nauplii of *Cyclops viridis, C. bicuspidatus,* and *Macrocyclops albidis* are more susceptible than adults. Development to the early third stage takes place in the body cavity of these hosts within 7 to 8 days following ingestion of second stage larvae. Experimentally, guppies are satisfactory second intermediate hosts. When encysted third stage larvae, recovered from naturally infected second intermediaries (snakes), are fed to raccoons, adult worms are recovered. After entering and before maturity is reached, the larvae leave the stomach or duodenum of the final host, enter the celom and musculature for a prolonged migration during which growth occurs. When sufficiently mature, they return to the lumen of the stomach by burrowing through the wall. Prepatency is about 4 months.

Family Spirocercidae

With well-developed stoma but no pseudolabia. The mouth is surrounded by 6 parenchymatous masses and 2 dorsal, 2 lateral, and 2 ventral papillae. The buccal capsule is smooth. Caudal alae are present but caudal papillae are few. Generally oviparous, and the vulva is usually near the posterior end of the body. Parasites in the esophagus, stomach, aorta, and lungs of mammals (Figure 189).

Spirocerca lupi

The esophageal worms of dogs have a wide distribution in warm parts of the world, including the United States. Adults occur in nodules in the esophagus of dogs and foxes, and migrating larval stages in the stomach wall and wall of the coronary, gastroepiploic and celiac arteries, and aorta, forward to the upper thoracic artery, and connective tissue between the upper aortic arch and anterior wall of the lower part of the esophagus.

Description

Adult worms are usually coiled in a spiral and when alive are red. The lateral pseudolabia are tri-lobed, each lobe consisting of a large internal pulpy mass with a broad basal and a pointed distal portion. These lobes are arranged so that there is a dorsal, median, and ventral one on each pseudolabium. Each dorsal and ventral lobe bears a distinct papilla on the outer margin of the base.

Males are up to 54 mm long, with ventrally curved tail bearing 4 pairs of lateral alae and a single median preanal papilla and 2 pairs of postanal papillae. In addition, a group of papillae is located near the tip of the tail. The left slender spicule is up to 2.8 mm long and the right, stouter one is up to 0.8 mm. The gubernaculum is rudimentary.

Females are up to 80 mm long with the vulva near the posterior end of the esophagus. Eggs have thick shells and measure 30-37 x 11-15 μ. They contain first stage larvae.

Life Cycle

The eggs are passed in the feces of the host and hatch only after they have been eaten by a suitable coprophagous Scarabaeidae beetle. Species of 3 genera of tumble bugs (*Canthon, Ontophagus,* and *Phanaes*) and 1 of earth-boring dung beetle (*Geotrupes*) are known suspectible hosts. The larvae develop to the infective third stage in 4 to 6 weeks and become encysted in the beetle, chiefly on the tracheal tubes. If beetles with infective larvae are eaten by an unsuitable vertebrate host, the larval worms promptly migrate from the intestinal lumen to the connective tissue surrounding the outer wall of the intestine where they reencyst. They are capable of reencystment in lizards, birds, and mammals, particularly species that regularly feed on dung beetles. The ability of an infective-stage larva to reestablish itself in an unsuitable vertebrate host, in which there is essentially no larval development, is known as paratenesis. The unsuitable host, which is normally inserted between the last intermediary and the final host, is known as a paratenic host. The final host may become infected by ingesting either beetles containing the infective larvae or paratenic hosts. On being liberated in the stomach, the larvae penetrate into the stomach wall and, upon reaching the arteries, migrate in the walls of the coronary and gastroepiploic arteries to the celiac artery and aorta, forward to the upper thoracic artery, and connective tissue between the upper aortic arch and anterior wall of the lower part of the esophagus. In experimental infections, the prepatent period is about 5.5 months.

Family Physalopteridae

The family is characterized by the large, simple, somewhat triangular, lateral lips armed with 1 or more teeth on the free margin, large caudal alae, long stalked papillae which appear to be supporting the alae, and sessile pre- and postanal papillae (Figure 197).

Physaloptera hispida

This species, which occurs in the pyloric region of the stomach in cotton rats *(Sigmodon hispidus)**
from the southeastern United States, is an excellent model of a Spirurina for anatomical studies of the
adults and biological investigations of the larval stages in an arthropod intermediate host. The life cycle
requires a long time to complete.

Description

Adult females are pink and the smaller males are white. There are 2 semicircular lips, each with a
lateral amphid, 1 externolateral tooth and 3 associated internolateral teeth on each lip, together with 1
subdorsal and 2 subventral papilla. Deirids and an excretory pore are at about the same level, slightly
posterior to the nerve ring. There is no cuticular collar extending forward over the head.

Males are up to 42 mm long by 1.4 mm in diameter. The tail is flexed ventrally and bears prominent caudal alae supported by 4 pairs of long, evenly spaced, pedunculated papillae; there are 3 sessile,
preanal papillae arranged like an inverted equilateral triangle, a transverse row of 4 papillae at the posterior margin of the anus, and 3 pairs of evenly spaced ones on the ventral side of the tail. The spicules are
slightly dissimilar, unequal, and short; the right one is up to 0.6 mm long and the left up to 0.5 mm.

Females are up to 64 mm long and 2.0 mm in diameter. The uterus is amphidelphic with the branches arising from the posterior end of the broad egg chamber which follows the slender vagina. The vulva
is near the junction of the first and second quarters of the body. The phasmids, midway between the
anus and the tip of the tail, appear clearly as a pair of lateral pores. Eggs 24-30 x 40-52 μ and are embryonated when laid.

Life Cycle

The eggs are passed in the feces of the host and hatch after they have been eaten by suitable insects.
Cockroaches *(Blattella germanica)*, earwigs *(Forficula auricularia)*, and ground beetles *(Harpalus* spp.)
are good hosts. The larvae become encysted in the wall of the colon or rectum and develop to the infective stage in 4 to 5 weeks. The final host becomes infected by ingesting intermediary insects containing
fully developed larvae. The prepatent period is 2.5 to 3 months. For life cycle details consult Schell
(1952).

Families Onchocercidae and Dirofilariidae

These two families, which are discussed here as representatives of the large group of filariae worms,
are characterized as slender, delicate parasites of the body cavities, air sacs, tissues, and circulatory system of amphibians, reptiles, birds, and mammals (Figure 148). Many species of this group of worms are
of great medical and veterinary importance, because of the serious and fatal diseases they produce in
their vertebrate hosts, including man and his domestic animals.

Females of both families are ovoviviparous, giving birth to larvae known as microfilariae (Figure
149), which in the Onchocercidae are sheathed and in the Dirofilariidae unsheathed. The microfilariae
may appear periodically with peak numbers in the blood at given times during the day or night, thus
being diurnal or nocturnal, or they may be present continuously. Such conditions are said to be of periodic or continuous occurrence in the blood.

Family Onchocercidae

Members of this family are characterized by a rudimentary stoma without lips or spines. There is no
ring around the mouth nor one connecting the buccal capsule and esophagus. The latter usually is not
divided into a short anterior muscular and a long posterior glandular part. Caudal alae generally are ab-

---

*The host rats of *P. hispida* as well as *Litomosoides carinii* (p. 174), may be trapped in the field or purchased from Rider Animal Co., Inc.,
Route 2, Box 270, Brooksville, Florida 33512.

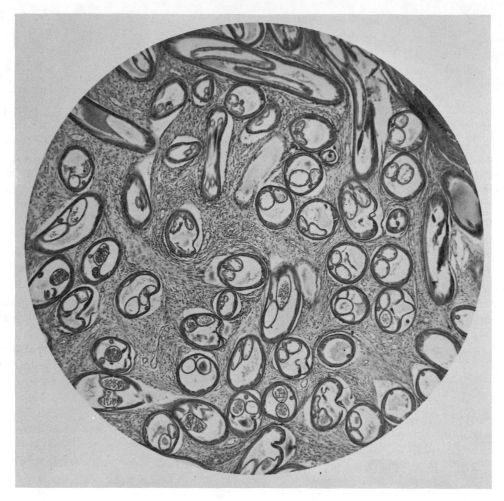

Figure 148. *Onchocerca volvulus,* section through nodule showing many adult worms.

sent, but if present, they are small and not supported by papillae (Figure 209). Microfilariae are periodic. Parasitic in the connective tissue, heart, arteries, or body cavities.

Wuchereria bancrofti

This is undoubtedly the most important medically of the Filariina. It is the cause of human bancroftian filariasis manifested by elephantine enlargement of pendent parts of the body such as the ear lobes, forearms, lower part of the legs, breasts, and male and female external genitalia. Its distribution includes the tropical regions of the world, both continental and insular.

Adults are not readily available for class use but they are described in detail in most texts on parasitology, from which information on the anatomy may be obtained. Blood smears with microfilariae are often available, as are sections of lymph nodules or vessels containing adult worms, and sections of mosquitoes with migrating larvae.

To find microfilariae of the periodic nocturnal variety, one must examine blood taken at night, preferably between 10 P.M. and 2 A.M. The best time to find the continuous-diurnal, subperiodic type in Polynesia is between 3 and 5 P.M. because the microfilariae are most numerous in the peripheral blood at this time. Examine well-stained specimens under high magnification, locating the basic anatomical landmarks characteristic of microfilariae, which are embryonic first stage larvae.

When born, the microfilariae are enclosed in a thin, transparent sheath which actually is the egg capsule stretched to a length greater than that of the larva, allowing the embryo to slide back and forth inside. The microfilariae measure up to 0.3 mm in length and up to 0.01 mm in diameter. A minute stylet may be seen at the anterior extremity. Delicate transverse cuticular striations occur along the entire length of the body, which is filled with nuclei except at each extremity (Figure 149).

The basic internal characters or landmarks and their relative location in the body provide the means for separating the species of microfilariae found in blood smears—hence the importance of knowing and recognizing them. They are located in percentage of total body length, beginning anteriorly. Locate the following structures. *Nerve ring:* it appears as a clear space extending across the body near the junction of the first and second fifths of it, or about 20% of the length of the larva from the anterior end. *Excretory pore:* it lies on the ventral side of the body behind the nerve ring slightly anterior to the union of the first and second thirds of the body, or about 30% of the length of the worm. *Excretory cell:* this is adjacent to the posterior side of the excretory pore. *Rectal cells:* these were formerly designated as G-cells and believed to be anlagen of the genital system but are now known to form the intestine and rectum. They consist of 4 cells. The first, R_1, is the largest and far caudad from the excretory pore, being near the union of the middle and last thirds of the body, or about 70% of the length of the worm. $R_{2, 3, 4}$ are small, close to each other, and arranged in a linear manner immediately anterior to the anal pore. *Anal pore:* this appears as a clear spot immediately behind R_4, or about 84% of the body length. A string of 10 to 12 body cells extends caudad from the anal pore, terminating short of the tip of the straight tail.

Upon entering the stomach of susceptible mosquitoes (*Culex quinquefasciatus*, *Aedes* spp., *Anopheles* spp.), the microfilarial sheath is shed quickly, the larvae penetrate through the stomach wall, migrate to the thoracic muscles, where they lie lengthwise between the muscle fibers, and begin growing into the sausage-shaped first stage larvae.

The fully developed sausage stage is 150 μ long by 10 μ in diameter. The bipartite esophagus, consisting of an anterior muscular and posterior glandular portion, is about one-half the length of the body and slightly longer than the intestine, which terminates near the end of the body in a large clear area, the anus. The excretory pore and cell appear as a large clear spot slightly anterior to the equatorial level of the esophagus. The genital primordium is located near the junction of the esophagus and intestine.

In the meantime, the larvae have molted twice and are infective. They now leave the thoracic muscles to make their way toward the head of the mosquito and into the proboscis. When the mosquito bites a warm moist skin, the larvae break free from the tip of the proboscis, escape head first, and enter the final host through the puncture made by the vector.

After entering the human host, the larvae migrate to their preferred sites of development, which are usually the main lymphatic vessels and in the sinuses of the lymphatic glands. The periodic form is most

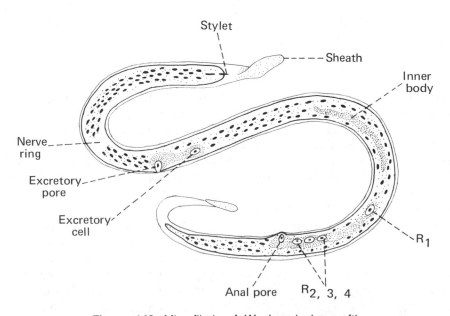

Figure 149. Microfilaria of *Wuchereria bancrofti.*

commonly located in the lymphatics of the lower limbs and the groin, whereas the subperiodic form is relatively more common in the upper limbs. After reaching the preferred site, they undergo 2 molts and the females, if accompanied by males, are fertilized and produce microfilariae which migrate into the bloodstream to await the appropriate vector.

The exact length of the prepatent period is not known, but circumstantial evidence suggests that it may be as much as a year. *Wuchereria bancrofti* is known to have survived for more than 10 years in persons who have moved to nonendemic areas. In bancroftian filariasis, it is unusual to see severe lesions except in patients with longstanding infections. The infection has usually been active for more than 20 years before the affected individuals develop severe elephantiasis (Nelson, 1966: 312).

Litomosoides carinii

This species is prevalent in cotton rats *(Sigmodon hispidus)* throughout their range in the southern half of the United States, especially in the southeastern states. It is the most useful model for studying filariae worms in the laboratory. Cotton rats are easily obtained, reproduce well in captivity, have a high incidence of natural infection, especially in the enzootic area, and are easily infected experimentally. The mite intermediate host is prevalent and easily reared in great numbers on white rats, which, if necessary, may be used in lieu of cotton rats as final hosts. Having a relatively short life cycle, all of the stages of a filarial worm with a bloodsucking arthropod intermediate host can be demonstrated in the laboratory.

Description

Adults are slender, cylindrical worms tapering to a bluntly rounded head and a slender conical tail. The buccal capsule is up to 23 μ long with cylindrical, heavily sclerotized walls encircled by thickened outer rings to which the esophageal muscles attach. The mouth is without lips or surrounding papillae. The esophagus is muscular throughout, up to 660 μ long, and is straight except for a bulbous anterior enlargement that surrounds the buccal capsule.

Males measure up to 28 mm long by 140 μ in greatest diameter. The tail, 180 μ long, is curled ventrally and tapers gradually in the posterior half of its length to a rounded tip. There are no preanal papillae, but 4 pairs of small, nearly equally spaced lateroventral papillae are located in the anterior half of the tail. Spicules are unequal and dissimilar. The shorter, right one is sclerotized and trough-shaped for up to 80 μ, about three-fourths its length, at which point there is an elbowlike, delicate, membranous projection up to 30 μ long; the longer left spicule has a total length of up to 295 μ; the anterior two-fifths is trough-shaped and sclerotized, followed by a portion partly sclerotized and partly membranous that ends in a delicate filament. The testis is located in the region of the posterior part of the esophagus and continues directly caudad as a straight vas deferens and seminal vesicle; the latter divided into 2 small tubes that empty into the cloaca.

Females are up to 100 mm long and have a tapering tail. The vulva is located a distance about twice the length of the esophagus from the anterior end of the body. The vulva is surrounded by a large muscular bulb, followed by the posteriorly directed, thick-walled muscular vagina (the single, expanded part of the anterior end of the uterus formed by the union of the 2 uteri), and 2 long, thin-walled, intricately looping uteri that extend to the posterior end of the body, and then loop forward to the anterior end, each terminating in a posteriorly directed threadlike ovary. The uteri are filled with eggs, and toward the proximal ends, with developing and sheathed microfilariae.

The anatomical landmarks of the microfilariae have not been described and located in percentages of the total body length. Obtain blood smears from infected cotton rats and stain some of them with Giemsa or other suitable blood stain (see p. 273). Locate the landmarks, using the pattern given for the microfilariae of *Wuchereria bancrofti*, and designate percentage-wise the position of (1) first body cells in the anterior part of the body, (2) nerve ring, (3) excretory pore, (4) excretory cell, (5) R_{1-4} cells, (6) anus, and (7) last body cell in tail.

Life Cycle

Adult worms lie free on the serosal surfaces of the pleural cavities of the rat host where the sheathed larvae are discharged. From the pleural cavities, the microfilariae enter the circulation and appear continuously in the peripheral blood vessels. When infected blood is ingested by the tropical rat mite (*Ornithonyssus bacoti*), the microfilariae burrow through the intestinal wall into the hemocel. By about the eighth day, the larvae become typically sausage-shaped and molt for the first time. The second molt occurs a day or 2 later and by the 14th to 15th days the slender third stage larvae are infective. After entering the rat, the larvae migrate to the pleural cavity. Here they undergo the fourth and fifth molts and mature. Prepatency is about 75 days.

Life Cycle Exercise

Animals required are some infected cotton rats that may be live-trapped in the southeastern United States or purchased from Florida where the parasite is enzootic. White rats may be used as nurse animals for developing colonies of mites. Cotton rats, wild brown rats, and often white rats are naturally infested with tropical rat mites.

To obtain mites to start a colony, place 1 or 2 infested rats in a cage provided with a tray beneath containing wood shavings. Engorged females drop off the host animals to lay their eggs. Twenty-four hours after putting the infested animals in the cages, examine the shavings by scattering and moving them over a large sheet of white paper to disclose the engorged female mites. Collect 50 to 100 mites by means of an aspirator bottle or a moist camel's hair brush to start a colony of uninfected mites. Adult tropical mites can be identified by consulting Baker *et al.* (1956).

A large number of mites can be reared for this experiment in an artificial nest constructed from a small, tight wooden box and homemade cages. Cover the bottom of the box with several layers of newspapers and then with a half-inch layer of loam soil (not sand) over which straw (not grass) is placed and bunched around the sides. The soil absorbs the urine voided by the rats and the straw provides a place for the mites to hide and oviposit, and young to develop.

Construct small but comfortable cages for individual rats from one-half-inch mesh hardware cloth bent and fitted appropriately. Put them on the straw and pack more about them (see Williams, 1946). Put an uninfected white rat in each small cage, provide it with food and water, and put the mites on them as culture animals. A thick ring of Vaseline or soft grease around the inside of the box above the cages and straw will discourage mites from crawling out. If the box is supported over a tray containing old lubricating oil, those mites negotiating the grease ring will be unable to escape into the room.

Within 10 to 14 days, the colony will contain hundreds of mites. It will consist of (1) bloodsucking females and nonbloodsucking males, (2) eggs, (3) hexapod nonfeeding larvae, (4) octopod bloodsucking protonymphs, and (5) nonfeeding deutonymphs that molt and develop into adults.

Engorged females are reddish and have a white Y-shaped mark on the dorsal side. Eggs are laid in 2 to 3 days after each engorgement and hatch in about 30 hours at room temperature. The whitish hexapod larvae about the size of the eggs do not feed, but molt when about 24 hours old to form 8-legged protonymphs. The protonymphs must engorge before being able to molt into deutonymphs 24 hours later. Deutonymphs do not feed but molt when 24 to 36 hours old into adults. The total time required to complete the life cycle of mites is 11.5 days at room temperature.

Uninfected mites are essential as a beginning point for experimental studies. They can be obtained easily by taking advantage of the developmental stages of mites. Place 50 or more adult females on a rat whose cage stands over a tray of water. Upon engorgement, the mites drop onto the surface of the water and are transferred to brood vials kept at room temperature. Oviposition takes place in them and the eggs hatch. The hexapod larvae soon molt to protonymphs which must have a blood meal. Use uninfected white rats for feeding the mites to continue their development. After the protonymphs have molted to the deutonymphs, they transform, without feeding, into the adult stage. The hungry and uninfected female mites are ready for infection.

To infect the mites, put adult females (protonymphs are not easily infected) on an infected cotton rat known to have microfilariae in the peripheral blood as demonstrated by stained blood smears. Allow the mites to engorge for 24 hours, at which time around 50% of them become infected; collect them from the surface of the water, and put in incubation vials. The mites must be allowed to feed on parasite-free rats on the 5th and 10th days following the initial engorgement. By the 15th day, the larvae have developed to the third stage. They are infective to rats and may be transmitted through the bites of the mites.

Dissect the mites in physiological solution, beginning 2 hours after they drop off the rat and every 4 hours thereafter during the first day to determine the sequence of events in the development of the larvae. Dissect other mites daily for the next 14 days and examine them for larvae, molts, and growth.

Larval worms in the mites fall into 3 separate and distinct stages. They are (1) first stage larvae, which include the microfilariae and finally the thick, stumpy, sausage forms which appear between the ninth and 13th days; (2) somewhat slender second stage worms appearing between the ninth and 13th days; and (3) third stage larvae, which are slender and longer than the preceding stages.

The first stage microfilariae in the stomach at the end of the blood meal are similar to those in the peripheral blood. After 6 hours, they are distributed randomly throughout the stomach. During the 24 hours residence in the stomach, the embryonic sheath is lost and by the end of this time most of the microfilariae have migrated through the stomach wall into the hemocel.

During the first week in the hemocel, the microfilariae gradually transform from the long, slender shape to the short, thick sausage form. At first, they become thickened equatorially. On the sixth day, the slender sickle-shaped tail appears along with other characteristics of the sausage form, which is attained on the seventh day. Molting begins on the eighth day, as indicated by the appearance of the loose cuticle at the ends of the body.

The first molt is completed by the ninth day while the body is still stumpy. By the end of the 13th day, nearly half of the larvae have molted and are in the third stage.

Third stage larvae are heralded by loosening of the cuticle as early as the ninth day and the second molt in some individuals is completed by the 10th day, producing third stage larvae. The majority of the larvae have completed the molt by the 13th day. The resultant infective larvae are up to 1 mm long.

Female mites that have fed on parasitized rats with microfilarae circulating in the peripheral blood harbor infective third stage larvae by the end of 2 weeks. Infection of parasite-free cotton rats can be accomplished easily and with a high degree of confidence by allowing these mites to feed on them. In order to follow the basic aspects of development in the vertebrate host, infect 5 clean, young, cotton rats by exposing each of them to 50 or more infected mites.

Four rats are to be examined postmortem for the developing stages of the worms in the pleural cavity. Keep the fifth one to determine the prepatent period.

After killing each infected rat, the skin is removed and the thoracic cavity opened. Carefully remove the lungs and heart to a large dish of physiological solution and wash them to remove any larvae from the serosal surfaces. Next wash out the thoracic cavity with the solution, catching all of the fluid. Sediment and decant the fluid until clear of blood. Any worms that have reached the pleural spaces from the site of inoculation by the mites can be recovered in this manner.

Within 18 hours after inoculation by the feeding mites, the third stage larvae are nearly 1 mm long and begin arriving in the thoracic cavity. Examine 1 rat on the second day after infection.

During the first week, the body attains a length of 1.5 mm. A well-defined group of cells in the anterior half of the body represents the ovary and vulva. The cloaca is distinct and the spicules show development. The buccal capsule which is characteristic of this stage is long and narrow with the outer surface of the wall bearing faint encircling thickened rings. By the end of the seventh day, evidence of the beginning of the third molt appears when the cuticle loosens and the third stage buccal capsule is hanging free. Examine another rat on the seventh day.

The third molt is accomplished as early as the eighth day after infection. Recently molted fourth stage larvae are around 1.2 mm long. Growth continues for 24 days during which time the worms develop sexually and grow in size. Males become 6.4 mm long and females 8.8 mm. The reproductive organs assume recognizable forms. Examine a rat on the 20th to 22nd days postinfection.

The final molt occurs about the 23rd to 24th days postinfection and the worms enter the preadult stage. They increase in length 1 to 2 mm during the molt and grow rapidly. The buccal capsule is thick-walled, cylindrical, and the outer surface of the wall has well-defined encircling rings. Examine a rat on the 25th to 26th days.

Examine a drop of blood prepared as a smear from the tail of the remaining rat at weekly intervals to determine the prepatent period. Microfilariae may appear in the blood between the 50th and 93rd days. When were they first seen in your rat?

Other species of Onchocercidae infecting man are *Brugia malayi* and *Onchocerca volvulus*. *Brugia* occurs in Indo-China, Malaya, Indonesia, India, and Ceylon. It is transmitted by species of *Mansonia* and *Anopheles* mosquitoes. *Onchocerca* is found in the tropical belt of west Africa, particularly the Congo area, and the Pacific slope of Guatemala and southern Mexico. It is transmitted by species of *Simulium* blackflies.

Family Dirofilariidae

Parasites of the heart and connective tissue of mammals. They are long, slender worms with a simple mouth devoid of lips. Caudal alae of the male are narrow and supported by large papillae. The vulva is located far anterior and microfilariae are without a sheath.

Dirofilaria immitis

This is the heart worm of dogs and occurs in the right side of the heart and in the pulmonary artery, often in large intertwined clusters (Figure 208). It is a dangerous parasite, as up to 100 of these large worms occur in a single animal, seriously interfering with the flow of blood. The infections occur in dogs over a wide area in the United States but they are especially prevalent along the Atlantic seaboard and southern half of the United States. In addition, foxes, wolves, coyotes, and cats are infected.

Description

They are very long, slender worms with a small circular mouth and 6 weakly developed cephalic papillae. The esophagus is short, narrow, and the anterior muscular and posterior glandular parts indistinctly separated.

Males are up to 200 mm long by nearly 1 mm in diameter, with a tapering, spirally coiled tail. The caudal alae are narrow and inconspicuous. Generally there are 5 pairs of large preanal papillae that support the alae and 6 pairs of smaller postanal papillae. The left spicule is up to 375 μ long and has a sharp elbowlike bend slightly posterior to the middle; the terminal portion ends in a narrow point; the right spicule is up to 229 μ long and basically arcuate in shape. A gubernaculum is lacking.

Females are up to 310 mm long by up to 1.3 mm in diameter. The tail is blunt and the vulva is located just behind the posterior end of the esophagus. The uteri and ovaries fill the body of gravid individuals. The vagina uterina and terminal parts of the uteri are packed with microfilariae.

Life Cycle

The unsheathed microfilariae are liberated in the blood of the heart and pulmonary artery and are carried throughout the body. They may, however, be subperiodic, showing a minor nocturnal peak. When the infected blood is ingested by *Aedes aegypti*, the yellow fever mosquito, several house mosquitoes, *e.g.*, *Culex pipiens*, *C. tarsalis* and *C. territans*, and the plasmodia-transmitting species *Anopheles maculipennis* and *A. quadrimaculatus*, the microfilariae migrate from the intestine to the Malpighian tubules, penetrate the cells at the tips, and develop to the sausage stage. After about 2 weeks, the larvae leave the Malpighian tubules and migrate anteriorly in the mosquitoes, going through the thoracic muscles and into the proboscis. In the meantime, they have developed to the third stage infective larvae. Transmission of larvae occurs while the vectors are feeding on the dog host. The young worms undergo development in various extravascular tissues before entering the blood vessels, migrating to the heart about 10 days postinfection. The prepatent period is 8 to 9 months.

A Dipetalonematidae, *Dipetalonema reconditum*, occurs in the connective tissues of dogs in this country and its microfilariae appear concurrently with those of *Dirofilaria immitis* in the peripheral blood. In general, they are smaller, have a small cephalic spine, and the landmarks have different locations percentage-wise in the body. The vectors are fleas (*Ctenocephalides canis*, *C. felis*) and lice (*Heterodoxus spiniger*). Consult Newton and Wright (1956) for the anatomical landmarks used in separating the 2 species of microfilariae.

Class ADENOPHOREA

The class is characterized by having an esophagus consisting of a long chain of individual cuboidal cells, or a cylindrical, muscular one in which case the males possess a fleshy, cup-shaped, rayless, copulatory bursa. Phasmids are absent.

Order TRICHINELLIDA

The esophagus consists of a short anterior muscular part followed by a long capillary tubule embedded in a long string of single cuboidal cells, the stichosome. It contains the single superfamily Trichinelloidea with 5 families.

Family Trichinellidae

This family contains a single species, *Trichinella spiralis*, known as the trichina or pork worm (Figure 214).

Trichinella spiralis

This species was discovered as calcified cysts in the muscles of a human cadaver in 1835 and in swine in 1846. A fatal case of human trichinellosis in 1850 was traced to the ingestion of ham and pork sausage, thus establishing the relationship between swine and human infection. The common hosts are, in addition to man and swine, rats, cats, dogs, and various wild rodents and carnivorous mammals, particularly foxes and bobcats. The incidence of trichinellosis in adult humans of the United States is 4.2%, according to Zimmermann et al. (1973), who examined more than 8,000 diaphragms from 48 States and the District of Columbia; adjustment for age distribution shows that approximately 2.2% of the population has detectable trichina infections.

Description

Adult worms are small, with the posterior end slightly thicker than the anterior part. The esophagus consists of an anterior muscular part, which terminates in a pseudobulb, followed by a constriction to a capillary tubule, that runs alongside a string of cuboidal cells, the stichosome, to enter the much larger intestine.

Males are up to 1.6 mm long and 50 μ in diameter. The posterior end of the body is blunt and bears a large, conical, copulatory papilla on each side of the terminal anus. The testis begins in the posterior fifth of the body as a large, elongate organ, continues forward to the posterior end of the esophagus, constricts, and bends backward, extending as a narrow tube to the cloaca. There is no spicule.

Females are up to 4 mm in length, with the greatest diameter exceeding that of the males. The posterior end is blunt and the anus terminal. The single ovary is in the posterior end of the body. It continues forward as a slender oviduct to about the junction of the middle and posterior thirds of the body, where it expands into a broad single uterus filled with eggs and hatched larvae. The vulva is in the esophageal, or stichosomal, region. These worms are ovoviviparous, with the larvae deposited in the host's intestinal mucosa.

Life Cycle

Infection of a new host occurs when another animal ingests: (1) meat containing viable encapsulated larvae, or (2) viable encapsulated or freed larvae, or adult worms contained in, or expelled from, the intestine of another host. Once excysted in the stomach, the larvae pass into the small intestine, where they develop to sexual maturity within 2 to 3 days. Mating occurs, after which the females penetrate the intestinal lining. The fertilized eggs hatch within the female and leave her as larvae. Larviposition begins with 4 to 5 days after mating and from 6 to 7 days postinfection.

The larvae escape from the parent and ordinarily enter directly into the lymph spaces. They may reach the thoracic duct, pass into the venous system and thus be carried via the liver, heart, lungs, back to the heart, and through the arterial circulation to the various parts of the body, where they leave the capillaries and enter the skeletal muscles. The highest concentrations of larvae usually occur in the muscles of the diaphragm, tongue, larynx, masseters, intercostals, and eyes.

There are indications, however, that the migration route of the larvae from the intestine to the muscles is not limited to the blood stream. In experiments with rats and mice, it has been observed that some, if not most, larvae reach the muscles by way of the abdominal and thoracic cavities, and serosal membranes (see Villella *in* Gould, 1970).

Once having penetrated the muscle fiber, the larva increases in size and becomes spirally rolled. Meanwhile the muscle fibers that have been invaded by the larvae undergo degenerative changes, and each parasite becomes surrounded by a membranous cyst formed from muscular connective tissue. The cyst begins forming about the third week and is completed within a month (Figure 150).

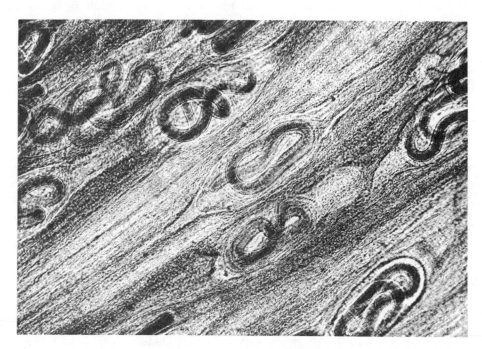

Figure 150. *Trichinella spiralis*, encapsulated larvae in skeletal muscle.

Investigators are not in agreement on the number, time and location of the molts, but it appears that *T. spiralis* has 4 molts and they occur in the intestine of the host. It is possible, as has been suggested, that the varying results have been influenced by : (1) strains of *T. spiralis* used, (2) species of experimental hosts, (3) intraspecific differences among experimental hosts, and (4) techniques employed.

Life Cycle Exercise

Infective larvae in muscle tissues, a few white mice, some cages, and a bit of ingenuity are necessary for demonstrating the life cycle. Larval stages from the intestine, body fluid, blood, serosa and muscles should be studied alive in physiological solution, using vital stains (see p. 279), as well as fixed and stained permanent mounts. There are 3 general developmental periods in the life cycle. They occur sequentially and in different parts of the body. The purpose of this study is to learn how to recognize them.

The first period begins in the uterus of the adults living in the small intestine, where larviposition occurs. The newly born larvae may enter the intestinal circulation and be carried to the heart, from which organ they are carried by the arterial system to the skeletal muscles, or they may reach the muscles via the body cavities and serous membranes.

The second period pertains to the development of the different parts of the reproductive system and other organs, and takes place in the muscles of the host. Encapsulation of the larvae terminates this period.

The third period is in the intestine of the host, beginning when the encapsulated larvae are ingested with meat and liberated from the capsule. The period ends with the appearance of the sexually mature worms.

Wild rats obtained on garbage dumps where organic refuse is deposited generally can be depended on to provide a source of larvae for beginning the study. Often, however, the department will have infected white rats on hand or be able to obtain one.

Because of their small size and low cost, white mice are preferred as experimental hosts in this study. The times given below for features in the life cycle are what occurs when mice are used. If white rats, which work equally as well, are substituted, consult Villella *in* Gould (1970); in rats, the timing of events, particularly molting, apparently differs from that in mice. Feed each of 30 hungry mice about 300 encapsulated larvae in small pieces of heavily infected muscle tissue from the donor rats.

Beginning 4 hours postinfection and continuing through the next 40 hours, examine a mouse every 4 hours. Remove intestine, cut into short lengths, slit open longitudinally, and place in physiological solution at 37 C. Occasionally agitate the intestinal segments with a dissecting needle to separate the emerging worms. After about an hour, remove the intestine and clean the solution by sedimentation and decantation preparatory to examination for larvae under the compound microscope. Compress the opened intestine between 2 glass plates and examine it with transmitted light under the high power of a dissecting microscope (or low power of a compound microscope) for worms that have entered the mucosa. Adult females will be present in the intestinal mucosa.

Some larvae fail to enter the intestinal mucosa and are passed alive in bits of undigested material or free in the feces. Feed fresh fecal pellets passed by mice during the first 2 days after a meal of trichinous meat to 3 trichina-free mice. About 10 days later, examine the diaphragm, tongue, masseter muscles, and intercostal muscles by the compression method. Presence of larvae in these muscles is evidence that they were in the feces.

Between the fifth and 14th days postinfection, examine the body fluid, blood, and serous membranes for larvae, to determine the time and route of migration. Beginning with the fifth day, examine a mouse every 24 hours for 10 consecutive days. Kill the animal, place it on its back, make a midventral and 2 lateral incisions, free the diaphragm, and fold back the body wall to expose the thoracic and abdominal cavities. With a Pasteur pipette transfer any body fluid to a slide and add a cover glass for examination under a compound microscope for young larvae. To collect the blood, ligate the hepatic portal vein near the intestine and the aortic arch above the heart. Cut the blood vessels so that the ligatures will keep the blood within the limits bounded by them and remove the lungs, heart, and liver as a unit to a dish of physiological solution where they should be opened to remove the blood. Sediment and decant until the solution is clear and examine for the larvae. Remove some serosa, transfer to a slide, smooth out in a drop of physiological solution, add a cover glass, and examine under a compound microscope.

The larvae begin entering the skeletal muscle bundles on about the sixth day postinfection and undergo a long period of development, which for convenience is divided into an early and a late period.

The early larval stage extends from the sixth to the 12th days. On the seventh day, they are shorter than the blood forms, there are 18 to 24 cuboidal cells in the esophagus, a syncytial cord extends to the posterior tip of the body, and a separate rectum has formed. The larvae are coiled on the ninth day, the intestine is a thin-walled tube, and the rectum differentiated. By the 12th day, the larvae are 270 μ long and 21 μ in diameter.

The late larval stage includes the period from the 14th day through the 29th day, during which time development has advanced markedly and the encapsulation has been completed. By the 14th day, the stichosome is 146 μ long; on the 16th day, the sexes are differentiated, with the rectum of the females 21 μ long and the males 28 μ; by the 17th day, the larvae are U-shaped or coiled. Encapsulation begins about the 21st day and is completed by the 29th. No molt occurs while the larvae are in the muscles, either free or encapsulated. On each of the 8 days indicated, kill a mouse, and examine one or more of the heavily infected muscles, compressed between slides, under a compound microscope, for the different developmental stages.

Four molts occur in the intestine of the host, following ingestion of infective larvae, as judged by the presence of a loose cuticular sheath and distinct anatomical changes.

First molt: the males molt in 9 to 10 hours after entering the intestine and the females in 12 to 13 hours. Are they in the lumen or intestinal wall? The body increases in length prior to the molt. The female genital primordium appears as a long chain of cells ventral to the stichosome and the male primordium is hook-shaped.

Second molt: this function occurs on about the 17th hour for males and the 19th for females. By this time, the cone-shaped copulatory papillae of the males are evident. The cuticular lining of the muscular part of the esophagus and the rectum is loose. Where are the larvae?

Third molt: this molt occurs about the 24th hour for males and the 26th for females. The sex organs are well developed, the copulatory papillae curved, and the vaginal plate present. Where are the larvae at this stage?

Fourth molt: males molt about the 29th hour and females the 36th hour. After molting, the worms are mature.

Examine mice at appropriate times in an attempt to identify the different larval stages.

Family Trichuridae

These, the whipworms, are medium to large in size. The esophageal part of the body is very slender and longer or shorter than the thick posterior part which contains the reproductive organs (Figure 211). They are parasites of both the intestine and cecum of amphibians, reptiles, birds, and mammals.

Trichuris trichiura

This, the human whipworm, also occurs in swine and monkeys. Adults commonly live in the cecum but may occur in the appendix, colon, and rectum. The long, slender portion of the body is buried in the mucosa.

Description

The anterior end of the body is slender and about twice as long as the very much thicker posterior end. The esophagus consists of a short anterior, muscular portion and a long posterior portion made up of a capillary tubule running alongside a row of closely packed cells, reaching to the union of the 2 parts of the body. The mouth has no lips but is provided with a tiny spear. A bacillary band (a band of fine, low, cuticular elevations) extends the length of the ventral side of the esophageal region.

Males are up to 45 mm in length. The hind end is coiled ventrally as much as 360 degrees at times. The testis begins near the posterior end of the body and extends forward as a straight tube to the esophago-intestinal juncture where it bends abruptly back to form the thick, muscular ejaculatory duct. A second constriction is followed by a broad, slightly shorter cloacal tube that empties into an almost equally long spicular tube. A single, 2.5 mm long spicule protrudes through a retractable sheath with a bulbous end.

Females are up to 50 mm long and bluntly rounded posteriorly. The ovary is a slender, sinuous organ, beginning near the middle of the enlarged portion of the body and extending to the posterior extremity where it sharply bends forward and quickly expands to form a narrow oviduct that opens into a broad uterus. The anterior end of the uterus narrows into a long vagina opening through the vulva located near the posterior end of the stichosome. The eggs are described under Helminth Eggs, p. 213.

Life Cycle

The life cycle is direct; the eggs reach the infective stage in about 3 weeks under favorable environmental conditions. They hatch in the alimentary canal and the larvae reach the cecum and mature in about a month. Eggs may remain viable in the soil for several years.

Other common species of whipworms include *T. ovis* of sheep, *T. opaca* of muskrats, *T. vulpis* of dogs and foxes, and *T. leporis* of rabbits. They are sufficiently similar to *T. trichiura* to replace it for anatomical studies.

Family Capillariidae

The members of this family are capillarylike in having a slender esophageal region about equal to or slightly shorter than the thin posterior part of the body, which is only slightly thicker than the anterior part. There is the stichosomal type of esophagus characteristic of the order. It is a large family with numerous species, some of which occur in all the classes of vertebrates.

Capillaria columbae

This species occurs in the small intestine of domestic and wild pigeons, mourning doves, fowl, turkeys, and other birds in many parts of the world (Figure 212).

Description

The body is threadlike with the esophageal part a little shorter and only slightly thinner than the posterior portion. The esophagus consists of the short anterior muscular portion, the long, posterior capillary tube, and the long stichosome (cell body) which is only slightly annulated except at the posterior end.

Males are up to 10.8 mm long by 0.06 mm in diameter. A narrow ventral and 2 well-developed lateral bacillary bands (bands of minute cuticular elevations) extend the length of the body. Each lateral one has a width equal to one-fourth to one-third the diameter of the body. The esophagus is up to 6 mm long. The posterior end of the body has a rounded, bursalike membrane supported by a pair of L-shaped processes, each of which has a terminal papilla. The spicule has an expanded base and a bluntly rounded tip; it is up to 1.6 mm long by up to 12 μ in maximum width. The spicule-sheath is transversely wrinkled but devoid of spines.

Females attain lengths up to 19 mm and diameters up to 0.09 mm. The posterior end is bluntly rounded and dark colored. The vulva is slightly behind the esophago-intestinal junction. The vagina is directed posteriorly and is heavily muscled. The anterior lip of the vulva is slightly prominent, but there appears to be no protrusible membrane.

Eggs are unembryonated when laid, yellowish in color, vary somewhat from symmetrical lemon-shaped to slightly asymmetrical, with a clear mucoid operculum at each end, and thick shells with a granular surface pattern. They measure 41-55 x 30-31 μ. The life cycle is direct and can be completed without the aid of an invertebrate intermediary.

Of the species with a monoxenous life cycle found in mammals, *Capillaria hepatica* is the most frequently encountered. It inhabits the liver of a variety of animals, primarily rodents, where its eggs accumulate without developing further. Since the eggs require exposure to air to embryonate, preliminary passage through the intestine of a predator animal, or death and decomposition of the original host, is necessary; infection does not result from eating the egg-burdened fresh liver. Infection results when rodents swallow the larvated eggs, which hatch in the small intestine and the larvae go to the cecum, enter the branches of the hepatic portal vein, and are carried to the liver where they mature. The eggs are described under Helminth Eggs, p. 213.

Some commonly encountered Capillariidae with a heteroxenous life cycle, requiring earthworms as intermediaries, include *Capillaria caudinflata* in the lower intestine of chickens; *C. annulata* in the wall of the esophagus and crop of chickens; and *C. plica* of the urinary bladder of dogs, foxes, and wolves.

Order DIOCTOPHYMATIDA

Members of the order are medium to large worms whose males have fleshy, bell-shaped bursa without rays and a single bristle-like spicule. The esophagus is muscular, cylindrical, and without a posterior bulb. Females have a single ovary.

Family Dioctophymatidae

Parasites of the kidney, occuring occasionally in the celomic cavity of carnivorous mammals.

Dioctophyma renale

This, the giant kidney worm, is a parasite normally of mink but occurs in dogs, foxes, and other mammals, including man (Figure 215). It occurs most commonly in the right kidney because the loop of the small intestine from which the larvae migrate lies close to or in contact with it. The kidney becomes a mere shell containing the coiled worm. Sometimes, they escape from the kidney and occur in the body cavity.

Description

These are large cylindrical worms with the cuticle transversely striated. The hexagonal mouth is devoid of lips but is encircled by 2 series of 6 papillae each. The body is attenuated at both ends and bears a series of papillae along each lateral line.

Males are up to 45 cm long by up to 6 mm in diameter. The copulatory bursa is cup- or bell-shaped;

its margin and inner surface are covered with minute papillae. The cloacal opening is near the center of the inside of the bursa. There is a single, slender spicule up to 6 mm long.

Females attain lengths of up to 100 cm and diameters up to 12 mm. The caudal end is rounded and the terminal anus crescent-shaped. The vulva is anterior, being up to 7 cm from the anterior end of the body.

Eggs are characterized by thick, grossly pitted shells, except the poles which are smooth, are brownish-yellow in color, and unembryonated when laid. They measure 64-68 x 40-44 μ.

Life Cycle

According to Woodhead (1950), 2 intermediate hosts are required in the life cycle. Hatching occurs when the embryonated eggs are swallowed by certain species of *Cambarincola* (Oligochaeta: Branchiobdellidae), which live as commensals in the gill chamber of crayfish. When a black bullhead *(Ictalurus melas)* ingests a crayfish with infected branchiobdellids, the second stage larvae migrate through the intestinal wall into the celom, where they molt twice, and encyst in the serous membranes of the fish. In the mammalian host, the fourth stage larvae excyst, migrate through the intestinal wall, and enter the kidney where the worm matures.

Karmanova (1962), working in the USSR, reexamined the life cycle. She found that the eggs hatch when they are swallowed by the oligochaete intermediary *(Lumbriculus variegatus)*. The larvae penetrate the intestinal wall and enter the ventral blood vessel, where they undergo 3 molts. When the infected oligochaete is eaten by the mammalian host, the larvae migrate to the kidney where maturity occurs. The experimentally determined life cycle took 8 to 9 months; 1 month for embryonation of the eggs, 4.5 to 5 months for larval development in the oligochaete, and 3 months for maturity in the final host.

Recently, investigators* working in Canada have confirmed Karmanova's findings. It seems, therefore, that only 1 intermediary is required, as in the case of other species of Dictophmatidae whose life cycles have been determined. If an infected oligochaete is eaten by a fish, the larvae penetrate the intestinal wall and re-encyst among the serous membranes. Thus, the fish is a paratenic host and not an obligate intermediary, as reported by Woodhead.

Key to Common Families of Parasitic Nematodes

Capital letters in parentheses refer to hosts: F-Fish, A-Amphibia, R-Reptiles, B-Birds, M-Mammals.

1 Esophagus club-shaped, cylindrical, or with posterior bulb and muscular, or with anterior muscular and posterior glandular portions; bilobed copulatory bursa when present always membranous and supported by 6 pairs of lateral rays; phasmids present; excretory canal lined with cuticle; males with paired genital papillae. Class SECERNENTEA 2

Esophagus with long posterior portion consisting of a chain of cuboidal cells, or cylindrical and muscular in which case males possess a fleshy, cup-shaped rayless copulatory bursa; phasmids absent; excretory duct not lined with cuticle; males usually without paired caudal papillae. Class ADENOPHOREA 60

2(1) Females only known from parasitic forms; esophagus of adult muscular, cylindrical and short or long. Order RHABDITIDA, Suborder Rhabditina 3

Both sexes parasitic and known 4

3(2) Esophagus short; buccal capsule present; vulva near middle of body; in lungs (Figure 151) (A) Superfamily Rhabditoidea .... Rhabditidae

Esophagus long; buccal capsule absent; vulva in posterior part of body; in intestine (Figure 152) (R, B, M) Superfamily Rhabdiasoidea .... Strongyloididae

4(2) Males with cuticular bilobed copulatory bursa supported by paired rays; esophagus more or less club-shaped. Order STRONGYLIDA 5

---

*Unpublished data kindly provided by Roy C. Anderson, Department of Zoology, University of Guelph, Guelph, Ontario, Canada.

Copulatory bursa absent; esophagus muscular or part muscular and part glandular 17

5(4) Buccal capsule large, with thick wall; body thick. Suborder Strongylina 6

Buccal capsule rudimentary or absent; filiform body 11

6(5) Mouth with ventral toothed or smooth cutting plates; head bent dorsally. Superfamily Ancylosto-matoidea 7

Mouth without cutting plates; head straight 8

7(6) Cutting plates with smooth margin; in intestine (Figure 153) . . . . (M) Uncinariidae

Cutting plates with teeth on margin; in intestine (Figure 154) . . . . (M) Ancylostomatidae

8(6) Buccal capsule with thickened rim, small teeth in base; mouth wide; small males often perma-nently attached to large females; in trachea (Figure 155) (B, M) Superfamily Syngamoidea . . . . Syngamidae

Buccal capsule without thickened rim; in alimentary canal or kidneys. Superfamily Strongyloidea 9

9(8) Buccal capsule short, ring-shaped; transverse groove on ventral side of body anterior to excretory pore; in cecum, colon (Figure 156) . . . . (M) Oesophagostomidae

No ventral transverse groove; buccal capsule globular or cup-shaped 10

10(9) Buccal capsule globular, usually with longitudinal dorsal groove and well-developed leaf crowns; copulatory bursa long; in intestine (Figure 157) . . . . (M) Strongylidae

Buccal capsule cup-shaped; copulatory bursa short; dorsal groove lacking; in kidneys (Figure 158) . . . . (M) Stephanuridae

11(5) Bursa and rays well developed, rays slender and not enlarged or fused; anterior end of the body usually inflated; body with longitudinal ridges; eggs thin-shelled, unembryonated in uterus; in alimentary canal and lungs. Suborder Trichostrongylina 12

Bursa less developed or absent, rays fused or abnormally large; body without anterior inflation or longitudinal ridges; uterus contains eggs with larvae or first stage larvae enclosed in a thin mem-brane; in respiratory tract. Suborder Metastrongylina 14

12(11) Genitalia of female single; spicules long, slender; vulva near anus; in intestine (Figure 159) (M) Superfamily Heligmosomatoidea . . . . Heligmosomatidae

Genitalia of female double; spicules short, stout; vulva not near anus. Superfamily Trichostrongyloidea 13

13(12) Buccal capsule small; spicules simple, sox-shaped; bursa short; in lungs (Figure 160) . . . . (M) Dictyocaulidae

Buccal capsule rudimentary or absent; spicules complex, with crests, or long and filiform; bursa with lateral lobes well developed; in alimentary canal (Figure 161) . . . . (A, R, B, M) Trichostrongylidae

14(11) Bursa reduced to 2 elongate lateral lobes projecting ventrally at almost right angles to body; spicules short, membranous; vulva near anus at extremity of tubular process; first stage larva in thin membrane in uterus; in bronchi, blood vessels, heart, frontal sinuses (Figure 162) (M) Superfamily Pseudalioidea . . . . Filaroididae

Bursa not reduced to 2 elongate lateral lobes, symmetrical or absent 15

15(14) Bursa well developed with large lateral lobes; dorsal ray double, externolateral ray lobed at tip; spicules long; buccal capsule small; in respiratory system (Figure 163) (M) Superfamily Metastrongyloidea . . . . Metastrongylidae

Bursa small, dorsal ray enlarged and single with 1 to several ventral papillae or with tip divided into 2 or more branches; spicules short; buccal capsule absent or rudimentary. Superfamily Pro-tostrongyloidea 16

16(15) Spicules long, slender, without alae; buccal capsule absent; gubernaculum present or absent; in circulatory or respiratory systems (Figure 164) . . . . (M) Angiostrongylidae

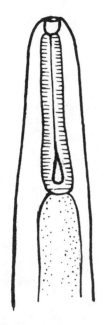

Figure 151. *Rhabdias bufonis.*

Figure 152. *Strongyloides stercoralis.*

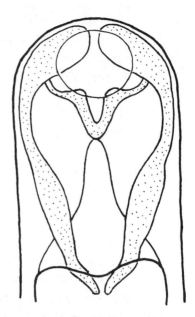

Figure 153. *Uncinaria stenocephala.*

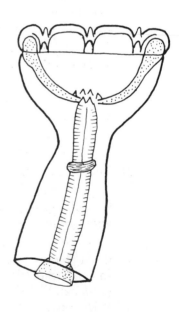

Figure 155. *Syngamus trachea.*

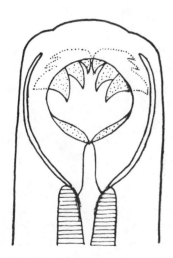

Figure 154. *Ancylostoma caninum.*

185

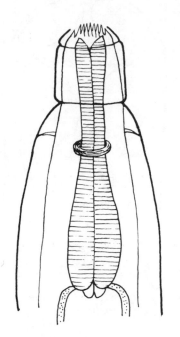

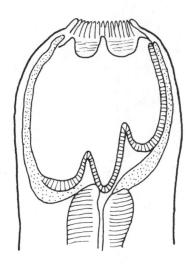

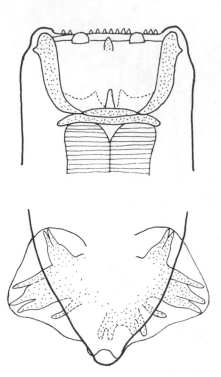

Figure 156. *Oesophagostomum columbianum.*

Figure 157. *Strongylus equinus.*

Figure 158. *Stephanurus dentatus.*

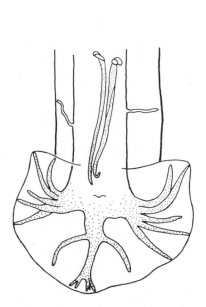

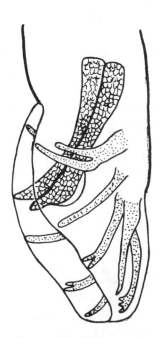

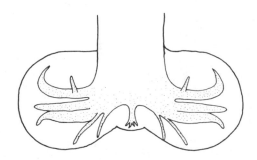

Figure 159. *Viannaia hamata.*

Figure 160. *Dictyocaulus filaria.*

Figure 161. *Trichostrongylus colubriformis.*

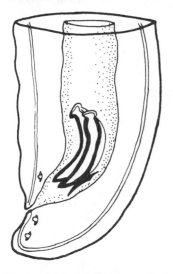

Figure 162. *Filarioides osleri.*

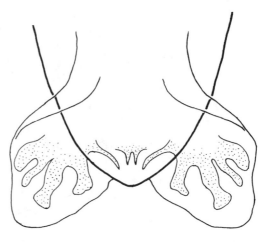

Figure 163. *Metastrongylus elongatus.*

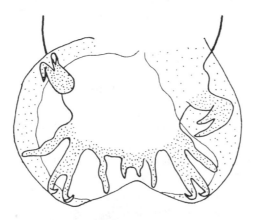

Figure 164. *Angiostrongylus gubernaculum.*

Spicules somewhat stout, long, tubular, and with striated alae in distal portion; buccal capsule absent or rudimentary; gubernaculum present (rarely absent); in respiratory and circulatory systems (Figure 165) . . . . (M) Protostrongylidae

17(4) Esophagus muscular throughout, cylindrical or with a basal bulb; spicules equal in length when present. Order ASCARIDA 18

Esophagus cylindrical, usually with short anterior, muscular and long posterior glandular parts (may be muscular throughout in Cucullanidae of the Suborder Camallanina); spicules usually different in shape and almost always unequal in size; esophageal glands uni- or multinucleate. Order SPIRURIDA 32

18(17) Esophagus more or less cylindrical. Suborder Ascaridina 19

Esophagus usually with distinct basal bulb. Suborder Oxyurina 23

19(18) Esophagus more or less cylindrical (sometimes with small posterior bulb), with unusual modifications such as a ventriculus with or without a cecum; a ventriculus and intestinal cecum, or only an intestinal cecum but no ventriculus; 3 lips, usually with interlabia; in alimentary canal (Figure 166) . . . . (B, M) Superfamily Heterocheiloidea. Heterocheilidae

Esophagus almost club-shaped, without ventriculus (except *Toxocara, Neoascaris)* or intestinal cecum; 3 lips without interlabia; large intestinal worms. Superfamily Ascaridoidea 20

20(19) Esophagus with ventriculus or glandular bulb 21

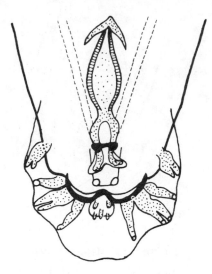

Figure 165. *Protostrongylus rufescens.*

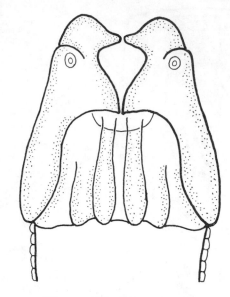

Figure 166. *Heterocheilus tunicatus.*

Esophagus without ventriculus 22

21(20) Lips greatly reduced; intestine with voluminous pyriform, cuticularized rectum; in intestine (Figure 167) ....(A, R) Oxyascarididae

Lips well developed; intestine without enlarged, cuticularized rectum; in intestine (Figure 168) ....(M) Toxocaridae

22(20) Lips well developed, esophagus never divided into different parts; vulva in anterior half of body: preanal sucker absent; in intestine (Figure 169) ....(R, B, M) Ascaridae

Lips small or absent; esophagus may be divided into narrow anterior and broader posterior parts; vulva in posterior half of body; preanal sucker usually present and without cuticularized rim; in intestine (Figure 170) ....(F) Quimperiidae

23(18) No preanal sucker. Suborder Oxyurina 24

Preanal sucker present. Suborder Heterakina 27

24(23) With a gubernaculum and a single spicule; in intestine (Figure 171).....................
....(M) Superfamily Syphacioidea. Syphaciidae

With or without single spicule but without a gubernaculum; with or without cuticular ornamentations. Superfamily Oxyuroidea 25

25(24) Single spicule; no cervical spine; in large intestine (Figure 172) ....(M) Oxyuridae

Without spicule (if present short) 26

26(25) Cuticular ornamentation in precloacal area; in intestine (Figure 173) . (M) Heteroxynematidae

Preanal ornamentation lacking; in intestine (Figure 174) ....(M) Aspicularidae

27(23) Preanal sucker circular, well formed, with cuticularized rim. Superfamily Heterakoidea ... 28

Preanal sucker (pseudosucker) formed by preanal muscles, without a cuticularized rim. Superfamily Subuluroidea 30

28(27) Cervical cordons present; precloacal muscles strongly developed; in intestine (Figure 175) ..
....(M) Aspidoderidae

Cervical cordons absent 29

29(28) Esophagus with well-defined posterior bulb, with short narrow anterior portion; male and female tails long; lateral caudal alae in males; small worms; in intestine (Figure 176) ....
....(A, R, B, M) Heterakidae

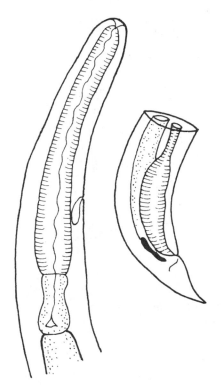

Figure 167. *Oxyascaris oxyascaris.*

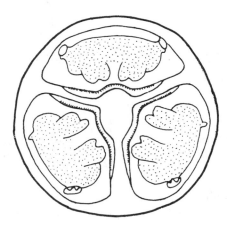

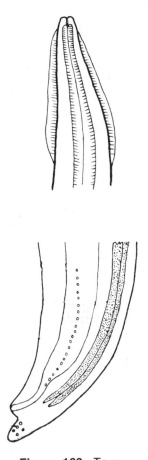

Figure 168. *Toxocara canis.*

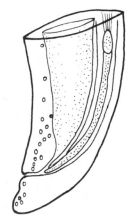

Figure 169. *Ascaris lumbricoides.*

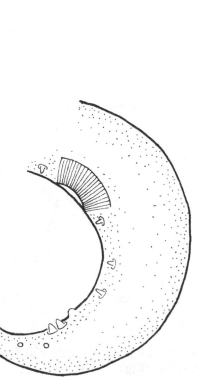

Figure 170. *Quimperia lanceolata.*

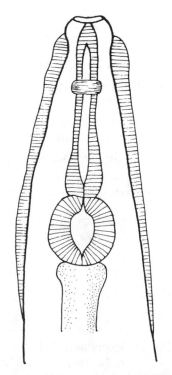

Figure 171. *Syphacia obvelata.*

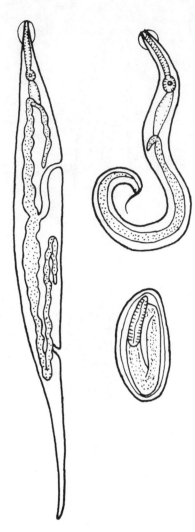

Figure 172. *Enterobius vermicularis.*

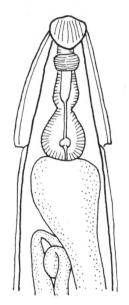

Figure 174. *Aspiculuris tetraptera.*

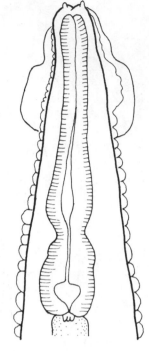

Figure 173. *Heteroxynema wernecki.*

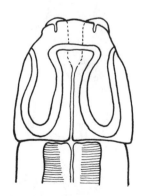

Figure 175. *Aspidodera scoleciformis.*

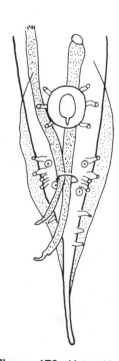

Figure 176. *Heterakis gallinae.*

190

Esophagus without bulb or short, narrow anterior portion; male and female tails conical; without caudal alae in males; medium-sized worms; in intestine (Figure 177) . . . .
. . . . (B) Ascaridiidae

30(27) Intestine with anterior diverticulum; in cecum, colon (Figure 178) . . . . . (A, R, B) Cruziidae
Intestine simple, without diverticulum 31

31(30) Preanal sucker circular, with cuticularized crown of about 70 leaflike elements and equal number of buttonlike basal elements; buccal capsule cylindrical, thin cuticularized walls; in intestine (Figure 179) . . . . (M) Parasubuluridae

Preanal sucker without cuticularized elements; precloacal muscles well developed, forming elongated pseudosucker; vestibule usually with basal teeth; in intestine (Figure 180) . . . .
. . . . (B, M) Subuluridae

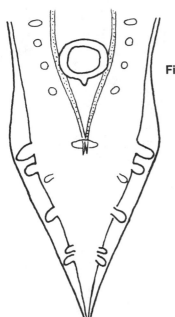

Figure 177. *Ascaridia sp.*

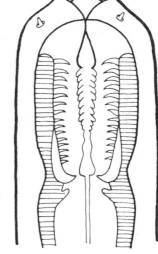

Figure 178. *Cruzia tentaculata.*

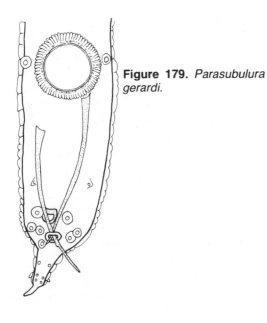

Figure 179. *Parasubulura gerardi.*

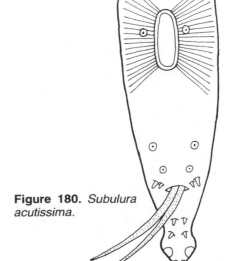

Figure 180. *Subulura acutissima.*

32(17) Esophageal glands uninucleate; internal and external circle of cephalic papillae present; mouth well developed or rudimentary; larvae with large pocketlike phasmids; intermediate hosts copepods. Suborder Camallanina 33

Esophageal glands multinucleate; stoma rudimentary or well developed, with or without pseudolabia; oviparous or ovoviviparous; intermediate hosts usually insects, occasionally copepods and acarines 36

33(32) Internal circle of papillae reduced in size, external circle partially fused; stoma well developed; oviparous or ovoviviparous. Superfamily Camallanoidea 34

Internal circle of papillae well developed; external circle of 8 separate papillae; stoma rudimentary; ovoviviparous. Superfamily Dracunculoidea 35

34(33) Buccal capsule globular or consisting of 2 large lateral cuticularized, shell-like parts; esophagus of anterior muscular and posterior glandular portions; vagina directed posteriorly; ovoviviparous; in alimentary canal (Figure 181) . . . . (F, A, R) Camallanidae

Esophagus muscular throughout with posterior club-shaped swelling and anterior dilation into a false buccal cavity; preanal sucker in males; vagina directed anteriorly; ovoviviparous; in intestine (Figure 182) . . . . (F) Cucullanidae

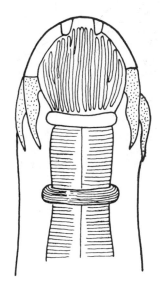

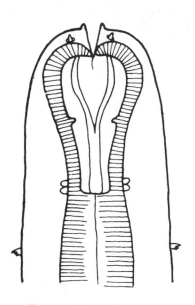

Figure 181. *Camallanus lacustris.*

Figure 182. *Cucullanus schubarti.*

35(33) Anterior extremity with cuticular thickening or shield; esophagus long and cylindrical with constriction at level of nerve ring; vulva immediately behind head; tail of female conical, slightly coiled; spicules equal; in tissues (Figure 183) . . . . (R, B, M) Dracunculidae

Anterior extremity without cuticular shield; esophagus short, cylindrical and without median constriction; tail of female rounded, straight; spicules equal; in tissues (Figure 184) . . . . (F) Philometridae

36(32) Stoma well developed with or without pseudolabia, or rudimentary and with pseudolabia; vulva usually near or posterior to middle of body; generally oviparous; intermediate hosts usually nonbloodsucking insects, occasionally copepods; in alimentary canal, orbital and nasal cavities, respiratory system. Suborder Spirurina 37

Stoma rudimentary without pseudolabia; vulva in anterior part of body; oviparous or ovoviviparous; in air sacs, peritoneal cavity, tissues, blood vessels; intermediate hosts bloodsucking insects and mites. Suborder Filariina 51

37(36) Pseudolabia absent (except *Physocephalus*) Superfamily Thelazioidea 38

        Two well-developed pseudolabia present 42

38(37) With cuticular plaques over anterior part of body; in wall of esophagus and stomach; slender worms (Figure 185) .... (B, M) Gongylonematidae

        Without cuticular plaques 39

39(38) With cuticular hooklike spines arranged in rows or circles over whole or anterior portion of body; in intestine (Figure 186) .... (R, B, M) Rictulariidae

        Without cuticular spines 40

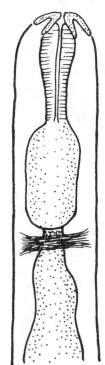
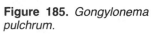

Figure 183. *Dracunculus medinensis.*

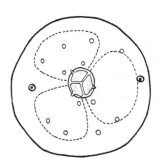

Figure 184. *Philometroides nodulosa.*

Figure 185. *Gongylonema pulchrum.*

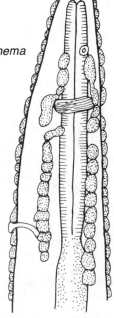

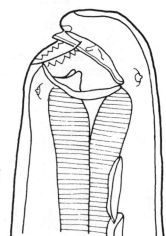
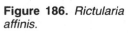

Figure 186. *Rictularia affinis.*

40(39) Buccal capsule long, cylindrical, composed of cuticular ridges in form of rings; in stomach (Figure 187) .... (M) Ascaropsidae

        Buccal capsule smooth.. 41

41(40) Caudal alae absent; in orbital and nasal sinuses, lungs (Figure 188) .... (B, M) Thelaziidae

        Caudal alae present; mouth surrounded by 6 masses of parenchyma; caudal papillae not numerous; in esophagus, stomach, aorta, lungs (Figure 189) .... (M) Spirocercidae

42(37) Pseudolabia lobed 43

        Pseudolabia not lobed 49

43(42) Cephalic papillae posterior to pseudolabia; interlabia present or absent. Superfamily Spiruroidea
.... 44

        Eight partially fused double cephalic papillae on pseudolabia. Superfamily Gnathostomatoidea. . . 47

44(43) Female fusiform or coiled to give fusiform shape; vulva near anus; males much smaller than females; females imbedded in crypts of proventriculus, males free in lumen (Figure 190) ....
.... (B) Tetrameridae

        Sexes more or less equal in size, females not fusiform 45

45(44) With 4 specialized lips, dorsal and ventral ones in form of isosceles triangles; male always rolled about female; festooned cuticularized ring at anterior extremity of apparently undivided esophagus; spicules equal; vulva near anus; tail of female invaginated to form groove with projecting hook; in intestine (Figure 191) .... (A, R) Hedruridae

        Without 4 lips, cuticular esophageal ring, or invaginated tail; spicules unequal; vulva near middle of body 46

46(45) Interlabia and cuticular flange absent; tail of male conical; in intestine (Figure 192) ....
.... (R, B, M) Spiruridae

        Interlabia usually present; cuticular flange present on 1 or both sides; tail of male coiled spirally; in stomach (Figure 193) .... (B, M) Habronematidae

47(43) Head with prominent backward pointing appendages of feathered or unfeathered appearance; in alimentary canal (Figure 194)................................... (R, B) Ancyracanthidae

        Head without appendages 48

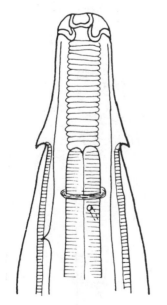

Figure 187. *Physocephalus sexulatus.*

Figure 188. *Thelazia rhodesi.*

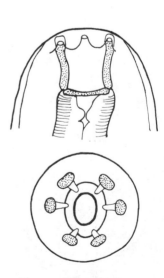

Figure 189. *Spirocerca sanguinolenta.*

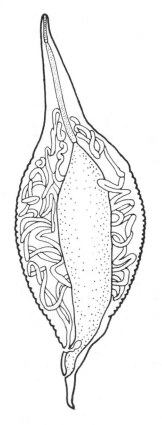

Figure 190. *Tetrameres fissipina.*

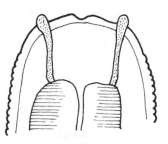

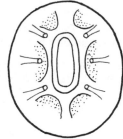

Figure 192. *Spirura talpae.*

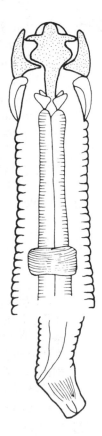

Figure 191. *Hedruris androphora.*

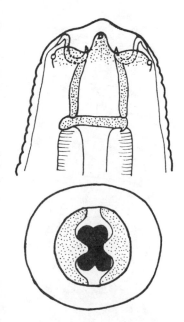

Figure 193. *Habronema muscae.*

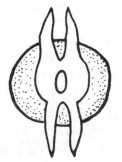

Figure 194. *Schistorophorus longicornis.*

48(47) Head inflated into prominent spined or striated bulb; vulva in posterior half of body; in alimentary canal (Figure 195) .... (F, R, B, M) Gnathostomatidae

Head without inflation; vulva near middle of body; in stomach (Figure 196) .... (R) Spiroxyidae

49(42) Anterior end without cordons; cuticle reflected forward over lips to form a cephalic collar; pseudolabia large, simple, triangular, and with 1 or more apical teeth; caudal alae wide, joined ventrally anterior to cloaca, supported by long papillae; in alimentary canal (Figure 197) ....(P, A, R, B, M) Superfamily Physalopteroidea. Physalopteridae

Anterior end of body with cordons (raised cuticular ridges), hood, or spined collarette extending caudad; buccal capsule long, cylindrical; caudal alae narrow, lateral. Superfamily Acuarioidea . . 50

50(49) Simple cordons present; in gizzard and intestine (Figure 198) ....(B, M) Acuariidae

Spined cephalic collar present; in intestine (Figure 199) ....(B) Seuratidae

51(36) Females oviparous; eggs thick-shelled, embryonated when laid. Superfamily Filarioidea 52

Females ovoviviparous; microfilariae in thin membranous sacs or free of membrane. Superfamily Onchocercoidea 56

52(51) With cuticularized trident on each side of anterior end of esophagus; in air sacs and body cavity (Figure 200) ....(B) Diplotriaenidae

Without tridents 53

53(52) Mouth simple without any surrounding structures 54

Mouth with cuticularized peribuccal ring with lateral expansions or surrounded by 4 cone-shaped elevations 55

54(53) Spicules unequal and dissimilar; vulva near mouth; caudal alae often present in male; in tissues (Figure 201) ....(M) Filariidae

Spicules equal and similar; vulva distinctly removed from region of mouth; caudal alae present or absent; in orbital and nasal cavities, subcutaneous tissues (Figure 202) ... (B) Aproctidae

55(53) Mouth surrounded by 4 conical cuticular elevations; 8 papillae in external circle and 4 small ones in internal circle; in air sacs and tissues (Figure 203) ....(B) Tetracheilonematidae

Figure 195. *Gnathostoma spinigerum.*

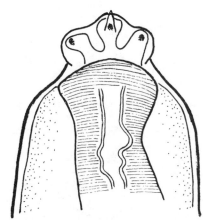

Figure 196. *Spiroxys contortus.*

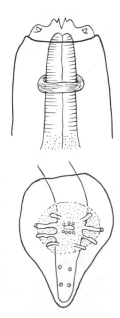

Figure 197. *Physaloptera clausa.*

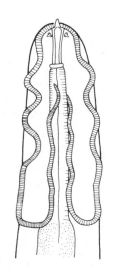

Figure 198. *Acuaria spiralis.*

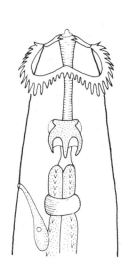

Figure 199. *Seuratia shipleyi.*

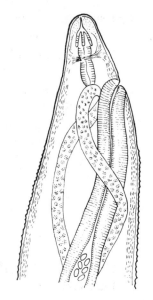

Figure 200. *Diplotriaena ozouxi.*

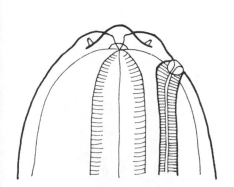

Figure 201. *Filaria martis.*

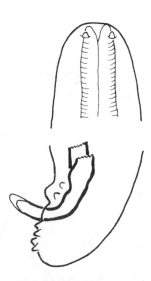

Figure 202. *Aprocta semenovi.*

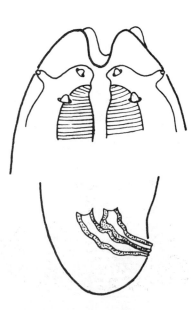

Figure 203. *Tetracheilonema tringae.*

Mouth surrounded by raised cuticularized peribuccal ring with a pair of lateral extensions; in air sacs and subcutaneous tissues (Figure 204) .... (R, B) Dicheilonematidae

56(51) Mouth surrounded by a ring of many minute cuticularized spines or with only 6 large papillae in external circle; amphids large; male tail with numerous cloacal papillae; in subcutaneous tissues (Figure 205) .... (M) Stephanofilariidae

Mouth not surrounded by ring of spines of 6 large papillae 57

57(56) Buccal capsule rudimentary or absent; with ring attaching it to anterior extremity of esophagus; in body cavity, tissues, heart (Figure 206) ....(M) Dipetalonematidae

No ring connecting buccal capsule and esophagus 58

58(57) Cuticularized peribuccal ring with medial or lateral protrusions, or mouth surrounded by lateral plates or stars; tail of female regularly long and ornamented with tubercules or 1 pair of lateral projections; in body cavity (Figure 207) .... (M) Setariidae

No peribuccal ring 59

59(58) Caudal alae well developed, supported by large cloacal papillae; male tail short (less than twice the diameter of body at level of anus); esophagus usually divided; in tissues, blood vessels, heart (Figure 208) .... (M) Dirofilariidae

Caudal alae absent, small, or well developed, but when well formed not supported by elongate papillae; esophagus usually not divided; tail long or short; in connective tissues, heart, arteries, body cavity (Figure 209)..(B, M) Onchocercidae

60(1) Small threadlike worms with esophagus composed of short, anterior muscular part and a long posterior chain of cuboidal cells with a capillary tubule running through them; caudal end of males without a bursa. Order TRICHINELLIDA, Superfamily Trichinelloidea............ 61

Large worms; males with fleshy, cup-shaped copulatory bursa without supporting rays; esophagus cylindrical, muscular. Order DIOCTOPHYMATIDA 65

61(60) Spicule and spicular sheath present, spicule only present, or, if spicule absent, spicular sheath present 62

Spicule and spicular sheath absent... 64

62(61) Spicule present, spicular sheath absent; small worms; vulva posterior to base of esophagus; ovoviviparous; in connective tissue (Figure 210).......................(M) Anatrichosomatidae

Spicule and spicular sheath present, or, if spicule absent, sheath present; large to medium-sized worms.. 63

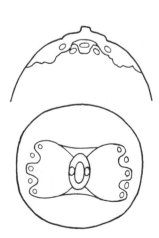

Figure 204. *Dicheilonema rheae.*

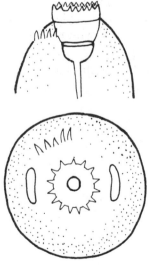

Figure 205. *Stephanofilaria stilesi.*

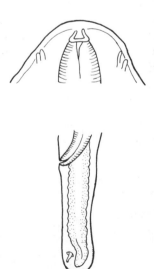

Figure 206. *Oswaldofilaria bacillaris.*

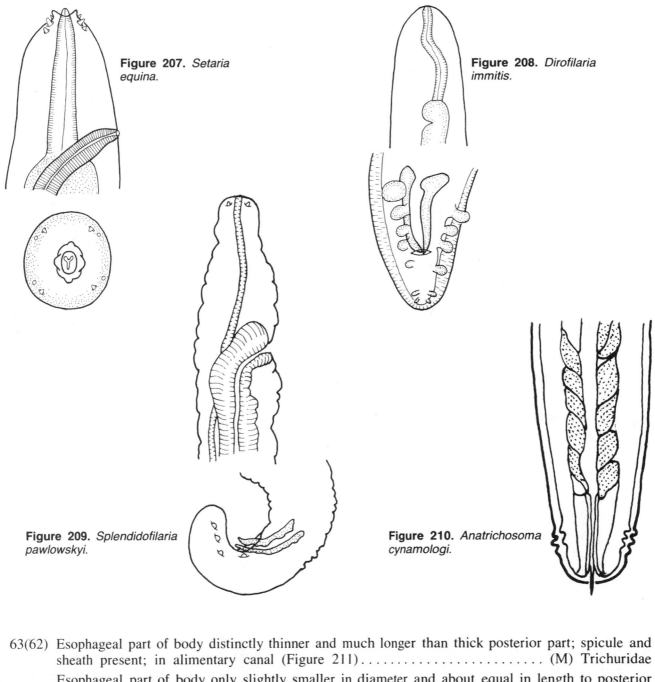

Figure 207. *Setaria equina.*

Figure 208. *Dirofilaria immitis.*

Figure 209. *Splendidofilaria pawlowskyi.*

Figure 210. *Anatrichosoma cynamologi.*

63(62) Esophageal part of body distinctly thinner and much longer than thick posterior part; spicule and sheath present; in alimentary canal (Figure 211) . . . . (M) Trichuridae

Esophageal part of body only slightly smaller in diameter and about equal in length to posterior part; spicule and sheath may be present, or spicule may be absent, in which case sheath is present; in alimentary canal (B, M), in liver and lungs (M) (Figure 212) . . . . . (M) Capillariidae

64(61) Males minute, in vagina or uterus of female; eggs with thick shell, bioperculate, fully embryonated when laid; in urinary tract (Figure 213) . . . . (M) Trichosomoididae

Male free, small but not minute, with 2 conical processes on caudal end; females embedded in intestinal mucosa, ovoviviparous; adults in intestine, larvae in muscles (Figure 214) . . . . (M) Trichinellidae

65(60) Six head papillae in 1 circle; vulva anterior; in kidney and body cavity (Figure 215) (M) Superfamily Dioctophymatoidea . . . . Dioctophymatidae

Twelve or more head papillae in 2 circles; vulva near anus; in forestomach (Figure 216) (B) Superfamily Eustrongyloidea . . . . Eustrongylididae

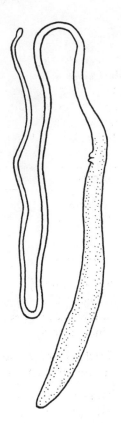

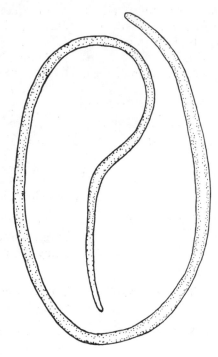

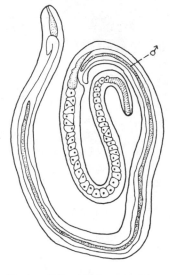

Figure 211. *Trichuris ovis.*

Figure 212. *Capillaria columbae.*

Figure 213. *Trichosomoides crassicauda.*

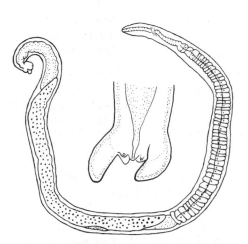

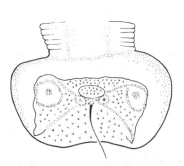

Figure 214. *Trichinella spiralis.*

Figure 215. *Dioctophyma renale.*

Figure 216. *Hystrichis tricolor.*

CLASSIFICATION

The classification followed is that of MayBelle Chitwood (1969) in which the phylum Nematoda is divided into 2 classes: the Secernentea (Phasmidia) and the Adenophorea (Aphasmidia).

The great majority of the Secernentea are parasitic. This class contains most of the important parasites of man and domestic and wild animals, being the cause of disease and death. Likewise, most of the species parasitic in plants are members of this class. Most of the Adenophorea are free-living in the soil and water, both fresh and salt. A few, however, are parasites of invertebrates, vertebrates, and plants.

Class	Order	Suborder	Superfamily	Family
SECERNENTEA	Ascarida	Ascaridina	Ascaridoidea	Ascaridae Toxocaridae
	Rhabditida	Rhabditina	Rhabditoidea Rhabdasoidea	Rhabditidae Strongyloididae
	Strongylida	Strongylina	Strongyloidea Syngamoidea Ancylostomatoidea	Strongylidae Syngamidae Ancylostomatidae Uncinariidae
		Trichostrongylina	Trichostrongyloidea Heligmosomatoidea	Trichostrongylidae Heligmosomatidae
		Metastrongylina	Metastrongyloidea Protostrongyloidea	Metastrongylidae Protostrongylidae
		Oxyurina	Oxyuroidea Syphacioidea	Oxyuridae Syphaciidae
	Spirurida	Spirurina	Spiruroidea Thelazioidea Gnathostomatoidea Physalopteroidea	Tetrameridae Spirocercidae Gnathostomatidae Physalopteridae
		Camallanina	Camallanoidea Dracunculoidea	Cucullanidae Dracunculidae Philometridae
		Filariina	Filarioidea Onchocercoidea	Diplotriaenidae Onchocercidae Dirofilariidae Dipetalonematidae
ADENOPHOREA	Trichinellida	Trichinellina	Trichinelloidea	Trichinellidae Trichuridae Capillariidae
	Dioctophymatida	Dioctophymatina	Dioctophymatoidea	Dioctophymatidae

REFERENCES

Bird, A.F. 1971. The Structure of Nematodes. Academic Press, New York, 318 p.

Chitwood, B.G., and M.B. Chitwood. 1950. An Introduction to Nematology. rev. ed. Monumental Printing Co., Baltimore, 213 p. Reprinted 1975, University Park Press, Baltimore.

Chitwood, MayBelle. 1969. The Systematics and Biology of Some Parasitic Nematodes. *In* Chemical Zoology, ed. M. Florkin and B.T. Scheer. Academic Press, New York, vol. 3, pp. 223-244.

Goodey, T. 1963. Soil and Freshwater Nematodes. rev. ed. John Wiley and Sons, New York, 544 p.

Lee, D.L. 1965. The Physiology of Nematodes. Oliver & Boyd, Edinburgh, 154 p.

Levine, N.D. 1968. Nematode Parasites of Domestic Animals and of Man. Burgess Publishing Company, Minneapolis, 600 p.

Thorne, G. 1961. Principles of Nematology. McGraw-Hill Book Company, Inc., New York, 553 p.

Yamaguti, S. 1961. Systema Helminthum. The Nematodes of Vertebrates. vol. 3, pts. 1 & 2. Interscience Publishers, Inc., New York, 1261 p.

Yorke, W., and P.A. Maplestone. 1926. The Nematode Parasites of Vertebrates. P. Blakiston's Son & Co., Philadelphia, 536 p. Reprinted 1962, Hafner Publ. Co., New York.

Ascaridae & Toxocaridae

Beaver, P.C. 1969. The nature of visceral larva migrans. J. Parasitol. 55: 3-12.

Douvres, F.W., F.G. Tromba, and G.M. Malakatis. 1969. Morphogenesis and migration of *Ascaris suum* larvae developing to fourth stage in swine. J. Parasitol. 55: 689-712.

Greve, J.H. 1971. Age resistance to *Toxocara canis* in ascarid-free dogs. Amer. J. Vet. Res. 32: 1185-1192.

Sprent, J.A.F. 1952-1953. On the migratory behavior of the larvae of various *Ascaris* species in white mice. J. Infect. Dis. 90: 165-176; 92: 114-117.

_____. 1955. On the invasion of the central nervous system by nematodes. Parasitology 45: 41-55.

_____. 1958. Observations on the development of *Toxocara canis* (Werner, 1782) in the dog. Parasitology 48: 184-209.

_____. 1961. Post-parturient infection of the bitch with *Toxocara canis*. J. Parasitol. 47: 284.

Stone, W.M., and M.H. Girardeau. 1967. Transmammary passage of infective-stage nematode larvae. Vet. Med./Small Anim. Clin. 62: 252-253.

Warren, E.G. 1969. Infections of *Toxocara canis* in dogs fed infected mouse tissues. Parasitology 57: 837-841.

Oxyurina

Beaver, P.C. 1949. Methods of pinworm diagnosis. Amer. J. Trop. Med. 29: 577-587.

Chan, K.F., and S. Kopilof. 1958. The distribution of *Syphacia obvelata* in the intestine of mice during their migratory period. J. Parasitol. 44: 245-246.

Hussey, Kathleen. 1957. *Syphacia muris* vs. *S. obvelata* in laboratory rats and mice. J. Parasitol. 43: 555-559.

Strongyloididae

Abadie, S.H. 1963. The life cycle of *Strongyloides ratti*. J. Parasitol. 49: 241-248.

Spindler, L.A. 1958. The occurrence of the intestinal threadworms, *Strongyloides ratti*, in the tissues of rats, following experimental percutaneous infection. Proc. Helminth. Soc. Washington 25: 106-111.

Rhabditidae

Walton, A.C. 1929. Studies on some nematodes of North American frogs. J. Parasitol. 15: 227-240.

Ancylostomatidae

Kalkofen, U.P. 1970. Attachment and feeding behavior of *Ancylostoma caninum*. Zeit. Parasitenk. 33: 339-354.

Miller, T.A. 1971. Vaccination Against the Canine Hookworm Diseases. *In* Advances in Parasitology, ed. B. Dawes. Academic Press, New York, vol. 9, pp. 153-183.

Nichols, R.L. 1956. The etiology of visceral larva migrans. II. Comparative larval morphology of *Ascaris lumbricoides, Necator americanus, Strongyloides stercoralis,* and *Ancylostoma caninum.* J. Parasitol. 42: 363-399.

Soh, C.T. 1958. The distribution and persistence of hookworm larvae in the tissues of mice in relation to species and to routes of inoculation. J. Parasitol. 44: 515-519.

Stone, W.M., and M. Girardeau. 1968. Transmammary passage of *Ancylostoma caninum* larvae in dogs. J. Parasitol. 54: 426-429.

Uncinariidae

Olsen, O.W., and E.T. Lyons. 1965. Life Cycle of *Uncinaria lucasi* Stiles, 1901 (Nematoda: Ancylostomatidae) of fur seals, *Callorhinus ursinus* Linn., on the Pribilof Islands, Alaska. J. Parasitol. 51: 689-700.

Syngamidae

Wehr, E.E. 1937. Observations on the development of the poultry gapeworm *Syngamus trachea.* Trans. Amer. Micr. Soc. 56: 72-78.

Heligmosomatidae

Ehrenford, F.A. 1954. The life cycle of *Nematospiroides dubius* Baylis (Nematoda: Heligmosomidae). J. Parasitol. 40: 480-481.

Fahmy, M.A.M. 1956. An investigation on the life cycle of *Nematospiroides dubius* (Nematoda: Heligmosomidae) with special reference to the free-living stages. Zeit. Parasitenk. 17: 394-399.

Dracunculidae

Hugghins, E.J. 1958. Guinea worms from carnivores in South Dakota and Minnesota, with a review of the distribution and taxonomy of dracunculid worms in North America. Proc. South Dakota Acad. Sci. 37: 40-46.

Mueller, R. 1971. *Dracunculus* and Dracunculiasis. *In* Advances in Parasitology, ed. B. Dawes. Academic Press, New York, vol. 9, pp. 73-151.

Philometridae

Dailey, M.D. 1967. Biology and morphology of *Philometroides nodulosa* (Thomas, 1929) n. comb. (Philometridae: Nematoda) in the western white sucker *(Catostomus commersoni).* Diss. Abstr. 28: 1266-1267.

Platzer, E.G., and J.R. Adams. 1967. The life history of a dracunculoid, *Philonema oncorhynchi,* in *Oncorhynchus nerka.* Canad. J. Zool. 45: 31-43.

Tetrameridae

Ellis, C.J. 1969. Life history of *Microtetrameres centuri* Barus, 1966. (Nematoda: Tetrameridae). I. Juveniles. J. Nemat. 1: 84-93.

_____. 1969. Life history of *Microtetrameres centuri* Barus, 1966. (Nematoda: Tetrameridae). II. Adults. J. Parasitol. 55: 713-719.

Spirocercidae

Bailey, W.S. 1972. *Spirocerca lupi:* a continuing inquiry. J. Parasitol. 58: 3-22.

Gnathostomatidae

Ash, L.R. 1962a. Development of *Gnathostoma procyonis* Chandler, 1942, in the first and second intermediate hosts. J. Parasitol. 48: 298-305.

———. 1962b. Migration and development of *Gnathostoma procyonis* Chandler, 1942, in mammalian hosts. J. Parasitol. 48: 306-313.

Physalopteridae

Schell, S.C. 1952. Studies on the life cycle of *Physaloptera hispida* Schell (Nematoda: Spiruroidea) a parasite of the cotton rat (*Sigmodon hispidus littoralis* Chapman). J. Parasitol. 38: 462-472.

Filariae

Hawking, F., and K. Gammage. 1968. The periodic migration of microfilariae of *Brugia malayi* and its response to various stimuli. Amer. Soc. Trop. Med. Hyg. 17: 724-729.

Kessel, J.F. 1965. Filarial infections of man. Amer. Zool. 5: 79-84.

Nelson, G.S. 1964. Factors Influencing the Development and Behavior of Filarial Nematodes in their Arthropodan Hosts. *In* Host-Parasite Relationships in Invertebrate Hosts. ed. Angela Taylor. Blackwell Scientific Publications, Oxford, pp. 75-119.

———. 1966. The Pathology of Filarial Infections. Helminthol. Abstr. 35: 311-336.

Onchocercidae

Duke, O.L. 1971. The Ecology of Onchocerciasis in Man and Animals. *In* Ecology and Physiology of Parasites. ed. A.M. Fallis. University of Toronto Press, Toronto, pp. 213-222.

Nelson, G.S. 1970. Onchocerciasis. *In* Advances in Parasitology. ed. B. Dawes. Academic Press, New York, vol. 8, pp. 173-224.

Scott, J.A., E.M. Macdonald, and B. Terman. 1951. A description of the stages in the life cycle of the filarial worm *Litomosoides carinii*. J. Parasitol. 37: 425-432.

WHO. 1966. WHO Expert Committee on Onchocerciasis, Second Rept. Wld. Hlth. Org.: Tech. Rept. Series, No. 335, 96 p.

Williams, R.W. 1946. The laboratory rearing of the tropical rat mite, *Liponyssus bacoti* (Hirst). J. Parasitol. 32: 252-256.

———. 1948. Studies on the life cycle of *Litomosoides carinii*, filariid parasite of the cotton rat, *Sigmodon hispidus litoralis*. J. Parasitol. 34: 24-43.

Dirofilariidae & Dipetalonematidae

Nelson, G.S. 1962. *Dipetalonema reconditum* (Grassi, 1889) from the dog with a note on its development in the flea, *Ctenocephalides felis* and the louse, *Heterodoxus spiniger*. J. Helminthol. 36: 297-308.

Newton, W.L. and W.H. Wright. 1956. The occurrence of a dog filariid other than *Dirofilaria immitis* in the United States. J. Parasitol. 42: 246-258.

Sawyer, T.K., E.F. Rubin, and R.F. Jackson. 1965. The cephalic hook in microfilariae of *Dipetalonema reconditum* in the differentiation of canine microfilariae. Proc. Helminthol. Soc. Washington 32: 15-20.

Taylor, A.E.R. 1960. The development of *Dirofilaria immitis* in the mosquito *Aedes aegypti*. J. Helminthol. 34: 27-38.

Trichinellidae

Denham, D.A. 1966. Infections with *Trichinella spiralis* passing from mother to filial mice pre- and post-natally. J. Helminthol. 40: 291-296.

Gould, S.E., ed. 1970. Trichinosis in Man and Animals. Charles C Thomas, Springfield, Ill., 540 p.

Robinson, H.A., and O.W. Olsen. 1960. The role of rats and mice in the transmission of the porkworm, *Trichinella spiralis* (Owens, 1835) Raillett, 1895. J. Parasitol. 46: 589-597.

Zimmermann, W.J., and E.D. Hubbard. 1969. Trichiniasis in wildlife of Iowa. Amer. J. Epidemiol. 90: 84-92.

_____, J.H. Steele, and I.G. Kagan. 1973. Trichiniasis in the U.S. Population, 1966-1970. Prevalence and epidemiologic factors. Health Service Repts. 88: 606-623.

_____, and D.E. Zinter. 1971. The prevalence of trichiniasis in swine in the United States, 1966-1970. Health Serv. Ment. Health Adm. Health Repts. 86: 937-945.

Capillariidae

Morehouse, N.F. 1944. Life cycle of *Capillaria caudinflata*, a nematode parasite of the common fowl. Iowa State College J. Sci. 18: 217-253.

Wehr, E.E. 1939. Studies on the development of the pigeon capillarid, *Capillaria columbae*. U.S. Dept. Agr. Tech. Bull. 679, 19 p.

Wright, K.A. 1961. Observations on the life cycle of *Capillaria hepatica* (Bancroft, 1893) with a description of the adult. Canad. J. Zool. 39: 167-182.

Dioctophymatidae

Hallberg, C.W. 1953. *Dioctophyma renale* (Goeze, 1782), a study of the migration routes to the kidneys of mammals and resultant pathology. Trans. Amer. Micr. Soc. 72: 351-363.

Karmanova, E.M. 1962. Development of *Dioctophyma renale* in the intermediate and final hosts. Trudy gel'mint. Lab. 12: 27-36. (In Russian) (see Helminthol. Abstr. 37: 303, No. 2309, 1968).

Woodhead, A.E. 1950. Life cycle of the giant kidney worm, *Dioctophyma renale* (Nematoda), of man and many other mammals. Trans. Amer. Micr. Soc. 69: 21-46.

Class Hirudinea

Hirudinea are the leeches, distinguished from other annelid worms by: a sucker or suckerlike depression at the anterior end of the body and a well-developed sucker at the posterior end, a constant number of body segments, no setae or parapodia, monoecious, and a body cavity largely filled with muscles and connective tissues.

Leeches vary in length from a few mm to more than 25 cm and are elongate, egg-shaped or leaflike in form. Since leeches have no caudal growth, the number of segments being fixed, increase in length results from the subdivision and elongation of the annuli into which the segments are divided. The number of annuli per segment in the midbody region is typical for any given species, 3 and 5 being commonest; but the number of annuli comprising the terminal segments is reduced. Eyes commonly vary from 1 to 5 pairs, placed anteriorly. The clitellar region, which secretes the egg cocoon, extends from segment X through XII and contains the male and female genital openings.

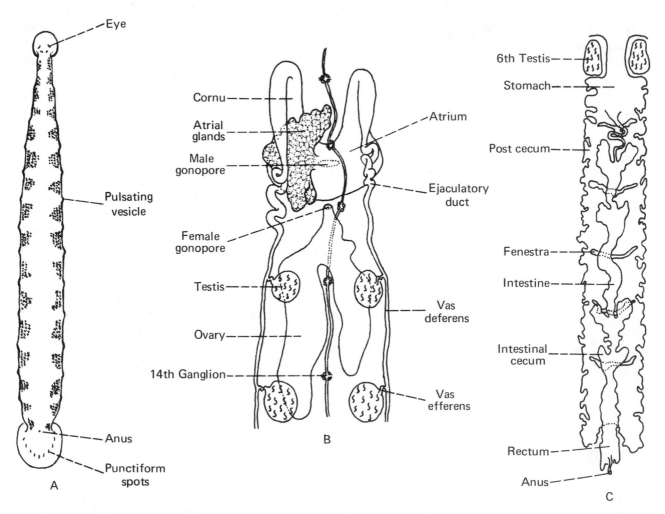

Figure 217. *Piscicola milneri:* (A) External; (B) Reproductive system; (C) Posterior portion of alimentary tract. All dorsal views.

206

The digestive tract is a tube from mouth to anus, and is divided into buccal chamber, pharynx, esophagus, stomach or crop, intestine, and rectum. Structure of the anterior portion constitutes 1 of the most important taxonomic characters for dividing the class into 2 orders: aproboscidal (Arhynchobdellae) and proboscidal (Rhynchobdellae) leeches. In the first group the buccal sinus is a restricted chamber housing the jaws (when present), and in the latter a space around the pharynx, which is thereby freed as a protrusible proboscis. The esophagus connects the pharynx with the stomach, which in predaceous species is tubular and simply chambered, without lateral ceca or with only a posterior pair. In bloodsucking leeches, the stomach becomes a spacious crop with lateral, paired ceca for the storage of food between feedings. A sphincter separates the storage stomach from the intestine, the principal digestive and absorptive region. The intestine, usually a straight, somewhat chambered tube, may in some species bear 4 paired ceca laterally. A simple rectum leads to the dorsally situated anus.

In all leeches, the internal sexual organs are paired, but the terminal parts and genital openings are single. In the male reproductive system, the testes, enclosed in celomic sacs, are multiple, usually in segmental pairs. Laterally from each testis, a small sperm duct enters a larger duct, which becomes enlarged and often much convoluted anteriorly to form a seminal vesicle or epididymis. Each of the 2 epididymides leads to a muscular, often bulbous ejaculatory duct, which opens into the median atrium, variously provided with glands. The female system consists of a pair of elongated ovisacs, which contain the true ovaries and unite anteriorly to form a common oviduct, which opens externally. In those forms in which fertilization occurs through copulation, this common passage serves as a vagina.

Reproduction in leeches is exclusively sexual. Fertilization occurs in 2 forms, depending upon whether a male copulatory organ is present. In the Rhynchobdellae and the Erpobdellidae, sperm transfer is accomplished by spermatophores; these are implanted by 1 leech onto the body of another. Sperms, forced out of the spermatophores, penetrate the tissues and enter the ovaries, resulting in fertilization. In the Hirudinidae and the Haemadipsidae direct copulation occurs, the sperm being transmitted by the cirrus of 1 animal into the vagina of another. Eggs are usually deposited in cocoons, soon after fertilization. Cocoons may be borne by the parent leech, attached to a host animal, to a foreign object, or deposited in damp earth. Development is direct.

Food relationships are varied. Leeches feed on oligochaete worms, crustaceans, insect larvae, molluscs, and all classes of vertebrates. Most subsist upon body fluids, involving all gradations from true predation to temporary and permanent parasitism.

Throughout the Mediterranean area and the Middle and Far East regions certain leeches are injurious to animals and man. Small, fasted leeches, taken in with drinking water, reach the laryngopharynx region, or may enter the nasopharynx, and become attached. By sucking blood they increase greatly in size, cause congestion and inflammatory swelling, and may occlude the passages involved. Other portals of body entry may involve the anus, external auditory canal, and urogenital sinus of both sexes. Attacks by land leeches (Haemadipsidae) are relatively common among pedestrian travelers through southern Asia and adjacent Sunda Islands.

The Piscicolidae occur as temporary or permanent parasites, chiefly on marine and freshwater fish, upon whose body fluids they feed. Not only may the host be killed or left in an emaciated condition as a result of heavy infestation, but the value of commercial and game fish may be materially impaired as a result of the inflammation developing where the leeches have attached. These breaks in the mucosal covering, resulting from attachment, may serve as a portal of entry for microbial organisms, which cause secondary infection. The glossiphoniid *Theromyzon* has a proclivity for the nares of waterfowl, often resulting in asphyxiation of the host bird.

Certain aquatic leeches serve as vectors of protozoan parasites such as *Trypanosoma, Cryptobia, Haemogregarina,* and probably others, of fish, amphibians, and reptiles. Leeches also serve as intermediate hosts for larval helminths, especially digenetic trematodes.

Order Rhynchobdellae

Marine and freshwater leeches with colorless blood, with an extensible proboscis, without jaws. The mouth is a small median opening situated within the anterior sucker, or rarely upon its front edge. Copulation by spermatophores.

Family Piscicolidae

Piscicola milneri

The anterior sucker, slightly narrower than the body at its widest point, has 2 pairs of distinct eyes arranged in the form of a trapezoid. Segments in the midregion of the yellowish colored body are characterized by large, brownish stellate flecks, arranged in 5 longitudinal rows, which extend dorsally, frequently meeting with those from the opposite side in the ganglionic region of the segment. Middorsally these flecks form a slightly acute angle and spread out laterally, where they are frequently continuous with those of the adjacent segments. Although these flecks are distributed over the entire length of the body, the triangular arrangement, so characteristic of the midregion, is lost anteriorly and posteriorly. The posterior sucker, equal to body width at its widest point, has a crown of 10 or 12 oculiform spots situated near the margin. Ratio of body length to width about 12:1. There are 11 pairs of lateral pulsating vesicles. Complete segments are divided into 14 annuli. The male gonopore is larger than that of the female and is located anterior to the latter (Figure 217).

Six pairs of metamerically arranged testes. The vasa deferentia continue anteriorly and dorsally slightly beyond the level of the male gonopore and then curve ventrally before uniting to form the muscular atrium. Ejaculatory ducts are located at the level of the gonopores. The paired ovisacs, which extend posteriorly to the second pair of testes, unite anteriorly to form a common oviduct, which opens 2 annuli posterior to the more conspicuous male gonopore. Gonopores are within segments XI and XII.

The stomach, the largest division of the alimentary canal, expands into segmental chambers which alternate between consecutive pairs of testes. Slightly posterior to the last pair of testes, in XIX, the alimentary canal divides into intestine (dorsally) and postcecum. The intestine, characterized by 4 pairs of lateral ceca, is followed by a rectum which empties posteriorly and dorsally through the anus. The postcecum, situated ventral to the intestine, has 4 fenestrae. Except for the occasional properly engorged specimen, serial sections are necessary for an understanding of the alimentation. Parasitic on various freshwater fish.

The life cycle of *Piscicola milneri* has not been studied, but with the possible exception of the time of reproduction it can be expected to be similar to that of *P. salmositica* which parasitizes Salmonidae throughout the Pacific Northwest.

The reproductive cycle of other freshwater leeches in temperate climates occurs during the spring and summer, but Becker and Katz (1965) found that *P. salmositica* breeds during the fall and winter months. Thus, the reproductive period of the leeches is adapted to the availability of the spawning salmon in fresh water, and not the usual rise in temperature.

Leeches appear in early fall attached to their hosts as they ascend into the hatchery and natural spawning areas. From late September through January, during a period of feeding, growth, and reproduction, the leech populations attain great abundance. By late winter, subsequent to the death of their hosts, the leeches have become virtually nonexistent. Only the small, immature leeches are believed to survive the summer and return attached to spawning salmon the following fall.

Other often available Piscicolidae with their hosts include *Piscicola punctata, Illinobdella moorei*, and *Cystobranchus verrilli*, all from various freshwater fish.

Key to Families of North American Hirudinea

1(4) Mouth a small pore on anterior sucker through which the proboscis protrudes; blood colorless
. . . . Order RHYNCHOBDELLAE. . .2

2(3) Body frequently divided into trachelosome and urosome, usually long and narrow, little flattened; complete segments usually more than 3-annulate (Figure 218) . . . . Piscicolidae

3(2) Body not divided into trachelosome and urosome, usually flattened and much wider than head; complete segments usually 3-annulate (Figure 219) . . . . Glossiphoniidae

4(1) Mouth large, opening from behind into entire sucker cavity; pharynx fixed, no proboscis; blood red
. . . . .Order ARHYNCHOBDELLAE. . .5

5(6) Pharynx a suction bulb, not reaching to clitellum; eyes 5 pairs, arranged in an arch; segments 5-annulate; muscular jaws usually present; testisacs large, metameric, usually 10 (Figure 220) .
. . . . Hirudinidae

6(5) Pharynx a crushing tube extending to clitellum; eyes 3 or 4 pairs in separate labial and buccal groups; segments 5-annulate but often further divided; 3 muscular pharyngeal ridges but no true jaws; testisacs small and numerous, in grape-bunch arrangement (Figure 221) . . . .
. . . . Erpobdellidae

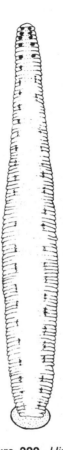

Figure 218. *Piscicola milneri.*

Figure 219. *Glossiphonia complanata.*

Figure 220. *Hirudo medicinalis.*

Figure 221. *Erpobdella punctata.*

REFERENCES

Mann, K.H. 1962. Leeches (Hirudinea), Their Structure, Physiology, Ecology, and Embryology. Pergamon Press, New York, 201 p.

Moore, J.P. 1959. Hirudinea. *In* Ward and Whipple's Fresh-water Biology, ed. W.T. Edmondson. 2nd ed. John Wiley and Sons, New York, pp. 542-557.

Piscicolidae

Becker, C.D., and M.J. Katz. 1965. Distribution, ecology and biology of the salmonid leech, *Piscicola salmositica* (Rhynchobdellae: Piscicolidae). J. Fish. Res. Bd. Canada, 22: 1175-1195.

Llewellyn, L.C. 1965. Some aspects of the biology of the marine leech *Hemibdella soleae*. Proc. Zool. Soc. London, 145: 509-528.

Meyer, M.C. 1940. A revision of the leeches (Piscicolidae) living on freshwater fishes of North America. Trans. Amer. Micr. Soc. 59: 354-376.

Helminth Eggs

TREMATODA

Clonorchis sinensis—Operculate; yellowish brown color; shape of an old-fashioned light bulb; abopercular end sometimes with a nipplelike protuberance; size 26-30 x 15-17 μ.

Fasciola hepatica—Operculate; delicate light brown in color; size 130-150 x 63-90 μ.

Fasciolopsis buski—Predominately oval in shape; clear, yellowish-tinged shell with a small and slightly convex operculum; size 130-140 x 80-85 μ. Eggs are practically identical with those of *Fasciola hepatica*.

Heterophyes heterophyes—With conspicuous conical operculum; size 20-30 x 10-20 μ.

Metagonimus yokogawai—With inconspicuous opercular shoulder; size 27-30 x 15-17 μ. Eggs practically identical with those of *H. heterophyes*, both of which closely resemble those of *Clonorchis sinensis*.

Paragonimus westermani—Broadly ovoidal, golden brown in color, somewhat flattened operculum; size 80-118 x 48-60 μ. As in eggs of *Clonorchis sinensis* some eggs of *P. westermani* have a nipplelike protuberance at the abopercular end.

Schistosoma haematobium—Oval at one end and conical at the other, tapering to a distinct spine; size 112-170 x 40-70 μ.

Schistosoma japonicum—Ovoidal, except for a shallow depression near one end, from which extends a short knoblike spine; pale yellow in color; size 65-90 x 50-65 μ.

Schistosoma mansoni—Oval at both ends and provided with a sharp lateral spine; size 114-175 x 45-68 μ.

CESTODA

Diphyllobothrium latum—Operculate; broadly ovoidal, evenly rounded at both ends, abopercular end sometimes with a nipplelike protuberance; contents granular, mulberrylike; shell thin and light straw colored; size 62-68 x 40-48 μ.

Dipylidium caninum—In gravid proglottids the eggs, measuring 35-60 μ, are bound together in egg-capsules, each containing 15-25 eggs.

Hymenolepis diminuta—Ovoidal; provided with two widely separated egg membranes, internal one with a thickening at each pole but lacking polar filaments. Between the membranes is a colorless, elastic, gelatinous substance; size 72-85 x 60-80 μ.

Hymenolepis nana—Spherical or subspherical; provided with two egg membranes closer together than are those of *H. diminuta*, internal one with two polar thickenings, from each arise 4 to 8 threadlike filaments lying in the space between the shells; size 40-60 x 30-50 μ.

Taenia spp. (indistinguishable)—Round or slightly ovoidal; embryophore commonly characterized by many radial striations; size 30-40 x 20-30 μ.

ACANTHOCEPHALA

Macracanthorhynchus hirudinaceus—Egg provided with 4 embryonic membranes, the outer one rugose; fully embryonated when voided; acanthor has 4 cephalic spines; ellipsoidal in shape; size 80-100 x 46-65 μ.

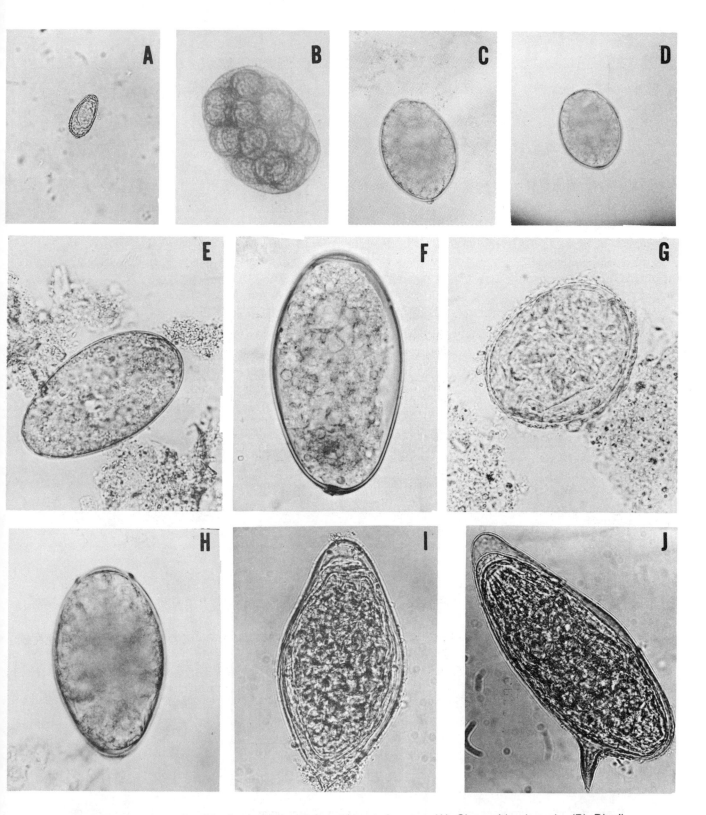

Plate IV. Photomicrographs of some trematode and cestode eggs: (A) *Clonorchis sinensis;* (B) *Dipyli-dium caninum;* (C) and (D) *Diphyllobothrium latum;* (E) *Fasciolopsis buski;* (F) *Fasciola hepatica;* (G) *Schistosoma japonicum;* (H) *Paragonimus westermani;* (I) *Schistosoma haematobium;* (J) *Schistosoma mansoni.*

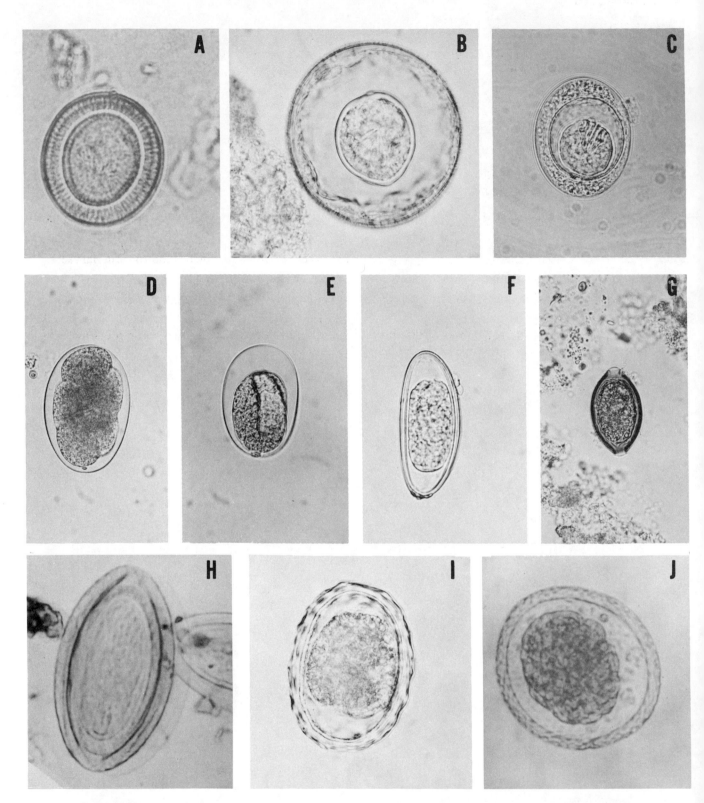

Plate V. Photomicrographs of some cestode, nematode and acanthocephalan eggs: (A) *Taenia* spp.; (B) *Hymenolepis diminuta;* (C) *Hymenolepis nana;* (D) *Necator americanus;* (E) *Strongyloides stercoralis;* (F) *Enterobius vermicularis;* (G) *Trichuris trichiura;* (H) *Macracanthorhynchus hirudinaceus;* (I) *Ascaris lumbricoides,* fertilized egg; (J) *Toxocara canis.*

NEMATODA

Ascaris lumbricoides—The fertilized egg is broadly ovoidal, with a thick transparent and an outer, rugose, albuminous covering; size 45-70 x 35-50 μ. Unfertilized egg longer and narrower (more elliptical), and usually has a thinner shell and an irregular coating of albumen; size 88-94 x 38-44 μ.

Capillaria hepatica—Eggs resemble those of *Trichuris trichiura* but have a velvety outer shell with radiating pores; size 51-67 x 30-35 μ.

Enterobius vermicularis—Egg in profile flattened on one side, rounded on the opposite side; a double membrane, containing a well-developed embryo; size 50-60 x 25-33 μ.

Necator americanus—Ovoidal, with bluntly rounded ends, shell thin and hyaline; at discharge normally in early stages of segmentation (2 to 8 celled stage); size 65-75 x 36-40 μ. Ova of *Ancylostoma caninum*, indistinguishable from those of *N. americanus*, may be used as a substitute. Ova of *Ancylostoma duodenale*, otherwise identical, measure 56-60 x 35-40 μ.

Strongyloides stercoralis—Ovoidal, thin-shelled, transparent; usually containing a larva; size 50-58 x 30-34 μ. Eggs rarely seen in feces; diagnosis usually based upon the typical rhabditiform larvae.

Toxascaris leonina—Slightly oval, with smooth shell; size 75-85 x 60-75 μ.

Toxocara canis—Subglobose to ovoidal, densely granular internally and a rather thick pitted shell; size 80-90 x 70-80 μ.

Trichuris trichiura—Barrel-shaped, with an outer and an inner shell, dark brown, with a distinct plug at each end; contents unsegmented; size 50-55 x 21-24 μ.

SECTION III

ARTHROPODA

Class Crustacea

The Crustacea comprise a group of mandibulate, usually aquatic, Arthropoda, typically possessing 2 pairs of antennae. Parasitologically, certain taxa are of dual concern: some as parasites and others as intermediate hosts in the life cycle of certain helminths. While the Ostracoda, Amphipoda, and Isopoda each contain a few parasitic species, they, along with the Decapoda, are certainly more important as intermediate hosts. The Branchiura, all of which are parasitic, infest principally marine and freshwater fish; the barnacles (Cirripedia) contain numerous species parasitizing various marine animals. But of the taxa of Crustacea, the Copepoda, both as intermediate hosts and as parasites, are of greatest importance.

Subclass Copepoda

The free-living copepods abound in both salt and freshwater, and may be collected with the aid of a plankton net in open waters throughout the year. The body is divided into 2 regions, a wider anterior division known as the cephalothorax and a rather narrow posterior one, the abdomen. The abdomen ends in a pair of caudal rami, which are provided with terminal, laterally situated setae. The cephalothorax can be further divided into 2 regions, the head with 2 pairs of antennae and 3 pairs of mouth parts, and a thorax with a pair of maxillipeds, 4 pairs of biramous swimming appendages, and a fifth uniramous, much reduced pair.

In the parasitic species the external body shape varies considerably, often showing little similarity to that of their free-living progenitors. The body metamerism is usually less prominent, and may be totally obscured. Thoracic appendages may be reduced in number or absent. Mouth parts are usually specialized for piercing and sucking, and the second antennae are usually specialized as prehensile organs. Thus, the cephalic region is often greatly modified in appearance.

A characteristic feature of both free-living and parasitic species is the presence of a pair of egg sacs in females. Most species are dioecious, with males usually smaller than females.

Family Lernaeidae

Lernaea cyprinaceae

Mature females attach to the skin of the host by a highly modified head buried in the epidermis, with the posterior part of the body hanging free. Because of the shape of the head and the elongate vermiform shape of the protruding body, species of *Lernaea* are commonly referred to as "anchor worms."

The body consists of a cephalothorax, free thorax, and abdomen (Figure 222, A). The head and thorax bear the diminutive appendages of these parts (Figure 222, B). There is a median tripartite eye. Anteriorly the cephalothorax expands into large cephalic horns, which are soft and leathery in texture. The cylindrical neck is soft and enlarges gradually to form the trunk, which bears on the ventral side the vulvae. The abdomen is short, rounded, and terminates bluntly. A pair of elongated to conical egg sacs is attached ventrally at the genital pore near the posterior end of the abdomen.

The life cycle and developmental stages of the parasitic copepods are basically similar to those of the free-living species. As 1 stage molts and metamorphoses into the next stage, the larvae are characterized by increase in size, in body segments, and in appendages.

Eggs hatch into nauplii which possess 3 pairs of appendages and a pair of spines posteriorly. The larval stages are motile, temporary parasites which feed on the host's body fluids. The nauplii molt and become metanauplii, which are characterized by an increase in the number of spines and body segmentation.

Males do not advance beyond the fifth copepodid stage and do not become permanent parasites; females do so only after copulation, which occurs during the fifth copepodid stage. After copulation the males disappear; the females penetrate the host tissue and attain a permanently fixed position. They then increase in length, develop cephalic horns, and become adults. After reaching maturity, they form egg sacs and eggs, completing the life cycle. The adult females are usually embedded at the base of the fins, particularly the dorsal fin, but they attach in the nares and over practically all other parts of the body (Figure 222, C). Under normal conditions, when the infestation is usually light, *L. cyprinaceae* and other parasitic copepods cause relatively little damage, but when abundant, as is likely to be the case under crowded hatchery conditions, they may become a serious menace to the life of their host. Not only may the fish be killed or left in an emaciated condition as a result of heavy infestation, but the value of the

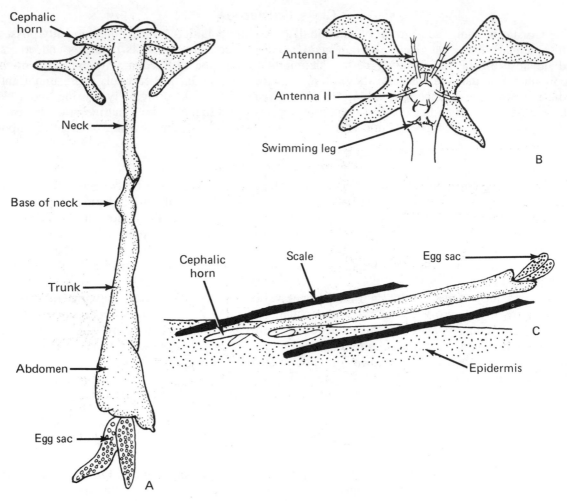

Figure 222. *Lernaea cyprinaceae:* (A) Adult female; (B) Cephalic horns with anterior end of cephalothorax and degenerated organs; (C) Adult embedded in fish epidermis between 2 scales.

commercial and game fish may be materially impaired as a result of the hemorrhagic lesions and tumor-like tissue at the site of parasite attachment. The breaks in the outer mucosal covering of the fish's skin may serve as a portal of entry for microbial organisms, often resulting in secondary infection.

Epizootics and mortalities of significant economic importance among fish due to massive infestations have been reported by Tidd (1934). There are also reports of heavy infestations in lakes and streams (Haley and Winn, 1959).

Subclass Branchiura

In this taxon of marine and freshwater crustaceans, the body is depressed dorsoventrally; the carapace extends laterally and posteriorly so that all or a portion of the segmented thorax and abdomen are free (Figure 223). There is a pair of ventral compound eyes and a dorsal median simple eye on the carapace; ventrally, each lateral lobe of the carapace contains 2 respiratory areas, whose arrangement and shape are useful in specific identifications. Ventrally, the first antennae are armed with claws and are prehensile; the second antennae are uniramous. In a longitudinal groove extending forward mesiad between the antennae is a preoral sting, which is retractile into a sheath. Posterior to and somewhat larger in diameter than the preoral sting is the proboscis, with the mouth. The first maxillae are modified as a pair of large sucking discs and the second pair consists of simple, 5-jointed leglike appendages terminating in a stout hook. There are 4 pairs of 3-jointed, setose walking legs. The abdomen is a single, bilobed

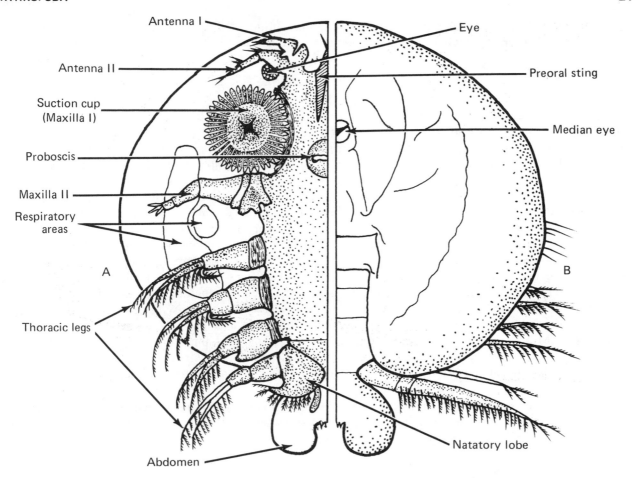

Figure 223. *Argulus catostomi,* ventral (A) and dorsal (B) sides of female.

segment. Elongate testes or rounded seminal receptacle, depending on the sex, are situated ventrally on the abdomen.

Branchiurans are periodic parasites on fish, occasionally on tadpoles of amphibians, with little host specificity. Specimens may be found anywhere on the body surface of the host and in the mouth and gill chamber. Attachment to the host is by the hooked antennae and sucking discs. As in the parasitic copepods above, in the case of heavy infestations the fish's vitality is weakened from the loss of blood, and the mechanical injury resulting to the host tissue may be equally, or even more, serious.

Both males and females can swim freely, and they leave their hosts regularly at the breeding season. Depending on the species, females deposit 30 to 600 eggs singly in rows, on plants and other fixed objects. Eggs are ovoidal in shape, about 450 μ long, and covered by a gelatinous capsule which secures each 1 firmly to the substrate.

The nauplius, metanauplius, and in some species the early copepodid stage occur in the egg, emerging in a late copepodid stage. Upon finding a host, the copepodid larva attaches and feeds for 5 to 6 weeks before becoming sexually mature. Adults may live free from the host for up to 15 days.

Depending on the species and temperature, completion of the life cycle requires from 40 to 100 days. The lifespan of females may be as long as 18 months, but males live for less than a year.

The subclass is represented in North America only by *Argulus,* the fish louse. Some common species with their usual fish hosts include: *Argulus alosae: Alosa, Salvelinus; A. appendiculosus: Stizostedion, Ictalurus; A. catostomi: Catostomus,* Cyprinidae; *A. flavescens: Amia, Micropterus.*

Class Arachnida

The members of this class, which includes the ticks, mites, spiders, and scorpions, are characterized by having no antennae, 4 pairs of legs in the adult stage, and the head and thorax usually fused into a cephalothorax. They are terrestrial, except for the aquatic mites. While many are parasitic and vectors of serious disease-producing agents, others are predatory feeders and are of no real parasitological concern.

Order Acarina

The Acarina includes the ticks and mites, which differ from true insects in normally having 3 pairs of legs as larvae and usually 4 pairs of legs in the nymphal and adult stages, and in lacking antennae. The head, thorax, and abdomen are fused into one body region, and the mouth parts are more or less distinctly set off from the rest of the body on a false head, the capitulum or gnathosoma.

Normally the life cycle includes 4 stages: the egg, larva, nymph, and adult. The eggs are usually deposited under the surface of the soil or in crevices or, in some parasites, under the skin of the host; some species are ovoviviparous. One or more nymphal stages occur between the larva and adult, and the nymph resembles the adult except in its smaller size and the absence of the genital opening.

The Acarina is important from the medical and veterinary point of view because some of its members are carriers of diseases and infections affecting man and his domestic animals, while others may cause dermatitis and allergic reactions in man. Still other species are important to the parasitologist because they occur as internal parasites in the lungs and air sacs of snakes, birds, and mammals.

Ticks and mites, although belonging to the same order, can be separated by the following characters. Ticks are usually 3 mm or more in length; have leathery body covering, with few short hairs; a hypostome toothed in some stage of life cycle (usually in adult); and a Haller's organ (sensory pore on tarsus of first leg). Mites, on the other hand, especially those of medical importance, are rarely more than 1 mm in length and when unfed have a white or pale yellow color. Their body texture is often membranous or heavily sclerotized, frequently with many long hairs; hypostome without teeth; and lacking a Haller's organ.

Baker and Wharton (1952) divided the Acarina into 5 suborders, one of which, the suborder Onychopalpida, is rare in the United States. The other 4 suborders are briefly characterized as follows, with a few of the more important species listed for each.

Ticks

Suborder Ixodides

With a pair of spiracles posterior or lateral to the third or fourth coxae, located in a stigmal or spiracular plate, without an elongated tube; Haller's organ present on the tarsus of the first leg; hypostome modified as a piercing organ.

The role of ticks in the human economy merits special attention, for not only are they annoying pests, but in temperate and tropical regions they surpass all other arthropods in the number and variety of disease-producing agents which they transmit to man and his domestic animals. As vectors of agents of human disease, they run mosquitoes a close second. Ticks are known to transmit 5 groups of microbial organisms: (1) spirochetal, such as relapsing fever; (2) rickettsial, such as Rocky Mountain spotted fever; (3) bacterial, such as tularemia; (4) viral, such as Colorado tick fever; and (5) protozoan, such as Texas fever. They also produce tick paralysis, which is probably caused by a neurotoxic substance in the tick's saliva. *Dermacentor andersoni* (*D. venustus* of some authors) has been referred to as a "veritable Pandora's box of disease-producing agents." It is a vector of the agents of Rocky Mountain spotted fever, tularemia, Colorado tick fever, and Q fever, and produces tick paralysis in both man and animals. It is also a vector of anaplasmosis in cattle.

Synopsis of Families of Ixodides

Scutum present; in males the scutum extends over the entire dorsal surface, in females only a portion of the anterior surface; capitulum terminal and visible from the dorsal surface . . . . Ixodidae

Scutum absent; capitulum ventral and usually concealed beneath the anterior margin . . Argasidae

In the United States, Ixodidae (hard ticks) are much more abundant than Argasidae (soft ticks), cause greater annoyance, and are far more important as vectors of disease-producing agents to man and animals.

Family Ixodidae

Dermacentor variabilis

This species, a vector of the agents of Rocky Mountain spotted fever and tularemia, is widely distributed east of the Rocky Mountains and also occurs on the Pacific Coast. The dog is the preferred host of the adult, although it feeds readily on other large mammals, including man. Adults are commonly found in the spring clinging to low vegetation, waiting to attach to a passing host. The male remains on the host for an indefinite time, alternately feeding and mating. The female feeds, mates, becomes engorged, and drops off to ovulate thousands of eggs.

Upon hatching, larvae remain close to the soil and mice runs. Larvae attach to and feed on mice and other small mammals, after which they drop off and seek a concealed niche for molting. Nymphs are similar in appearance to the larvae, but have 4 instead of 3 pairs of legs and a pair of spiracular plates. They attach to small mammals, feed, drop to the ground, and molt. The entire life cycle, typical for 3-host ticks, requires from 4 months to a year.

The adult is oval in contour and somewhat flattened dorsoventrally (Figure 224). Anteriorly there is a deeply concave region of the scutum or dorsal shield into which is fitted the capitulum, comprising the following 6 mouth parts (2 of which are paired): a median hypostome which has 6 rows of large, backwardly directed teeth on its ventral surface; a pair of chelicerae lying close together immediately above the hypostome, each chelicera consisting of a cylindrical shaft, which projects for a considerable distance into the body of the tick, and armed distally with a pair of serrated digits which can be moved by muscular action, and each is protected by a chitinous sheath into which it can be withdrawn when not in use; and a pair of 4-segmented palpi, inserted anterolaterally on the basis capituli. The basis capituli is the dense basal part articulating with the body (Figure 225).

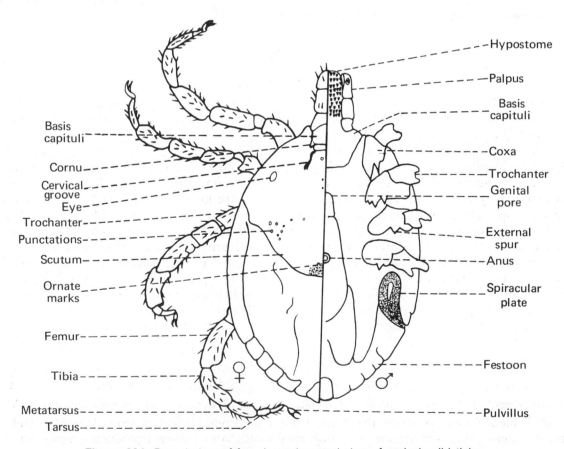

Figure 224. Dorsal view of female and ventral view of male ixodid ticks.

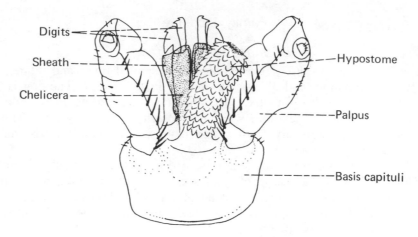

Figure 225. Capitulum of *Dermacentor variabilis,* with hypostome slightly displaced to left to show chelicerae, ventral view.

Directly back of the head lies the scutum or dorsal shield. This is small in the female but covers or almost covers the entire body in the male. Note the following scutal structures: the scapulae, the anterior angles or shoulders projecting on either side of the emargination; the cervical grooves, a pair of furrows running backward from the inner angles of the scapulae. Lateral grooves, dorsally situated furrows more or less parallel with the edge of the body, absent in female *Dermacentor,* but in the male they may extend into 1 or more of the festoons on either side. The festoons are the uniform rectangular areas into which the posterior margin of the body is divided.

On the venter note the following: the genital pore, situated just posterior to the basis capituli; genital grooves, starting at either side of the gonopore and running posteriorly; the anus, situated in the median line, posterior to the coxae; postanal groove, a curved transverse furrow immediately posterior to the anus. The spiracular plates or spiracles, paired respiratory organs, are situated posteriolaterally, posterior to coxae IV. These plates vary greatly in shape and configuration in the different species. In *D. variabilis,* the plate has a dorsal prolongation and the goblets (beadlike structures under the plate) are very numerous and small; in *D. albipictus,* it lacks a dorsal prolongation and the goblets are few and large.

The 6-segmented legs consist of the coxa, the immovable segment to which the rest are attached, trochanter, femur, tibia, metatarsus, and tarsus bearing a stalk with claws.

Other widely distributed hard ticks with their common hosts include *Dermacentor albipictus,* deer, elk, horse, moose; *Haemaphysalis leporispalustris,* rabbits; and *Rhipicephalus sanguineus,* dogs.

Mites

Suborder Sarcoptiformes

Without spiracles, or a few with a system of tracheae opening through stigma and porose areas on various parts of the body; chelicerae usually scissorslike, for chewing; palpi simple; anal suckers often present in the male.

These are minute, rounded, short-legged, flattened mites, causing itch, mange or scabies in man and various animals. They spend their entire life on the host. Transmission is usually by direct contact, either with infested animals or with objects with which the hosts have been in contact.

The burrowing and feeding of the itch mite, *Sarcoptes scabiei,* of man causes intense itching, inflammation, and swelling. The fertilized females cause the most trouble due to their tunneling just beneath the surface of the skin, laying eggs as they progress. The burrows ooze serum which hardens into scabs. Different varieties attack livestock, dogs, and rabbits (Figure 226).

Notoedres cati causes a very severe and occasionally fatal mange in cats. The mites burrow into the skin of the face and ears and sometimes spread to the legs and around the genitals. The burrowing of the mites causes a constant intense itching, and scratching and rubbing done by the host will often produce open sores which frequently become infected. Scabs are formed as with *S. scabiei.*

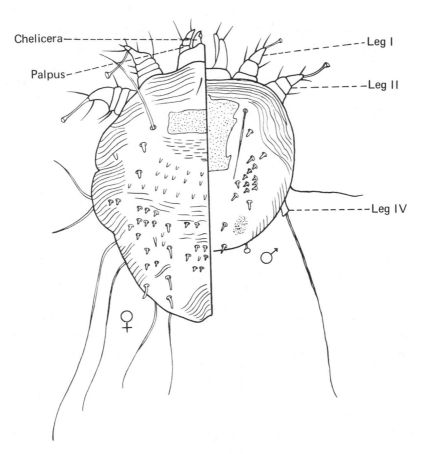

Figure 226. *Sarcoptes scabiei,* dorsal view of female and male.

The ear mange mite, *Otodectes cynotis,* infests dogs and cats. The mites do not burrow into the skin, but are found deep in the ear canal near the eardrum, where they feed upon tissue fluids. Intense irritation results and the canal becomes filled with inflammatory products and modified ear wax as well as mites. The host animal scratches and rubs its ears and shakes its head or holds its head to 1 side.

Psoroptes mites cause mange or scab in livestock and rabbits. All the mites of cattle, sheep, goats, horses, and rabbits are considered as varieties of *P. equi.* The mites live on the surface of the skin on any part of the body that is thickly covered with hair. They feed upon serum or lymph and continually migrate to the periphery of the lesions as they become larger. Scabs are formed from serum, dirt, and blood.

A condition known as scaly-leg in chickens and turkeys is caused by *Knemidokoptes mutans.* The mites apparently get onto the feet of the birds from the ground, since the lesion usually develops from the toes upwards. They pierce the skin beneath the scales, causing the scales to separate from the skin and the feet and legs to swell and become deformed. Closely related to the scaly-leg mite is the depluming mite of chickens and other birds *(K. laevis* var. *gallinae).* These mites, which are embedded in the tissues or scales at the base of the quills, cause a falling out of feathers over the back and sides.

In addition to the above parasitic mites, the oribatid, or soil, mites are of parasitological concern because they serve as intermediate hosts for Anoplocephalidae cestodes. The eggs of the tapeworms are ingested by the mites, in which they develop to the infective cysticercoid stage in several months. Animals feeding on the vegetation on which infected mites are crawling easily acquire the infection, and the tapeworms develop to the adult stage within a few weeks.

Suborder Trombidiformes

A pair of spiracles on or near the capitulum, but occasionally the stigmata may be absent; palpi usually free; chelicerae usually modified for piercing. These include, among others, *Demodex folliculo-*

rum, the hair follicle mite, and species of *Trombicula*, the chiggers or redbugs, and other Trombiculidae.

The hair follicle mite, *D. folliculorum* (Figure 227), an almost wormlike form, is found in the pores of man, especially around the nose and eyelids. The short and stumpy legs are 3-jointed. Setae are lacking on the body. The palpi are closely appressed to the tiny rostrum. It has a typical trombidiform life cycle, the egg, larva, protonymph, deutonymph, and adult. The entire life cycle takes about 14 days and is spent on the host. The deutonymph is the distributive stage (Spickett, 1961), and transmission is by direct contact. Numerous forms from various animals have been described as different species, but they are very similar and their specificity is questionable.

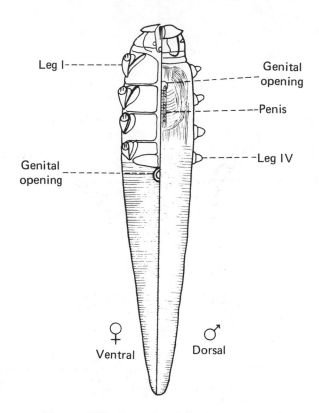

Figure 227. *Demodex folliculorum,* ventral view of female and dorsal view of male.

While the attacks of a number of mites are responsible for dermatitis and allergic reactions in man, the most important ones are the chiggers of the genus *Trombicula*. In the United States, these are common in the southern states, the Mississippi valley, and certain areas of California. They are most abundant in woody and grassy areas, swamps, and particularly where wild rodents and birds occur, such as berry patches. Only the larval stage feeds on vertebrates, the nymphal and adult stages being free-living. Chiggers attach especially in areas where the clothing fits tightly, such as the tops of shoes and socks, the waist region where belts and underwear are fastened, and under the armpits.

The larvae inject saliva into the host tissue, forming a stylostome, or feeding tube, through which the larva sucks up the semidigested tissue debris. After engorgement, the larva drops off and molts.

Chiggers are not known to transmit any agents of diseases in the United States, but their attacks produce violent itching, which when scratched may result in secondary infections. In Southeast Asia and the Southwest Pacific, larval chiggers transmit the rickettsiae causing scrub typhus, also known as tsutsugamushi fever. This disease is often fatal, and took a heavy toll among troops during WW II. Since the larvae feed on man or other vertebrates only once in their life, the rickettsiae are transmitted transovarially from infected adults through the egg to the "chigger" larval stage which feeds on man and rodents. The postlarval stages are predatory, feeding on small arthropods and their eggs in the soil.

Suborder Mesostigmata

Body well chitinized, with dorsal and ventral plates; capitulum small, anterior; 1 pair of spiracles lateral to legs, usually associated with an elongate tube or peritreme, or if absent, highly specialized parasites of the respiratory tract of vertebrates. Representatives with their hosts include: *Dermanyssus gallinae*, chicken mites (Figure 228); *Ornithonyssus bacoti*, tropical rat mite, the vector of *Litomosoides carinii*, a filarial nematode parasitic in rodents (p. 174); *O. bursa*, tropical fowl mite; *O. sylviarum*, northern fowl mite; and *Echinolaelaps echidninus*, spiny rat mite, which serves as the final host for *Hepatozoon muris*, a haemogregarine of the rat. Mites become infected by feeding on infected rats, but cannot transmit the infection in this manner, as rats acquire the infection only by ingesting infected mites.

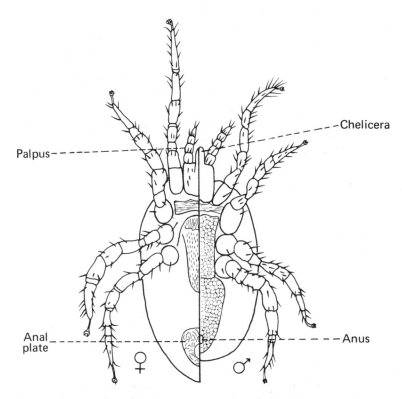

Figure 228. *Dermanyssus gallinae,* ventral view of female and male.

Class Pentastomida

The Pentastomida, commonly referred to as tongue worms, are of uncertain taxonomic position. Some workers consider them as a class of the phylum Arthropoda; others consider them closely related to the phylum Annelida; while still others recognize them as an independent phylum with the belief that they share a common ancestry with the annelids and arthropods (see Self, 1969).

Pentastomes are parasitic, both as larvae and as adults, in the respiratory organs usually of reptiles and carnivorous mammals. The adults are vermiform, with a short cephalothorax and an elongate, annulate abdomen. On the ventral surface, the cephalothorax bears 2 pairs of hooklike claws on each side of the mouth.

Life cycles are not well known. Usually an intermediate or paratenic host harbors the immature stage, but development may be completed in a single host. Late larval stages are referred to as nymphs. The pentastome egg contains a fully developed larva when oviposited by the female. The eggs are ingested by a vertebrate, and the larva is liberated in the digestive tract. The larva bores through the intestinal wall and encapsulates in the viscera and associated mesenteries. In the capsule, it undergoes metamorphosis and attains the adult form, except that it is not sexually mature.

Family Porocephalidae

Porocephalus crotali

In the United States, *P. crotali* adults occur commonly in the lungs of Crotalidae snakes, especially rattlesnakes and water moccasins; immature stages occur in rodents (muskrats, wood mice). Probably a wide range of wild rodents serve as intermediate hosts, because white mice, rats, and hamsters are satisfactory intermediaries in experimental infections.

Adult females measure up to 7 cm long (Figure 229, A) with males somewhat smaller. A pair of hooks occurs on each side of the mouth (Figure 229, B). The genital opening in the male is near the anterior end of the body and in the female near the posterior end.

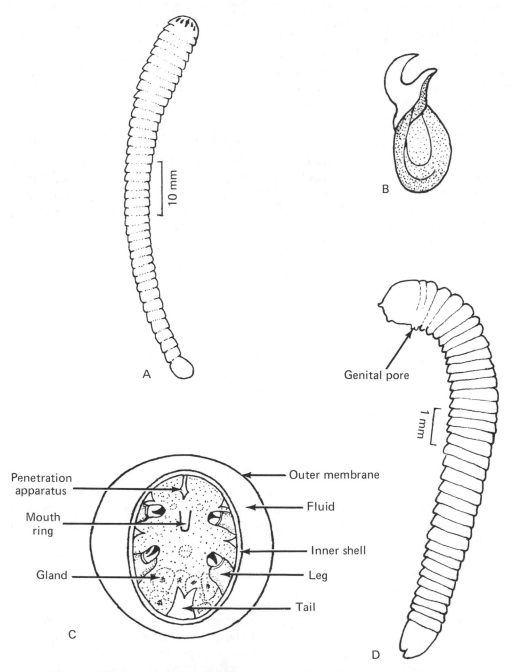

Figure 229. *Porocephalus crotali:* (A) Adult female; (B) Hook from anterior end of adult; (C) Mature egg, intact larva in ventral view; (D) Sixth stage of infective male nymph, from rodent.

Eggs containing fully developed 4-legged larvae (Figure 229, C) are deposited in the lungs, pass up the trachea, are swallowed, and voided in the feces. The egg proper consists of a thin, permeable, outer membrane and a thick, inner impermeable one. The space between the 2 membranes is filled with fluid.

Larvae bear a spearlike anterior penetration apparatus and a short, broad bifurcated tail. There are 2 pairs of short, stumpy legs that terminate in a pair of sharp, slender claws. Dorsally on the body surface are a pair of stigmata and a median dorsal organ. On the ventral side, between the anterior pair of legs, is a U-shaped, heavily chitinized mouth ring. Several large uninucleate cells, a large circumesophageal ganglion, and a blind gut appear internally.

Upon being swallowed by a rodent, the eggs hatch quickly in the duodenum, and the larvae are in the body cavity within an hour. They wander about in the celom until the seventh to eighth days, when the first molt occurs. At this time, they become tightly encapsulated in the host tissues.

Once encapsulated, 6 molts occur, each nymphal stage being followed by a period of growth and differentiation. The sixth stage nymph (Figure 229, D) is infective to snakes.

Snakes become infected by ingesting infected rodents. When freed sixth stage nymphs are fed snakes, they penetrate the gut wall and are in the body cavity within 24 hours. Since the alimentary canal and lungs are juxtaposed, passage from one to the other is easy and quick. The nymphs attach to the lungs in a few days.

Class Insecta

The members of this class are air-breathing arthropods with distinct head, thorax, and abdomen, except in some larvae and some modified adults. They have a single pair of antennae, 3 pairs of thoracic legs, and usually 1 or 2 pairs of wings in the adult stage. As chiefly terrestrial animals, insects have become adapted to the greatest variety of conditions and have become so successful that they outnumber all other metazoan animals both in species and in individuals. Practically every possible relationship of organic beings is found in the list of the habits of insects. Both as predators and as parasites upon plants and other animals, and in their relations to the spread of etiologic agents of disease, they occupy a position of economic importance not excelled by any other animal group.

Order Siphonaptera

Fleas are wingless, laterally compressed insects. They are armed with backwardly directed spines, long stout spiny legs, and short clubbed antennae which lie in grooves along the side of the head. The mouth parts are elongated, adapted for piercing the skin and sucking blood.

Their life cycle includes a complete metamorphosis, with egg, larva, pupa, and adult stages. The female lays her eggs in the host's nest or bedding, or in the host's fur or feathers, whence they drop into the nest or onto the ground. The full quota of eggs is not laid all at once but singly or in batches over a considerable period of time, punctuated by blood meals which are necessary for their development. The length of the entire life cycle may be as short as a month for some individuals in hot, moist climates, but may be much longer, even in other individuals of the same brood. The adults infest mammals and birds and feed only on blood, although they may spend considerable time in the burrow or nest of their host instead of on its body. Of the known species, about 95% occur on mammals and only about 5% on birds.

Most fleas have preferred hosts, but many of them will feed upon a variety of animals and will bite man readily in the absence of their normal host species. Thus, man is always liable to flea infestations from his dogs or cats, or even from rats living in the house or adjacent structures. In early summer fleas on their hosts lay large numbers of eggs, the eggs dropping off into cracks of the flooring or into debris. Here the larvae hatch and feed, mature, and transform into pupae. Then about midsummer, often about the time the occupants of the infested dwelling return from their summer vacation, large numbers of the second generation of adults appear. In the summer, fleas of cats and dogs, particularly if stray pets are around, will breed out of doors in vacant lots, under buildings, and similar situations.

Fleas are of importance as pests and in connection with the transmission of disease-producing agents. By their insidious attacks on man and his domestic animals, they cause irritation, blood loss, and discomfort. In addition to their role as pests, fleas are of greater importance because of their connection with the transmission, both among their reservoir hosts and to man, of the etiologic agents of two human diseases of outstanding importance—plague and murine or endemic typhus.

Plague, caused by the bacterium *Yersinia pestis* (formerly *Pasturella pestis*), has been a scourge of mankind since early times. Of the 2 chief clinical types of plague, bubonic and pneumonic, the former is the commoner and much less severe. It is characterized by swellings of the lymph glands, especially those of the groin and armpits. When untreated, the fatality rate may exceed 50%. The pneumonic form is a highly contagious and usually fatal pneumonialike disease, spread directly among people in sputum or droplets coughed up by the sick. It occurs secondarily to the bubonic type when the plague bacteria become localized in the lungs.

Bubonic plague ordinarily results from the bite of an infected flea, but the disease may be contracted by direct contact or the bite of an infected rodent. Fleas become infected when feeding on the blood of a diseased host. The bacteria multiply so rapidly in the proventriculus and stomach of the flea that an obstruction is formed. As these "blocked" fleas attempt to feed on man or a rodent, the blood which cannot pass beyond the obstruction, becomes mixed with the plague bacilli. When the contaminated blood is regurgitated into the bite-puncture or other skin abrasion, infection results. All rodent fleas are not equally susceptible to stomach blockage, and hence different species are not equally dangerous as plague vectors. The oriental rat flea, *Xenopsylla cheopis*, is the most important vector of urban plague because of its great tendency to become blocked, its ability to feed on both rodents and man, and its great abundance in urban centers. It occurs in most of the coastal ports of the United States, and has been found in a number of inland states.

There are 2 epidemiological types of plague: urban plague, the classical form of the disease and the type usually contracted in cities where people have close contact with domestic rats and fleas; and sylvatic plague, which is more often contracted by people in rural areas who have contact with wild rodents or their fleas.

Murine typhus is primarily a disease of murine rodents, which in North America include only rats and mice, and is caused by *Rickettsia typhi*. It is transmitted among rodents by their fleas, lice and possibly mites, and occasionally to man from rats and mice, presumably by their fleas. The most likely mode of transmission is by contamination of the bite-wound or abraded skin with the infected feces or crushed fleas, especially *X. cheopis*. According to Jellison, transmission by direct bite remains a possibility, and infection by inhalation and ingestion is probable. In the southern United States, murine typhus is common among Norway and roof rats without evidence of ill effect. Fleas are not harmed by the rickettsia and presumably remain infected for life but do not transmit the organism to their progeny. While 5,400 cases of murine typhus were officially reported in the United States as recently as 1944, the disease has since all but disappeared in this country as a result of the vigorous control programs of domestic rodents and their fleas.

In addition to the 2 above-mentioned familiar pathogens which are transmitted by fleas, they transmit a nonpathogenic rat hemoflagellate, *Trypanosoma lewisi* (see p. 11), and serve as intermediate hosts for certain tapeworms which occasionally occur in man, especially children. One of these is *Dipylidium caninum* (see p. 114) of dogs and cats, the cysticercoids of which occur in *Ctenocephalides canis*, *C. felis*, and *Pulex irritans*. The worm eggs are ingested by larvae, but the cysticercoids finish their development in the adult fleas. Two other tapeworms, *Hymenolepis diminuta* and *H. nana* of rats and mice, which occasionally are found in man, may utilize certain fleas as intermediate hosts. A nematode, *Dipetalonema reconditum*, found in the subcutaneous tissues of dogs, is transmitted by species of *Ctenocephalides*.

Family Pulicidae

Ctenocephalides felis

This species, the cat flea, and the closely allied dog flea, *C. canis*, are now almost cosmopolitan, having followed man and his domestic pet animals over most of the world. The cat flea is usually more abundant and generally more widely distributed than the dog flea. In both species, the eyes are large and deeply pigmented; both genal and pronotal combs are heavily pigmented and sharply pointed. Typical specimens of the 2 species of *Ctenocephalides* can be readily separated by the shape of the head and the character of the genal combs. In *C. felis* (Figures 230; 231, D), the head is quite long and pointed, and the first and second genal spines are of approximately equal length; in *C. canis* the head is relatively short and rounded, and the first genal spine is shorter than the second.

Xenopsylla cheopis

This species, the oriental rat flea, has been introduced throughout much of the world with the Norway and roof rats. It is established throughout most of North America. When rats are killed, these fleas leave their hosts and bite man readily. Medically, it is undoubtedly the most important flea in the world, for a large percentage of the human cases of bubonic and pneumonic plague that have occurred during the many historical pandemics and epidemics of this disease may be attributed to the presence of this flea. Also, it is the chief vector of the agent of murine typhus. Eyes well developed; genal and pronotal combs lacking; ocular bristle situated anterior to eye (cf. *Pulex*). Females can be easily distinguished by the large deeply pigmented spermatheca (Figure 231, C).

Pulex irritans

This species, the human flea, though found on a wide variety of other hosts in nature, is widespread throughout the warmer regions of the world. Of all the species of fleas, *P. irritans* is probably the most nearly cosmopolitan in distribution. In addition to being a serious pest of man and serving as an intermediate host of *Dipylidium caninum*, the human flea has been found infected with plague bacteria in nature. Eyes large and deeply pigmented; genal comb absent, or represented by 1 inconspicuous tooth; pronotal comb absent; ocular bristle situated ventral to eye (cf. *Xenopsylla*) (Figure 231, A).

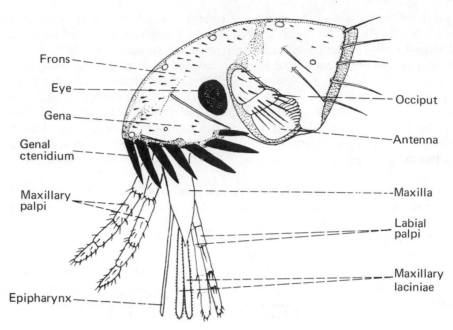

Figure 230. Head of *Ctenocephalides felis*, male, lateral view.

Family Ceratophyllidae

Nosopsyllus fasciatus

This species, the northern rat flea, is found commonly on domestic rats and house mice throughout North America. It does not bite man readily and is found more commonly in temperate regions, where plague is less likely to be present. This species may be of importance in the transmission of the plague bacterium among rats. While relatively unimportant in the transmission of plague to man, it is an efficient vector of murine typhus. Eyes well developed; genal comb lacking; pronotal comb with long, dark brown spines (Figure 231, B).

Key to Common Families of Siphonaptera

1(2) The 3 thoracic segments combined, shorter than the first abdominal tergite (Figure 232) . . . . .
. . . . Tungidae

2(1) The first 3 thoracic segments combined, longer than the first abdominal tergite 3

3(4) Abdominal tergites II through VII, each with only 1 row of setae (Figure 231, A) . Pulicidae

4(3) Abdominal tergites each with more than 1 row of setae 5

5(6) Lacking genal comb; eyes usually well developed (Figure 231, A) . . . . Ceratophyllidae

Eyes present and pigmented (Figure 233) . . . . Amphipsyllinae

Eyes usually present (vestigial in *Dactylopsylla* and *Foxella*) (Figure 231, B) . . Ceratophyllinae

Eyes absent (Figure 234) . . . . (*Dolichopsyllus*, only genus) Dolichopsyllinae

6(5) Usually with genal comb; eyes not well developed 7

7(10) Usually with genal comb; with 2 spines (on each side) 8

8(9) Genal comb with 2 spines, situated at the tip; eyes vestigial (on bats) (Figure 235) . . . .
. . . . Ischnopsyllidae

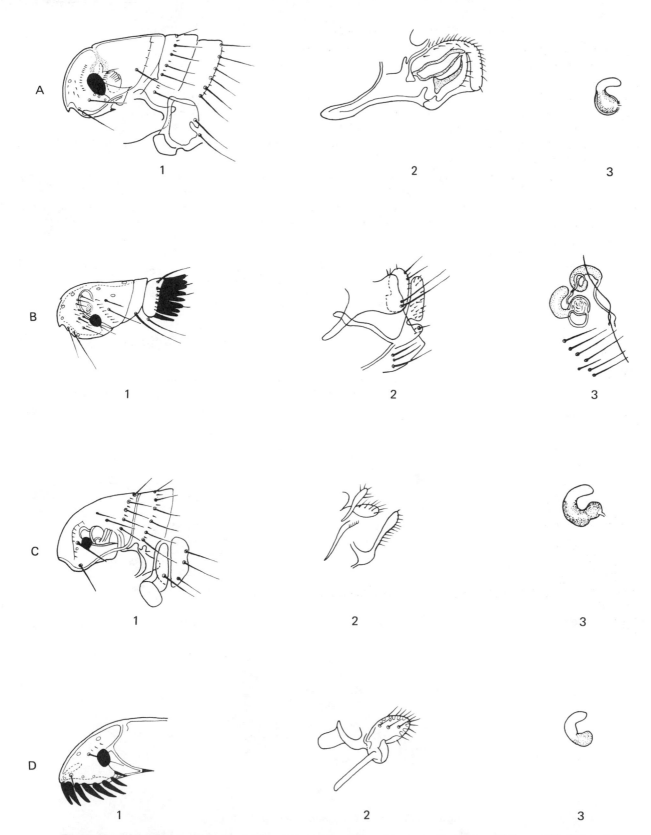

Figure 231. Taxonomic characteristics of four common and important species of fleas. (A) *Pulex irritans*; (B) *Nosophyllus fasciatus*, showing pronotal comb; (C) *Xenopsylla cheopis*; and (D) *Ctenocephalides felis*, showing genal comb. (1) Head, (2) Clasper of male, and (3) Spermatheca of female.

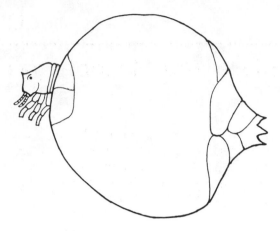

Figure 232. *Tunga penetrans,* engorged female.

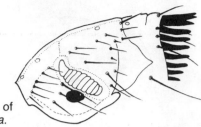

Figure 233. Head of *Amphipsylla sibirica.*

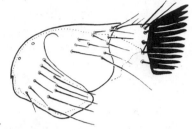

Figure 234. Head of *Dolichopsyllus stylosus.*

9(8) Genal comb with 2 or 4 spines, distinct, not overlapping, situated behind the eyes; eyes much reduced (Figure 236) . . . . Leptopsyllidae

10(7) Usually with genal comb, with more than 2 spines (if only 2, overlapping and not distinct, and situated near the eyes); eyes absent or vestigial, not large and heavily pigmented (Figure 237) . . . . Hystrichopsyllidae[1]

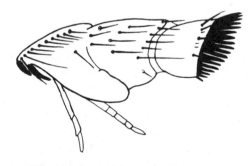

Figure 235. Head of *Eptescopsylla vancouverensis.*

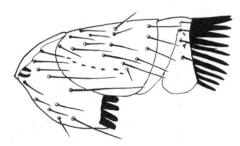

Figure 236. Head of *Leptopsylla segnis.*

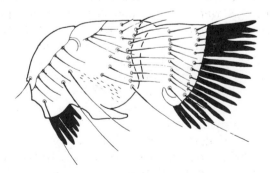

Figure 237. Head of *Hystrichopsylla dippiei.*

1. Included are a few genera lacking genal comb, e.g., *Atyphloceras, Callistopsyllus, Catallagia, Conorhinopsylla, Delotelis, Megarthroglossus, Saphiopsylla, Trichopsylloides.*

Lice (Anoplura and Mallophaga)

Lice are wingless, dorsoventrally flattened insects, ectoparasitic on birds and mammals. They are parasitic throughout their entire life cycle, even their eggs being firmly attached to the hairs or feathers of their hosts. The immature stages have the same habit of life as the adults and resemble them closely, so that there is little metamorphosis. There are 3 nymphal instars before reaching the adult stage. They have only short 3 to 5 segmented antennae, sometimes hidden in a recess on the head; reduced or no compound eyes; and no ocelli. The thoracic segments are sometimes indistinct; the legs are short but stout; and the abdomen has from 5 to 8 distinct segments.

Authorities are not in agreement as to whether the sucking lice of mammals and the chewing lice of birds and mammals should be placed together in the order Phthiraptera, or the former in the suborder Anoplura and the latter in the suborder Mallophaga, or the suborders be given ordinal status. Since the 2 taxa are anatomically distinct and have different habits and host preferences, they are here considered as separate orders.

Order Anoplura

The Anoplura have highly specialized mouth parts, which when not in use are retracted into a pouch, adapted for piercing the skin of their hosts and sucking blood. The head is narrower than the thorax, the 3 segments of which are fused together, without any clear division of segments. The number of spiracles has been reduced to not more than 1 pair on the thorax and 6 pairs on the abdomen. The single tarsal joint bears only 1 claw and in most species the tip of the tibia is widened and drawn out into a "thumblike process," opposable to the claw. All are ectoparasitic on mammals and very specific in their host requirements.

Man is infested with 2 species of sucking lice, i.e., the crab louse, *Pthirus pubis*, and the human louse, *Pediculus humanus*. The latter is commonly regarded as comprising 2 biological races, anatomically similar and known to interbreed, but occurring on different body locations of the host. Those normally clinging on man's inner garments, except when feeding on the body, are sometimes referred to as *P. humanus humanus*, and the others, inhabiting the head, are known as *P. h. capitis*. But Ferris (1951) prefers to use the appellation *P. humanus* to cover the population as a whole, with the addition of the vernacular names body louse and head louse for those forms when the occasion demands. Except for the body louse, which attaches its eggs to the fibers of the underwear, lice attach their eggs to the hairs of the host. Sucking lice have been intimately associated with man from time immemorial. They are most common in times of stress, such as war, famine, and disaster, when people are forced to live under crowded conditions and facilities for cleanliness are inadequate. Anoplura are of importance in 2 ways: their attacks on man and his domestic animals result in irritation and loss of blood, and they are involved in the transmission of microbial organisms causing human disease.

They inject an irritating saliva into the host during feeding, resulting in considerable itching. A person's first exposure to lice results in little or no discomfort, but sensitivity occurs after about 7-10 days, when the average individual develops an intense itching from the feeding of the lice. Heavy infestations may lead to scratching and secondary infections.

The body louse is the vector of epidemic typhus, which is highly fatal during epidemics, and relapsing fever of Europe, Africa, and Asia. The first is caused by *Rickettsia prowazeki* and the second by the spirochete, *Borellia recurrentis*. Lice become infected with the rickettsial agent by ingesting blood from a diseased person; the parasites multiply in the midgut epithelium of the louse and are voided with the feces. Man usually acquires the infection by the contamination of the bite wound or abraded skin with the infected feces or crushed lice. Lice frequently leave the typhus patient when high fever develops and attack other human hosts, causing a rapid spread of the disease. In the case of the spirochete causing relapsing fever, which is likewise acquired by the louse with the ingested blood of a positive patient, it multiplies throughout the body of the louse. Neither the feces nor the bites of the lice are infective, transmission occurring only by crushing lice and inoculating the bitten or abraded skin with the infective body fluid.

Family Pediculidae

Pediculus humanus

The human louse bears a pair of prominent eyes, a pair of 5-jointed antennae, and mouth parts which are withdrawn and not visible. The 3 thoracic segments, each with a pair of legs terminating in a hooklike claw, are fused together. The abdomen is elliptical and consists of 7 segments. The margins are festooned and chitinized to form deeply pigmented plates on which 6 pairs of spiracles are situated. In the male, the abdomen is rounded posteriorly and the genital organ, the aedeagus, is easily visible and usually extruded; in the female the terminal portion of the abdomen is deeply cleft. The life cycle may be completed in about 18 days (Figure 238).

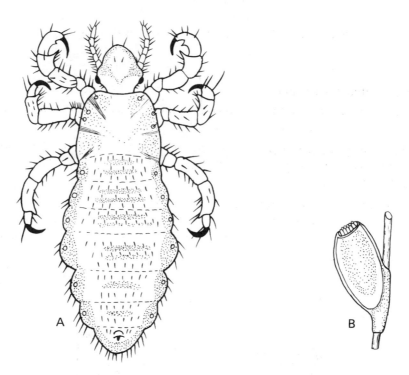

Figure 238. Body louse, *Pediculus humanus;* (A) Adult male; (B) Egg attached to body hair.

Pthirus pubis

The crab louse is characterized by its relatively short head which fits into a broad depression in the thorax. The latter is broad and flat and fused with the abdomen. The first pair of legs is slender and terminated by a straight claw, the second and third pairs are thicker and provided with powerful claws for clinging to hairs. The first 3 abdominal segments are fused into 1, bearing 3 pairs of spiracles, followed by 3 tandem arranged segments with as many pairs of spiracles. The last 4 abdominal segments bear wartlike processes laterally, known as tubercles, the last of which is the longest. The life cycle takes about 15 days (Figure 239).

Key to Common Families of Anoplura

1(2) Abdomen with paratergal plates 3

2(1) Abdomen without paratergal plates. 7

3(4) Paratergal plates with their apical parts projecting free from body (on rodents, a few on insectivores and primates) (Figure 240). . . . . Hoplopleuridae

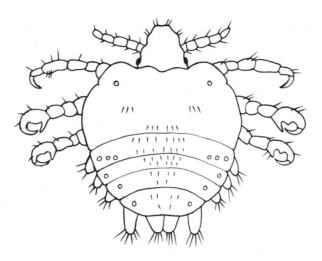

Figure 239. Adult crab louse, *Pthirus pubis*, female

4(3) Paratergal plates without free apices 5

5(6) Eyes present, pigmented; head not retracted deeply into thorax (on primates, including man) (Figure 238). . . . . Pediculidae

6(5) Eyes absent, but with prominent ocular lobes posterior to the antennae (*Pecaroecus* from peccaries, an aberrant genus, lacks ocular lobes); head deeply retracted into the thorax (on Artiodactyla and Equidae) (Figure 241) . . . . Haematopinidae

7(8) Body thick and stout, with heavy spines and, in some cases, scales; legs, at least 2 pairs, with stout undivided tibiotarsus (on marine mammals) (Figure 242). . . . . Echinophthiriidae

8(7) Body with setae but not scales, first pair of legs smaller than the others (on Artiodactyla and Canidae) (Figure 243) . . . . Linognathidae

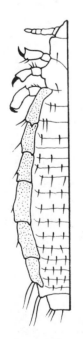

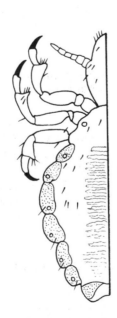

Figure 240. *Hoplopleura hesperomydis*, female.

Figure 241. *Haematopinus suis*, female.

Figure 242. *Antarcophthirus callorhini*, female.

Figure 243. *Linognathus setosus*, female.

Order Mallophaga

These chewing or biting lice, like their sucking counterpart, are wingless, dorsoventrally flattened insects which are parasitic throughout their entire life cycle, even the eggs being firmly attached to the feathers or hairs of their hosts. But they differ from the Anoplura in several respects: the mandibles are heavily chitinized; the head is very broad, usually as wide or wider than the thorax; the thorax is divided into at least 2 distinct segments; the last of the 2 tarsal joints bears 1 or 2 claws, and only rarely is the tip of the tibia so formed that it is opposable to the claw; and most are ectoparasites of birds, but several genera infest mammals only.

The "biting" ascribed to the Mallophaga refers to their manner of feeding and not to any wound they may inflict on the host, since in most species the food consists of feathers and hairs, though some are found to have an admixture of blood in their diet. Whatever their feeding habits may be, however, they are most annoying pests because of the irritation produced by their constant crawling and nibbling. Nearly all birds are infested with these lice, and domestic poultry and pigeons suffer particularly from them, owing to the crowded conditions under which these birds live. While less widely distributed on mammals, domestic animals are attacked by various species of *Trichodectes*. Dogs, cats, horses, cattle, sheep, and goats may suffer considerable loss of condition if badly infested with chewing lice.

Suborder Amblycera   Family Menoponidae

Menopon gallinae

The common shaft louse of chickens is a small pale species, about 1.5 mm long. The head is broadly triangular, strongly enlarged on the temples and evenly enlarged behind. Head without ventral sclerotized processes situated lateral and posterior to the mandibles. Prothorax smaller than the head, meso- and metathorax united. This is a large genus and is well represented on galliform birds (Figure 244).

Suborder Ischnocera   Family Trichodectidae

Trichodectes canis

The dog biting louse, which occasionally also infests other canines, is most troublesome on puppies. It is characterized by a broad head, greater in width than length, and a broadly rounded preantennary region. The antennae are sexually dimorphic; in the male they are shorter than the head. Six pairs of abdominal spiracles (Figure 245).

This species may serve as the intermediate host of the double-pored tapeworm, *Dipylidium caninum*, of dogs, cats, and occasionally man, especially children (see p. 114).

Key to Common Families of Mallophaga

1(4) Maxillary palps present; antennae arising from ventral portion of head, usually concealed in grooves
. . . . .Suborder AMBLYCERA. . .2

2(3) Tarsi with 1 claw or none (on guinea pigs) (Figure 246). . . . . Gyropidae

3(2) Tarsi with 2 claws (on birds) (Figure 244) . . . . .Menoponidae

4(1) Maxillary palps absent; antennae arising from or near lateral margin of head, not concealed in grooves. . . . . .Suborder ISCHNOCERA. . .5

5(6) Antennae 5-segmented, tarsi with 2 claws (on birds) (Figure 247) . . . . Philopteridae

6(5) Antennae 3-segmented, tarsi with 1 claw (on mammals) (Figure 245) . . . . .Trichodectidae

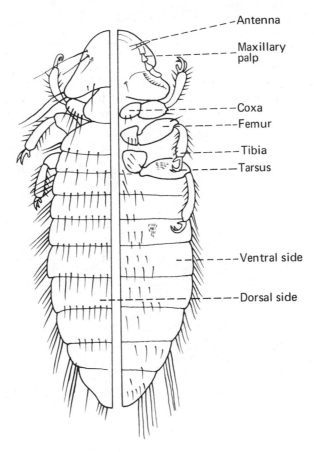

Figure 244. *Menopon gallinae*, dorsal and ventral side of female.

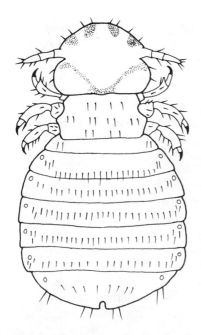

Figure 245. *Trichodectes canis*, dorsal view of female.

Figure 246. *Gyropus ovalis*, female.

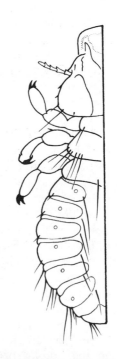

Figure 247. *Philopterus dentatus*, female.

Order Diptera

The members of this order possess 1 pair of functional wings, and rudiments of a second pair in the form of short, knobbed organs known as halteres. Mouth parts are modified for piercing and sucking, or for lapping. Metamorphosis is complete. The larvae are legless and often wormlike, and frequently they are found in the tissues or cavities of man and animals, causing myiasis. There are 3 suborders, based on the antennal characters of the adults.

Of all the taxa of arthropods which are involved as vectors of disease and infectious organisms the Diptera, or true flies, are of paramount importance. The order includes a great variety of bloodsucking forms responsible for the transmission of organisms that cause such important diseases as malaria, yellow fever, dengue, trypanosomiasis and leishmaniasis, encephalitis, and filariasis, as well as a number of nonbloodsucking forms of sanitary interest, such as the ordinary houseflies and blowflies.

Suborder Cyclorrhapha

Antennae are very short, generally 3-segmented; the first 2 segments are short, the third is much larger and with an arista, a bristle or featherlike structure, usually dorsal in position. Palpi are 1-segmented. This suborder, by far the largest of the 3 suborders of Diptera, is variously divided by taxonomists; the diagnostic characters can only be used with certainty by specialists familiar with the group.

Family Muscidae

The family is such a large and heterogeneous group that it is not possible to write a description which would be generally applicable to the species of parasitological importance. Because of the great diversity in form, for convenience the family is divided into groups based upon medical importance: (1) the "biters," the most famous of which are the tsetses (*Glossina*), (2) the "nonbiters," of primary importance as agents in the occasional mechanical transmission of certain infections, and (3) the myiasis-producing larvae.

The tsetses, all of which belong to *Glossina*, are the most important group of bloodsucking muscids directly affecting man and animals. They are distinguished from other flies by the slender, forward-projecting proboscis; the spiny, long, slender palpi; and the branched dorsal hairs on the dorsal side of the antennal arista (Figure 248). The living fly can be recognized when at rest by the way in which it folds its wings scissors-like above the abdomen.

The female is ovoviviparous and produces 1 mature larva each 10 to 12 days during her lifetime of about 40 days. These larvae pupate almost immediately, the puparia occurring in dry, loose soil in shaded, protected areas. The flies are diurnal feeders, and both sexes are bloodsuckers. Tsetses are confined to tropical and subtropical Africa, where they hinder settlement and development. They occur in "fly belts"; the type of area infested varies with the species.

Among biting flies, the tsetses rank second only to the mosquitoes as vectors of disease-producing agents of man and animals. They are obligate vectors of the trypanosomes causing sleeping sicknesses. *Glossina palpalis* is the most important vector of *Trypanosoma gambiense*, causing Gambian sleeping sickness in man, chiefly in West Africa. The most important vector of *T. rhodesiense*, the organism of Rhodesian sleeping sickness in man and nagana in animals (horses, mules, camels, and dogs), chiefly in East Africa, is *G. morsitans*. Nagana, or African trypanosomiasis of animals, is a disease caused by *T. brucei* and closely related species. In addition, *G. morsitans*, *G. longipalpis*, and *G. pallidipes* relate to nagana in nearly the same way that other *Glossina* species relate to African human sleeping sickness in that the flies are infective for a day or 2 after feeding, then become noninfective for about 3 weeks, when they again become infective and remain so for life.

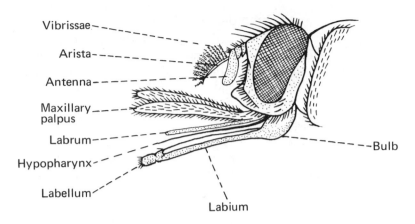

Vibrissae
Arista
Antenna
Maxillary palpus
Labrum
Hypopharynx
Labellum
Labium
Bulb

Figure 248. *Diagrammatic figure of head of a tsetse (Glossina),* showing mouth parts.

Family Hippoboscidae

These, the louse flies, are degenerate Diptera which have adapted themselves to more or less continuous existence on mammals and birds. The body is broad and flattened dorsoventrally; the abdomen is indistinctly segmented, and usually rather leathery in texture. Wings are present in some and absent in others. In the winged species, the eyes are large and the antennae exposed, but in the wingless ked or "sheep tick," *Melophagus ovinus* (Figure 249), the eyes are small and the reduced antennae lie in pits on the forehead. Hippoboscids give birth to mature larvae which are ready to pupate. Usually the larvae are deposited in dry soil or humus, in the nests of birds, or other suitable places where they transform into black pupae in a few hours. In the case of the ked, however, the females attach their larvae to the wool of the host by means of a gluelike substance.

The 2 species of greatest importance in North America are *M. ovinus* and the pigeon fly, *Pseudolynchia canariensis*. The ked is especially injurious to young lambs, to which it migrates in large numbers when the older sheep of the flock are sheared. It serves as a vector for *Trypanosoma melophagium*, a nonpathogenic hemoflagellate of sheep. The pigeon fly, which has a pair of transparent wings, in addition to taking blood, especially on squabs, when the feathers begin to form and afford protection, transmits *Haemoproteus columbae*, a blood protozoan of pigeons.

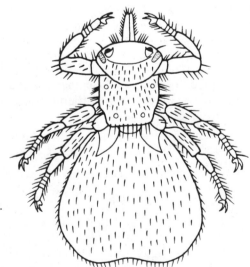

Figure 249. Adult *Melophagus ovinus*.

Suborder Brachycera

Antennae shorter than thorax, usually 3-segmented, variously formed and usually held horizontally erect. The last segment may be ringed to form several smaller units. Arista, when present, terminal or nearly so. The palpi are 1- or 2-segmented. The only family of parasitological importance is the Tabanidae, which includes the horseflies *(Tabanus)* and the deerflies *(Chrysops)* (Figure 250).

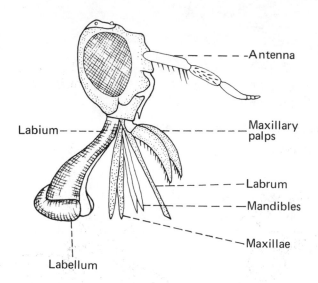

Figure 250. Diagrammatic figure of head of *Chrysops*, showing mouth parts.

Family Tabanidae

The adults are sturdy flies with large heads and eyes. The females are bloodsuckers; their mouth parts are developed for cutting skin and sucking blood which oozes from the wound. The males feed on nectar. Typically, the eggs are deposited in masses on aquatic plants growing over water. The newly hatched larvae drop to the water, or make their way to damp earth, and live on snails, insect larvae, and other aquatic animals. The larval stage lasts for several months or a year, and the pupal period ranges from 1 to 3 weeks.

Chrysops dimidiata and *C. silicea* are important vectors of the filarial worm *Loa loa* in various parts of Africa, particularly in the Belgian Congo, and *C. discalis* is an important transmitter of the etiologic agent of tularemia in the western United States. Species of *Tabanus* and *Hybomitra* transmit the filarioid *Elaeophora schneideri* to sheep, deer, and elk in western and southwestern United States. It causes blindness and death in elk and sorehead in sheep. Several species of *Tabanus* are known to transmit trypanosomes, causing surra in horses, cattle, camels, dogs, etc.

Suborder Nematocera

Antennae many segmented, each segment similar and of approximately equal diameter throughout its length, often quite long, longer than the length of the head and thorax combined. There is no arista; the palpi are 4- or 5-segmented.

Family Psychodidae

The members of this family, commonly known as sand flies, are small, mothlike flies found in nearly all warm and tropical climates. Their bodies and wings are hairy and the antennae are long, consisting of 12 to 16 segments (Figure 251).

The eggs of *Phlebotomus*, laid in batches of about 50, are deposited in dark, moist crevices of rocks or concrete walls, damp cracks in shaded soil, caves, or other such places where there is sufficient moisture. In about a week, the eggs hatch into larvae, which after about a month transform into naked pupae. The pupal stage lasts about 10 days. Under favorable conditions, the life cycle is completed in about 2 months. The flies are active at night only and hide during the day in cool, damp places, such as masonry cracks, stone walls, animal burrows, and hollow trees.

Sand flies are vicious biters and carriers of at least 3 serious disease agents of man, although none of the species of *Phlebotomus* occurring in the United States is a carrier. The dangerous verruga or Oroya fever of South America is transmitted principally by *Phlebotomus verrucarum*, and sand-fly fever of the Mediterranean region, Near East, southern China, Ceylon, and India is transmitted by *P. papatasii*. Kala-azar, a leishmaniasis endemic to the Mediterranean region, Near East, southern China, Ceylon, India, and Central and South America, is transmitted by various species of *Phlebotomus*. Only the females suck blood.

Family Heleidae

These are minute flies, seldom exceeding 2 to 3 mm in length, that are called punkies, biting midges, or no-see-ums. The wings are more or less covered with erect hairs and, in many species, are marked with pale spots on a dark background (Figure 252).

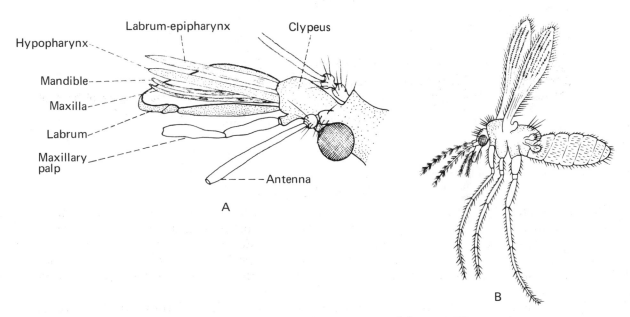

Figure 251. *Phlebotomus,* a sand fly. (A) Mouth parts of female; (B) Adult female.

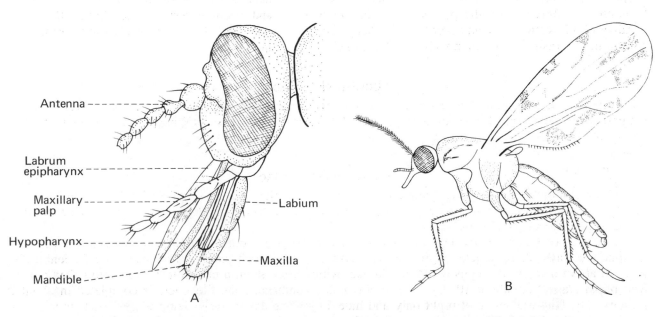

Figure 252. *Culicoides,* a biting midge. (A) Head showing mouth parts; (B) Adult female.

The eggs of *Culicoides,* several hundred in number, are laid in gelatinous masses and are usually anchored to some underwater object. The larvae, which emerge after a few days, are aquatic or semiaquatic, whitish, with a small brown head and 12 body segments. The slender brown pupae have 2 short breathing tubes on the thorax. They float nearly motionless in a vertical position, the breathing tubes in contact with the surface film. The life cycle requires from 6 to 12 months.

The great majority of the species which attack man belong to *Culicoides.* Only the females suck blood. They become active in the evening and early morning, but if disturbed many will bite in the shade, even on bright days. The bites can be extremely irritating, and often they produce delayed reactions. They have been shown to be quite aggressive on occasion, crawling into the hair and beneath articles of clothing to obtain a blood meal. However annoying, they are not known to carry human disease agents in this country. Various species of *Culicoides* serve as vectors of 3 filarial worms of man: *Dipetalonema perstans* in parts of Africa and eastern South America; *D. streptocerca* in Africa; and *Mansonella ozzardi* in the British West Indies. In addition, at least 3 filariae of animals, *Onchocerca reticulata* of horses, *O. gibsoni* of cattle, and *Icosiella neglecta* of amphibians, are transmitted by species of *Culicoides* (see Nelson, 1964; 1966).

Family Simuliidae

These are the small (1 to 5 mm long) dark colored diptera, known as blackflies. The thorax is humped over the head and the proboscis is short. The antennae are many jointed (10 or 11), but are short, with beadlike segments instead of being long and filamentous as in some other Nematocera. The wings are broad and unspotted. They have no scales and they are not hairy, except for bristles on the anterior veins (Figure 253).

The shiny eggs are laid in masses of several hundred on stones and plants in flowing streams. They usually hatch in a few days and the larvae attach themselves to objects in the water, from which they gather food with a set of motile brushes around the mouth. After about a month the larvae spin weblike cocoons, open at the anterior end for the extrusion of the branching gill filaments used for respiration under the water, and pupate in them. The adults emerge in a few days to a week or more and are carried to the surface by a bubble of air which has been collected inside the pupal case. The life cycle requires from 60 days to 15 weeks or longer, and the number of generations per year, in temperate regions, is

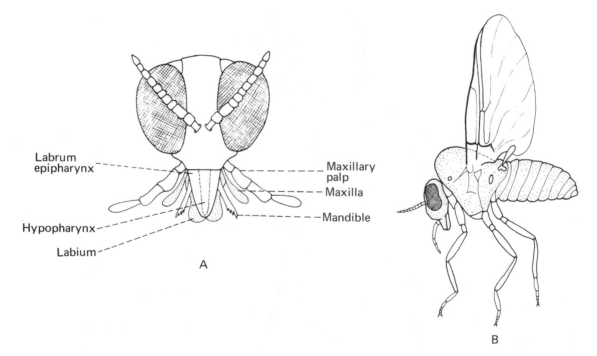

Figure 253. *Simulium,* a blackfly. (A) Head showing mouth parts; (B) Adult female.

from 1 to 5 or 6, depending upon the species and climatic conditions. The flies are active mainly during the daylight hours.

Few insects are more capable tormentors of man and animals than some species of blackflies. One who has never been the victim of their bites cannot fully appreciate the suffering which accompanies and follows them. The flies often appear suddenly in great clouds, and on such occasions are able, within a few hours, to cause the death of large animals and even man if he is unprotected. Only the females suck blood. Certain species of *Simulium* are vectors of the filarial worm *Onchocerca volvulus* which infects man in certain parts of Mexico, Central America, and Africa. Blackflies also transmit the blood protozoans, *Leucocytozoon simondi* of young ducks and geese in northern United States and Canada, and *L. smithi* of turkeys in various parts of the United States. The chief vectors are *Simulium rugglesi* and *S. occidentale*.

Family Culicidae

This family includes the mosquitoes, the Diptera with scales along the wing veins and a prominent fringe of scales along the margin of the wings. The antennae, of 14 or 15 segments, are conspicuous and can usually be used to separate the sexes. In females, they are long and slender with a whorl of a few short hairs at each joint, whereas in the male they have a feathery appearance due to tufts of long and numerous hairs at the joints. Mosquitoes are such a diversified group that it is not possible to write a short general description of the adults which would be generally applicable (Figure 254). Foote and Cook (1959) give excellent descriptions and figures of larvae and adults of mosquitoes of medical importance. The family may be divided into 2 subfamilies, the Anophelinae (including *Anopheles*) and the Culicinae (including *Culex*, *Aedes* et al.).

The eggs are usually laid on or in water or in areas subject to flooding at some later time, and each species has its special requirements, which are usually very restricted. *Aedes* and *Psorophora* lay their eggs singly out of water, in places likely to be flooded later; *Anopheles* deposit them singly on the surface of the water; and *Culex*, *Culiseta*, *Mansonia*, and *Uranotaenia* lay them in small boat-shaped rafts on the surface of the water, or attach them to submerged objects.

Eggs laid on water hatch in a few days; but those deposited out of water or subjected to desiccation

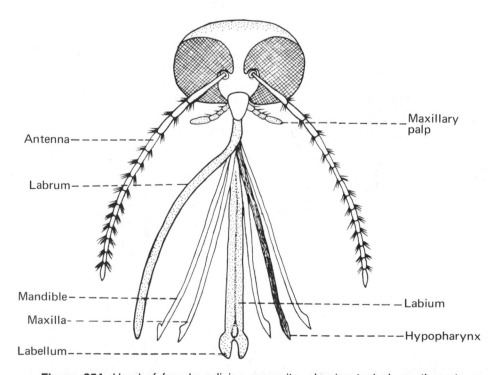

Figure 254. Head of female culicine mosquito, showing typical mouth parts.

do not hatch until submerged, which may be weeks or months; even then only a few of them normally hatch at the first flooding, but some remain for the second and subsequent inundations.

The larvae, which are always aquatic, are well known as wrigglers. As they feed and grow, they pass through 4 instars, shedding their skins between each. Since the hairs and other structures which are used in identifying the larvae usually differ between instars, descriptions of mosquito larvae and keys for identifying them usually refer only to the full grown or fourth stage larvae. Anophelinae larvae are readily recognized in life by the fact that they have no elongate air tube and lie just under and parallel to the water surface, whereas in Culicinae larvae the air tube is well developed and the body hangs downward at an angle of about 45 degrees, touching the surface film only with the end of the air tube. Species of *Mansonia* absorb air from the air-carrying tissues in the roots of certain aquatic plants, piercing them with the tip of the breathing tube.

After 5-14 days or longer, the fourth-instar larva molts and transforms into a pupa or tumbler. The pupal stage is quite short, usually 2 or 3 days. Within the pupa, a marked transformation is occurring; new structures are forming which will adapt the mosquito to terrestrial life. Eventually the adult within emerges through a slit on the back of the thorax, spreads and dries its wings while clinging to the old pupal case on the surface of the water, and then flies off to mate and feed.

Some mosquitoes, particularly those which breed in relatively permanent water, have several generations each year, the number depending upon water temperatures and the length of the summer season. Others, which breed in temporary rain pools, appear each time their breeding places are flooded; while still others, such as the early spring *Aedes*, produce but one brood each year.

Some species bite at all hours of the day and night; others, such as the Anophelinae, are principally night biters; still others, such as *Aedes aegypti*, usually bite in the daytime. Only the females suck blood; the males feed on plant nectar or other sweet or fermenting substances. Although the females of most species feed on warm-blooded animals, others select cold-blooded animals, such as frogs and snakes. The life span of adult mosquitoes is not well known, but for most species in the southern United States it apparently is only a few weeks during the summer months. In the northern latitudes of this country, the females of *Culex* and *Anopheles* usually hibernate, and may live for 6 months or more.

Mosquitoes probably have had a greater influence on human health throughout the world than any other insect. This is not due wholly to the important human diseases and infections they transmit, but also to the severe annoyance they cause man and animals. They are vectors of the causitive organisms of malaria, yellow fever, dengue, and filariasis, 4 of the most important diseases of the tropical and subtropical regions of the world. *Anopheles, Aedes,* and *Culex* are genera of particular medical importance. Plasmodia, the cause of human malaria, are transmitted only by species of *Anopheles,* and the virus causing dengue only by species of *Aedes,* primarily *A. aegypti* and *A. albopictus. Aedes aegypti* also transmits the yellow fever virus from man to man under urban conditions, but in tropical rain forests, species of other genera are also involved in the transmission of the infection from animal to animal, and incidentally to man. The filarial worms, *Wuchereria bancrofti* and *Brugia malayi,* are transmitted by species of several genera. Today these diseases have been reduced to minor or historical importance in the United States. But epidemics of 3 types of human encephalitides continue to occur in many parts of this country and are the most important mosquito-borne disease in the United States. Species of several genera, among them *Culex tarsalis, C. pipiens,* and *Culiseta melanura,* serve as vectors.

Order Hemiptera

The members of this order have a jointed fleshy beak attached anteriorly; when not in use it is flexed backwards beneath the head. The mouth parts are of the piercing-sucking type. The winged members of the order have each of the front wings modified into a thickened basal portion, and a membranous distal portion. The membranous portions often overlap. The second pair of wings are membranous and folded beneath the front wings. Metamorphosis is incomplete. Two families are of interest to parasitologists.

Family Reduviidae

Members of the family have a short, 3-segmented beak which is attached to the tip of the head. The antennae are 4-segmented and the anterior portion of the head is elongate and coneshaped (Figure 255).

The reduviids usually inhabit the burrows or nests of animals, but a few feed on man and his domestic animals. The eggs are glued into cracks and crevices in houses or in the nests of their hosts. They are deposited singly or in small batches, each female ultimately laying about 20. After an incubation period of 8 days to a month, the first instar nymphs emerge. A blood meal is necessary before each molt, and the time between instars is about 40 to 50 days. The life cycle requires a year or more.

Species of several genera are important vectors of *Trypanosoma cruzi*, the hemoflagellate causing Chagas' disease, which occurs throughout South and Central America, especially Brazil, Argentina, and Mexico. In addition to *Panstrongylus megistus*, the most important natural vectors are *Triatoma infestans* and *Rhodnius prolixus*. The adults of both sexes attack man.

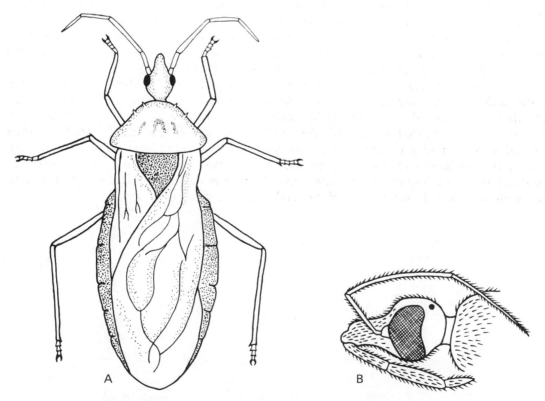

Figure 255. *Panstrongylus megistus*, a reduviid bug. (A) Adult, dorsal view; (B) Head, lateral view.

Family Cimicidae

In this family the wings are vestigial, the hind wings being absent while the fore wings are reduced as small pads. The body is broad, the prothorax large; its concave anterior border receives the head. The antennae are 4-segmented; the beak and the tarsi are 3-segmented (Figure 256).

The eggs are laid at the rate of about 2 to 8 each day, glued by secretion in cracks and crevices and under wall paper. Each female may average 100 to 250 eggs during her lifetime. In warm weather, the eggs usually hatch in about 8 days; at lower temperatures it may take as much as 30 days. The young nymphs, which must have a blood meal between instars, molt 5 times before becoming adults. The life cycle requires from 7 to 10 weeks.

Man is attacked by the common bedbug *Cimex lectularius*, which may become an important pest in living quarters of all kinds. They are most frequently encountered in sleeping quarters, hiding during the day in cracks and crevices of woodwork, furniture, and debris and emerging at night to suck blood from their victims. Despite the fact that *C. lectularius* can transmit experimentally the microbial agents of certain diseases, there is no convincing evidence that it is a vector of any human or animal diseases in nature.

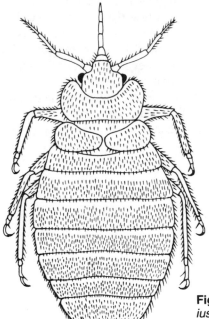

Figure 256. Dorsal view of adult female bedbug, *Cimex lectularius*.

CLASSIFICATION

The list includes only the commonly used higher taxa of the phylum Arthropoda containing species of parasitological importance treated herein.

Class	Subclass	Order	Suborder	Family
CRUSTACEA	Copepoda	Caligidea		Lernaeidae
	Branchiura	Argulidea		Argulidae
ARACHNIDA		Acarina	Ixodides	Ixodidae
				Argasidae
			Sarcoptiformes	Sarcoptidae
				Psoroptidae
			Trombidiformes	Demodicidae
			Mesostigmata	Dermanyssidae
PENTASTOMIDA		Porocephalida		Porocephalidae
INSECTA		Siphonaptera		Ceratophyllidae
				Leptopsyllidae
				Pulicidae
				Tungidae
		Anoplura		Hoplopleuridae
				Pediculidae
				Haematopinidae
				Linognathidae
		Mallophaga	Amblycera	Menoponidae
				Gyropidae
			Ischnocera	Trichodectidae
				Philopteridae
		Diptera	Cyclorrhapha	Muscidae
				Hippoboscidae
			Brachycera	Tabanidae
			Nematocera	Psychodidae
				Heleidae
				Simuliidae
				Culicidae
		Hemiptera		Reduviidae
				Cimicidae

Some Vertebrate Diseases and Infections Together With the Arthropods Important in Their Transmission

Diseases/infections	Hosts	
	Vertebrates	Arthropods
SPIROCHAETES		
Louse-borne relapsing fever	Man	ANOPLURA
		Pediculus humanus
Tick-borne relapsing fever	do.	ACARINA
		Ornithodoros
RICKETTSIAE		
Q fever	Man, cattle et al.	*Dermacentor* et al.
Rocky Mountain spotted fever	Man	do.
Rickettsial pox	Man, mice	*Allodermanyssus*
Scrub typhus	Man, rodents	*Trombicula*
Epidemic typhus	Man	ANOPLURA
		Pediculus humanus
Murine typhus	Man, rodents	SIPHONAPTERA
		Xenopsylla cheopis
BACTERIA		
Plague	do.	*X. cheopis* et al.
Tularemia	Man, rabbits	DIPTERA
		Chrysops
		ACARINA
		Haemaphysalis
		Dermacentor
VIRUSES		
Colorado tick fever	Man	*Dermacentor andersoni*
		DIPTERA
St. Louis encephalitis (SLE)	Man, birds	*Culex*
Western encephalitis (WE)	Man, horses, birds	do.
Eastern encephalitis (EE)	do.	*Culiseta melanura* et al.
Yellow fever	Man, monkeys	*Aedes aegypti* et al.
Sand-fly fever	Man	*Phlebotomus*
PROTOZOA		
Leishmania spp.	Man et al.	do.

Parahaemoproteus nettionis . . . . Ducks, geese . . . . *Culicoides* sp.

Haemoproteus spp. . . . . Birds . . . . *Lynchia* et al.

Leucocytozoon spp . . . . .do. . . . . *Simulium*

Plasmodium spp . . . . Man et al. . . . . *Anopheles* et al.

Trypanosoma spp . . . . .do. . . . . *Glossina* et al.

HEMIPTERA

Triatoma

Babesia spp . . . . Cattle et al . . . . ACARINA

Haemaphysalis

Dermacentor

Boophilus

TREMATODA

Prosthogonimus macrorchis . . . . Chickens et al . . . . ODONATA

Leucorrhinia et al.

Haematoloechus medioplexus . . . . Frogs . . . . *Sympetrum*

Paragonimus spp . . . . Man, mink . . . . DECAPODA

Cambarus et al.

Dicrocoelium dendriticum . . . . Sheep et al. . . . . HYMENOPTERA

Formica fusca

CESTODA

Anoplocephalidae . . . . Chiefly mammals . . . . ACARINA

Oribatid mites

Diphyllobothrium spp . . . . Man et al. . . . . COPEPODA

Diaptomus

Cyclops

Spirometra spp . . . . Cat et al. . . . . do.

Proteocephalus ambloplitis . . . . Bass et al. . . . . do.

Choanotaenia infundibulum . . . . Chickens et al. DIPTERA

Musca domestica et al.

Hymenolepis carioca . . . . Domestic fowl et al. . . . . COLEOPTERA

Scarabaeidae

Dipylidium caninum . . . . Man, dog, cat . . . . MALLOPHAGA

Trichodectes

SIPHONAPTERA

Ctenocephalides et al.

Hymenolepis diminuta . . . . Man, mice . . . . *Xenopsylla cheopis* et al.

MALLOPHAGA

Trichodectes

LEPIDOPTERA

Pyralis et al.

COLEOPTERA

Asopis

ACANTHOCEPHALA
Macracanthorhynchus hirudinaceus......................Swine.......................*Cotinus*
                                                                                 Phyllophaga
Leptorhynchoides thecatus.................................Fish..................AMPHIPODA
                                                                                 Hyalella
Pomphorhynchus bulbocolli ....Sucker et al.............................do.
Prosthorhynchus formosus...............................Birds .... ISOPODA
                                                          Armadillidium vulgare et al.
Moniliformis moniliformis .... Rat ....ORTHOPTERA
                                                                                 Periplaneta
Mediorhynchus grandis .... Birds ..... *Schistocerca americana* et al.
Neoechinorhynchus cylindricum ....Fish ....OSTRACODA
                                                                                 Cypria

NEMATODA
Dracunculus medinensis ....Man .... COPEPODA
                                                                                 Cyclops
Philonema oncorhynchi ....Salmonid fish ....do.
Ascarops strongylina ....Swine ....COLEOPTERA
                                                                                 Aphodius et al.
Litomosoides carinii................................Cotton rat ....ACARINA
                                                                                 Ornithonyssus bacoti
Habronema megastoma .... Horse .... DIPTERA
                                                                                 Musca
Wuchereria bancrofti ....Man .... *Anopheles*
                                                                                 Aedes
                                                                                 Culex
Brugia malayi................................. Man et al.................... *Anopheles*
                                                                                 Mansoni
Loa loa ....Man .... *Chrysops*
Onchocerca volvulus .... do .... *Simulium*
Mansonella ozzardi ....do. .... *Culicoides*
Dirofilaria immitis .... Dog ....*Anopheles* et al.
D. scapiceps... Hare ....*Aedes*
Dipetalonema arbuta .... Porcupine ....do.
D. perstans............................... Man, monkey .... *Culicoides*
D. reconditum ....Dog .... SIPHONAPTERA
                                                                                 Ctenocephalides

REFERENCES

Beesley, W.N. 1973. Control of Arthropods of Medical and Veterinary Importance. *In* Advances in Parasitology, ed. B. Dawes. Academic Press, New York, vol. 11, pp. 115-192.

Horsfall, W.R. 1962. Medical Entomology: Arthropods and Human Diseases. Ronald Press Company, New York, 467 p.

James, M.T., and R.F. Harwood. 1969. Herms' Medical Entomology, 6th ed. Macmillan Company, New York, 484 p.

Matheson, R. 1950. Medical Entomology, 2nd ed. Comstock Publishing Associates, Ithaca, New York, 612 p.

Philip, C.B., and W. Burgdorfer. 1961. Arthropod Vectors as Reservoirs of Microbial Disease Agents. *In* Annual Review of Entomology, ed. E.A. Steinhaus and R.F. Smith. Annual Reviews, Inc., Palo Alto, Cal., vol. 6, pp. 391-412.

Smart, J. 1956. A Handbook for the Identification of Insects of Medical Importance. British Museum (Natural History), London, 303 p.

Yamaguti, S. 1963. Parasitic Copepoda and Branchiura of Fishes. Interscience Publishers, New York, 1104 p.

Copepoda

Bird, N.T. 1968. Effects of mating on subsequent development of a parasitic copepod. J. Parasitol. 54: 1194-1196.

Grabda, J. 1958. Developmental cycle of *Lernaea cyprinacea* L. Wiad. Parazytol. 4: 634-636.

Haley, A.J., and H.E. Winn. 1959. Observations on a lernaean parasite of freshwater fish. Trans. Amer. Fish. Soc. 88: 128-129.

Tidd, W.M. 1934. Recent infestations of goldfish and carp by the "anchor parasite," *Lernaea carassii.* Trans. Amer. Fish. Soc. 64: 176-180.

Branchiura

Cressey, R.F. 1972. The genus *Argulus* (Crustacea: Branchiura) of the United States. Biota of Freshwater Ecosystems, Washington, D.C., No. 2, 14 p.

Meehean, O.L. 1940. A review of the parasitic Crustacea of the genus *Argulus* in the collections of the United States National Museum. Proc. U.S. Nat. Mus. 88: 459-522.

Wilson, C.B. 1944. Parasitic copepods in the United States National Museum. Proc. U.S. Nat. Mus. 94: 529-582.

Acarina

Arthur, D.R. 1962. Ticks and Disease. Row, Peterson and Company, Evanston, 445 p.

_____. 1965 & 1970. Feeding in Ectoparasitic Acari with Special Reference to Ticks. *In* Advances in Parasitology, ed. B. Dawes, Academic Press, New York, vol. 3, pp. 249-298; vol. 8, pp. 275-292.

Baker, E.W., and G.W. Wharton. 1952. An Introduction to Acarology. The Macmillan Company, New York, 465 p.

Baker, E.W., T.M. Evans, D.J. Gould, W.B. Hull, and H.L. Keegan. 1956. A Manual of Parasitic Mites of Medical or Economic Importance. Nat. Pest Control Assoc. Tech. Publ., New York, 170 p.

Cooley, R.A. 1938. The Genera *Dermacentor* and *Otocentor* (Ixodidae) in the United States, with Studies in Variation. Nat. Inst. Health Bull. No. 171, 89 p.

Spickett, S.G. 1961. Studies on *Demodex folliculorum* Simon (1842). Parasitology. 51: 181-192.

U.S. Department of Agriculture. 1965. Manual on Livestock Ticks. Agr. Res. Service 91-49, 142 p.

Pentastomida

Esslinger, J.H. 1962. Development of *Porocephalus crotali* (Humboldt, 1808) (Pentastomida) in experimental intermediate hosts. J. Parasitol. 48: 452-456.

Self, J.T. 1969. Biological relationships of the Pentastomida; a bibliography on the Pentastomida. Exptl. Parasitol. 24: 63-119.

_____, and B.F. McMurray. 1948. *Porocephalus crotali* Humboldt (Pentastomida) in Oklahoma. J. Parasitol. 34: 21-23.

Siphonaptera

Ewing, H.E., and I. Fox. 1943. The fleas of North America. U.S. Dept. Agr., Misc. Publ. No. 500, 142 p.

Holland, G.P. 1949. The Siphonaptera of Canada. Canadian Dept. Agr., Pub. 817, Tech. Bull. 70, 306 p.

Jellison, W.L. 1959. Fleas and Disease. *In* Annual Review of Entomology, ed. E.A. Steinhaus and R.F. Smith. Annual Reviews, Inc., Palo Alto, Cal., vol. 4, pp. 389-414.

_____, B. Locker, and R. Bacon. 1953. A synopsis of North American fleas, north of Mexico, and notice of a supplementary index. J. Parasitol. 39: 610-618.

Lice (Anoplura and Mallophaga)

Ferris, G.F. 1951. The sucking lice. Pacific Coast Ent. Soc. Mem. 1: 1-320.

Matthysse, J.G. 1946. Cattle lice, their biology and control. Bull. No. 832, Cornell Univ. Agric. Exp. Sta., Ithaca, New York, 67 p.

Rothschild, M., and T. Clay. 1952. Fleas, Flukes and Cuckoos: A Study of Bird Parasites. Collins, London, 304 p.

Weyer, F. 1960. Biological Relationships Between Lice (Anoplura) and Microbial Agents. *In* Annual Review of Entomology, ed. E.A. Steinhaus and R.F. Smith. Annual Reviews, Inc., Palo Alto, Cal., vol. 5, pp. 405-420.

Muscidae

Glasgow, J.P. 1963. The Distribution and Abundance of Tsetse. Pergamon Press, New York, 241 p.

_____. 1967. Recent fundamental work on tsetse flies. *In* Annual Review of Entomology, ed. R.F. Smith and T.E. Mittler. Annual Reviews, Inc., Palo Alto, Cal., vol. 12, pp. 421-438.

Hippoboscidae

Bequaert, J. 1942. A monograph of the Melophaginae, or ked-flies, of sheep, goats, deer and antelopes (Diptera, Hippoboscidae). Entomol. Americana 22: 1-64.

Bishopp, F.C. 1929. The pigeon fly, an important pest of pigeons in the United States. J. Econ. Entom. 2: 974-980.

Tabanidae

Anthony, D.W. 1962. Tabanidae as disease vectors. *In* Biological Transmission of Disease Agents, ed. K. Maramorosch. Academic Press, New York, pp. 93-107.

Hibler, C.P., and J.L. Adcock. 1971. Elaeophorosis. *In* Parasitic Diseases of Wild Mammals, ed. J.W. Davis and R.C. Anderson. Iowa State Univ. Press, Ames, Iowa, pp. 263-278.

James, M.T. 1947. The Flies that Cause Myiasis in Man. U.S. Dept. Agr. Misc. Publ. 631, 175 p.

Psychodidae

Adler, S., and O. Theodor. 1957. Transmission of disease agents by phlebotomine sandflies. *In* Annual Review of Entomology, ed. E.A. Steinhaus and R.F. Smith. Annual Reviews, Inc., Palo Alto, Cal., vol. 2, pp. 203-226.

Heleidae

Foote, R.H., and H.D. Pratt. 1954. The *Culicoides* of the Eastern United States (Diptera, Heleidae), a Review. Public Health Monogr., No. 18, 53 p.

Kettle, D.S. 1965. Biting ceratopogonids as vectors of human and animal diseases. Acta Trop. 22: 356-362.

Simuliidae

Dalmat, H.T. 1955. The black flies (Diptera, Simuliidae) of Guatemala and their role as vectors of onchocerciasis. Smithsonian Misc. Collections, (Publ. 4173), vol. 125, No. 1, 425 p.

Fallis, A.M. 1964. Feeding and related behavior of female Simulidae (Diptera). Exptl. Parasitol. 15: 439-470.

Culicidae

Foote, R.H., and D.R. Cook. 1959. Mosquitoes of Medical Importance. U.S. Dept. Agr., Agr. Handbook No. 152, 158 p.

Horsfall, W.R. 1955. Mosquitoes: Their Bionomics and Relation to Disease. Ronald Press Company, New York, 723 p.

Mattingly, P.F. 1970. The Biology of Mosquito-Borne Diseases. American Elsevier Publishing Co., New York, 184 p.

Reduviidae

Usinger, R.L. 1944. The Triatominae of North and Central America and the West Indies and their public health significance. U.S. Public Health Bull. No. 288, 83 p.

Cimicidae

Usinger, R.L. 1966. Monograph of Cimicidae (Hemiptera-Heteroptera). Thomas Say Foundation (Ent. Soc. Amer.), vol. 7, 585 p.

SECTION IV

TECHNIQUE PROCEDURES

Standardization of the Microscope

The calibration of the microscope is required for the measurement of microscopic objects. In reality, however, it is the calibration of the ocular scale, used in the microscope, combined with the different objectives. In the calibration, 2 scales are necessary; the ocular micrometer, which is a glass disc bearing an arbitrary scale of usually 100 divisions, and a stage micrometer which is a slide etched with a known scale of 2 mm usually subdivided into units of 0.1 mm and 0.01 mm. For a few remaining microscopes having an adjustable draw tube, care should be taken to see that the tube is set at 160 mm when doing the calibration and making measurements. When working with the microscope, both eyes should be kept open. With a little practice, one soon learns to ignore what the unused eye sees. This method avoids eye strain.

Remove the regular microscope ocular and substitute an ocular equipped with an ocular micrometer disc. If an extra eyepiece with a disc is not available, unscrew the ocular mount and place the ocular micrometer with the engraved side down (figures should appear erect) over the eyepiece diaphragm. Replace the mount and reinsert the ocular in the microscope tube. With the aid of the plane mirror and the iris diaphragm adjust the microscope so that optimum lighting is obtained, for only then will the scales be sharp and clear. Place the stage or slide micrometer on the microscope stage and adjust the 2 scales, with the zero points of both coinciding, so that they are parallel. Reading is easier if 1 scale is slightly to 1 side of the other rather than being superimposed. A similar point where lines coincide should now be found at the extreme right. The farther the matching lines are from the zero point, the less will be the error. Count the number of lines (hundredths-of-mms) on the stage micrometer (SM) and the number of lines on the ocular micrometer (OM) between the zero point and where the lines coincide exactly. Then divide the number of SM lines by the number of OM lines and you will have the value of one OM unit, with that objective. There are 1,000 micra in a millimeter.

STAGE MICROMETER

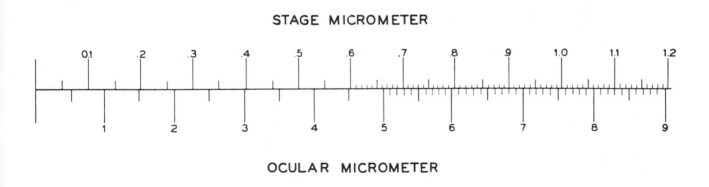

OCULAR MICROMETER

Figure 257. Example of microscope calibration.

In the accompanying example, 83 divisions on the ocular micrometer scale are equal to 110 small divisions or 1.10 mm on the stage micrometer. By dividing the number of SM lines by the number of OM lines the value of OM unit is obtained, e.g.,

$$\frac{SM}{OM} = \text{value for each OM division, or } \frac{1.10}{83} = 0.0132 \text{ mm.}$$

257

Inasmuch as there are 1,000 micra in one mm, it follows that 1 ocular unit equals 13.2 micra (0.0132 x 1,000). Two ocular units = 26.4 micra, five ocular units = 66.0 micra, etc.

Following the above procedure, calibrate in order the ocular with the low, high dry, and oil immersion objectives. To lessen the error, make 5 different readings with each objective and take the average. It must be remembered that the calibration thus obtained is accurate only when the same microscope, the same ocular, and the same objectives are used.

Table II. MICROSCOPE CALIBRATION

Microscope No._____   Ocular Micrometer No._____   Oil Imm. Obj. No._____

Objectives		No. OM spaces	No. SM spaces in mm	μ value of one OM space	Average
low 16 mm (10 X)	1				
	2				
	3				
	4				
	5				
high 4 mm (43 X)	1				
	2				
	3				
	4				
	5				
oil 1.8 mm (97 X)	1				
	2				
	3				
	4				
	5				

Suggestions for Obtaining Fresh Material

The use of fresh material which may be available locally is too frequently neglected in favor of prepared slides and other fixed materials. Nothing will serve to stimulate the student and vitalize the course more than actual contact with living specimens. Fixed and stained material is essential in the identification of certain parasites (notably plasmodia), to illustrate forms unavailable in the living state, and to supplement the study of viable material. It seems unwise, however, to use only fixed and stained material to illustrate certain stages which the student will rarely, if ever, see in such condition thereafter. However valuable a stained smear of *Entamoeba histolytica* may be to illustrate the morphology of the trophozoite, it is much less important than the viable trophozoites of any parasitic amebae in the training of students. Actually the student, in his future contacts, will be called upon to recognize the forms in feces, probably without the facilities or opportunity for making stained preparations. One should be mindful that a purely anatomical approach to parasitism is not only sadly stereotyped, often becoming tiresome to the student, but it fails to stimulate students to think in terms of basic parasitological principles and encourage further elucidation of the adaptive properties of parasites and host-parasite relations. Therefore, in conjunction with laboratory studies on the digenetic trematodes it is always interesting and advisable to allow students to observe miracidia hatching from the egg, infecting a snail, and, if possible, recover the sporocysts, rediae, and cercariae. As an alternative project, students may examine local or imported snails for infections. Instructions for examination begin on p. 278.

What has been emphasized here for *E. histolytica* and the larval stages of trematodes holds with varying degrees for other forms studied. Emphasis should be placed on the fact, however, that all forms encountered cannot always be correctly identified in the living state and, if possible, stained preparations should be made for the purpose of positive identification. More individual effort is required on the part of the instructor to provide living material, but the reward is worth the extra effort. This is not to say that living materials should supplant prepared mounts; on the contrary, they are necessary to satisfy basic technical needs of medical technology, fisheries, and wildlife students.

In view of the proven value of fresh material, several generally available hosts which normally harbor a rich parasitic fauna will be used as examples. Some of the parasites discussed here are listed again along with others under their respective taxa, together with their hosts.

Frogs (*Rana* spp.), generally available locally or from dealers, and routinely used in several zoology courses, are particularly useful for this purpose since they harbor a variety of parasites. Following the instructions given below under Internal Examination (p. 261), remove the lungs and tease open in physiological solution. Two helminths are commonly found here. One, a pale grey trematode, is a species of *Haematolechus*; the other is the nematode *Rhabdias ranae*. If removed to a slide in distilled water, both forms will be observed to discharge eggs in enormous numbers. In the case of *R. ranae*, the eggs will frequently be seen to hatch, releasing small, writhing larvae. Through manipulating the iris diaphragm to reduce the light of the compound microscope, one may readily study the internal anatomy of both larvae and adults. An examination of a blood smear may reveal the hemoflagellate *Trypanosoma rotatorium*, although seldom present in great numbers. The blood from kidney and liver furnishes more specimens than the peripheral blood.

Another likely source of parasites is the intestine, which should be removed, slit open, and examined in physiological solution. A drop of the solution examined under the compound microscope will often show many larvae and adult organisms. If *Rhabdias ranae* is in the lungs, its larvae and eggs will probably be present in the intestine, as will the eggs of flukes. The eggs can be readily distinguished by the color, shape, and the presence of larvae in the nematode eggs. Moving swiftly about may be seen a small, ovoid protozoan with a sharply pointed end, an undulating membrane, and several flagella which are actively lashing about. This is probably *Tritrichomonas batrachorum*. Three large ciliate protozoa are commonly present: *Opalina obtrigonoidea*, flattened in cross section, roughly spindle-shaped and curved, with many cilia; *Balantidium entozoon*, ovoid, with terminal oral groove, smaller than *Opalina* and of interest because of its relationship to the human *B. coli*; and *Nyctotherus cordiformis*, which resembles *Balantidium*, but may be distinguished from it by the prominent oral groove which traverses the body obliquely.

In addition, the following distome trematodes occur in the indicated locations: *Glyphthelmins quieta*, duodenum and adjacent intestine; *Loxogenes arcanum*, bile duct and adjacent region; *Cephalogonimus americanus*, small intestine; a species of *Halipegus*, the eustachian tube and oral cavity; and *Megalodiscus temperatus*, an amphistome, the large intestine. *Cylindrotaenia americana*, a tapeworm, is also occasionally found in the small intestine.

Two kinds of distomes belonging to the genera *Gorgodera* and *Gorgoderina* occur in the urinary bladder. These are characterized by a much-enlarged ventral sucker, with which they may adhere firmly to the bladder wall. A monogenetic trematode, *Polystoma nearcticum*, with 6 muscular suckers and a pair of large hooks on the opisthaptor, occurs in the urinary bladder of the tree frog *(Hyla versicolor)*.

Another widely distributed, generally abundant, expendable, and fairly easily obtained host which yields a variety of parasitic species is the white sucker *(Catostomus commersoni)*. On the gills, one can expect to find the polystome *Octomacrum lanceatum* and the leech *Actinobdella triannulata*, and attached in the branchial chamber or on the body surface the fish louse *Argulus catostomi*. The nematode, *Philometroides nodulosa*, may be observed beneath the subcutaneous tissue of the cheek and fins. An examination of the opened intestine is likely to yield 1 or more species of trematodes belonging to the genus *Triganodistomum*, monozoic tapeworms of the genus *Glaridacris*, and the acanthocephalans *Pomphorhynchus bulbocolli*, *Neoechinorhynchus cylindratum*, and *Octospinifera macilentus*.

The muskrat *(Ondatra zibethicus)*, which also has a wide distribution, is unexcelled as a mammalian host for class use. Fresh carcasses, usually available in season from trappers, are small and clean, and yield a wide variety of helminths very suitable for student study and preparation. The following species, among others, in the intestine unless otherwise indicated, from the various groups are commonly recovered: TREMATODES—(1) monostomes, *Catatropis fimbriata*, *Notocotylus filamentis*, *Nudacotyle novica*, *Paramonostomum echinum*, *Quinqueserialis quinqueserialis*; (2) echinostomes, *Echinoparyphium recurvatum*, *Echinostoma revolutum*; (3) amphistome, *Wardius zibethicus*; (4) schistosome (blood fluke) *Schistosomatium douthitti* in the mesenteric veins; and (5) distomes, *Opisthorchis tonkae* in the gallbladder and bile ducts, and *Plagiorchis proximus*; CESTODES—*Hymenolepis evaginata* and strobilocerci of *Hydatigera* in the serous membranes; and NEMATODES—*Trichuris opaca*.

Host Autopsy and Recovery of Parasites

Whether the animal is a piscine, amphibian, reptilian, avian, or mammalian host, the process of autopsy is essentially the same. Animals should be autopsied as soon after death as possible since some species of worms migrate upon death of the host, with the result that their normal location becomes uncertain. Others undergo changes in the dead host. If the autopsy must be delayed, animals should be kept in a refrigerator until it can be accomplished. The instructor will advise as to the disposal of the carcass following autopsy. Under no circumstances, however, should remains be left in the laboratory waste jar.

1. External Examination

When examining fish, the entire outer surface should be searched carefully, especially the oral region, the gills and opercula, and the fins. On the gills and often on the fins, one finds monogenetic trematodes, larvae of freshwater clams (glochidia), leeches, and fish lice. On the outer surface, especially the fins, may be found encysted digenetic trematodes (metacercariae), as well as leeches and fish lice. Species of certain metacercariae give the entire surface a salt-and-pepper appearance, known as "black spot." Instructions for the recovery of the smaller, microscopic Monogenea are found on p. 278. When examining birds and unskinned mammals, be on the lookout for mites, ticks, fleas, lice, and their eggs. Once collected, small mammals and birds should be promptly placed individually in a paper bag, since some arthropods, especially fleas, leave their host soon after death. The top of the bag is then folded over a few times and secured with paper clips. Upon returning to the laboratory the bag may be opened, a small wad of chloroform- or ether-saturated cotton dropped in, and the bag closed again until the vapors have inactivated the parasites. The bag is then torn open and laid out flat, and the host is examined carefully by brushing anteriorly through the hair or feathers with a pair of forceps. Parasites are picked up, using a toothpick or camel's hair brush previously dipped in alcohol, and transferred to a vial containing preserving fluid. The nares and the oral cavity of waterfowl should be dissected open and examined for "duck" leeches and mites.

2. Internal Examination

Following the external examination, make an incision through the midventral body wall, cutting around the anus and urogenital opening. When working with mammals, prior to making the abdominal incision, wet down the ventral surface with waste alcohol as an aid in preventing excessive contamination of the area with hair. Before removing the viscera, look for parasites that may be free in the body cavity, beneath the peritoneum, or in the mesenteries, and examine the liver surface for spots or nodules (cysts) which might be due to parasites. Encysted flukes can be distinguished by their thin hyaline appearance, through which the larva is often obvious. A cyst that is more or less opaque, nonhyaline, can be suspected of being that of a metacestode. These cysts resemble the host tissue, of which they are composed. Furthermore, metacestodes can be distinguished from those of trematodes which contain numerous, glasslike granules, known as calcareous bodies. Cysts too small to open with dissecting needles should be pressed between slides and examined under the low power of the compound microscope to determine, if possible, the contents. Cnidosporidian cysts may be mistaken for those of flukes or tapeworms unless carefully observed. But the presence of polar capsules, which may be expressed by pressure on the glass cover, and minute pyriform spores, identifies them as Cnidospora. When examining fish, in order to facilitate separating the organs, it is suggested that 1 side be cut away. Now remove the internal organs, 1 at a time, and place them in separate dishes of physiological solution. A temporary label giving relevant data should accompany each organ.

Once the urinary bladder and the segments of the alimentary canal are opened (by making a longitudinal slit with blunt-tipped scissors or scapel) and the other organs teased apart in the physiological solution, the parasites, by their own active movements, usually free themselves from the mucus. If pyloric ceca are present, they should be opened. Tapeworm-infected ceca, due to their whitish, opaque appearance, can be distinguished from worm-free ceca. If the infected ceca are opened at the distal end, by cutting off the tips with scissors, the worms will usually protrude enough so that one can disengage them by placing a dissecting needle or forceps through the exposed loop and pulling gently. The remaining organs should likewise be opened and torn apart by teasing. Not only should every organ and each opened alimentary segment be examined under the dissecting microscope, but the mucosal wall of the intestine should be scraped to loosen small, embedded worms. For the examination of intestinal contents under the dissecting microscope, a dark background reveals the light-colored worms better than a light background. When a parasite is recovered, call it to the attention of the instructor.

Acanthocephala are occasionally found free in the intestinal lumen, but usually the proboscis is embedded in the intestinal wall. Unless removed with care, the proboscis will be broken off and left in the intestinal wall, or its hooks, which are essential for species determination, may be torn loose. A pair of fine-pointed dissecting needles may be inserted into the craterlike opening in the host tissue, and by careful manipulation of the needles the crater may be spread wide enough to free the proboscis of the worm. Or the intestine can be cut into small squares, each holding an embedded proboscis. After these have been preserved, the host tissue may be dissected away before the worms are stained.

If pressed for time or when dealing with a large number of animals, the intestine, having first been opened as described above, should be placed together with its contents in physiological solution and shaken for a minute or so in an appropriately sized jar. The contents should be allowed to settle; then the upper, clearer portion of the liquid is poured off. This process should be repeated several times, with the decanted liquid being replaced each time with fresh solution. The parasites, which are heavier than the organic debris, are the first to settle to the bottom and are concentrated in the remaining liquid. By examining only a small portion of the concentrate at a time in a Syracuse watch glass or petri dish under a dissecting microscope in good light, any parasites remaining will be found. These should be transferred with a spatula, bulbed pipette, or, tweezers if caution is exercised so as not to injure the worms, to a Syracuse watch glass filled with physiological solution.

Preparation of Specimens for Study

1. Fixation

While killing, fixation, and hardening are commonly referred to as separate processes in histology and cytology, when applied to materials to be made into whole mounts nothing is gained by considering them separately since killing and hardening are usually features of fixation. In its broad sense, fixation consists of arresting the life processes, preserving and hardening the animal in as nearly as possible its condition in life. This is accomplished through the use of certain chemicals which are known accordingly as fixing reagents. Fixation is probably the most important step in the preparation of materials for microscope slides, for here the elements are set and cannot be changed.

Preparation—Worms should be thoroughly washed in the physiological solution and cleansed of mucus prior to fixation. After the mucus has been removed, it often proves advantageous to leave digenetic trematodes and cestodes in distilled water for some minutes prior to killing. This serves to relax the specimens further and in some species of flukes causes them to void most of the eggs. Since an egg-filled uterus usually obscures much of the internal anatomy, voidance of eggs is highly desirable. Trematodes and cestodes give considerable trouble by contracting and thickening upon fixation unless precautions are taken. Some workers prefer to place flukes and tapeworms between 2 slides, after which the fixing solution is added. But this is objectionable since the pressure distorts the shape of the animal and disturbs the normal position of the internal organs.

Killing and fixing—Fairly satisfactory results can be obtained if the mucus-free trematodes are dropped into 10% formalin, A-F-A, or Bouin's fluid that is heated until it just starts to bubble (but not boil). Trematodes expand and die instantly upon striking the hot fixing fluid. Cercariae are killed and fixed in a similar manner.

Tapeworms should be plunged into a dish of water heated to about 70 C and agitated with a camel's hair brush or a dissecting needle, or suspended over a hooked needle and dipped repeatedly and rapidly. Since the water is cooled 5 or more degrees when poured into the dish from the beaker, the original temperature must be adjusted accordingly. In the case of specimens with a thick body wall, e.g., *Ligula*, *Schistocephalus*, *Taenia* et al., the water should be heated 10 or more degrees higher. They must be plunged into it quickly so that the whole worms come into contact with water simultaneously, thus killing them in a uniformly extended condition. Worms are then promptly transferred to the fixing solution. For plerocercoids and other delicate material, the water temperature should be 10 or more degrees lower, to avoid overheating and excessive distention. Using a bulbed pipette, transfer such specimens to the hot water, agitate gently and transfer promptly to the fixing solution. Whether dealing with larvae or adults, the proper temperature of the water for killing is determined by trial.

Living Acanthocephala should be placed in water in a petri dish or Syracuse watch glass. For a time, specimens retain the ability to invert the proboscis but ultimately the worms take up water and distend so that the proboscis remains fully extended even when stimulated. After the worms have become immobile, the excess water in the dish is poured or pipetted off, leaving only a thin film covering the specimens. The fixing reagent, A-F-A, 10% formalin, or 70% alcohol to which a few drops of glacial acetic acid have been added, is heated until it just begins to bubble, and then poured over the worms. If the specimens are small enough to enter a bulbed pipette, they may be pipetted directly into the steaming fixing reagent.

Large nematodes should be dropped into steaming acetic alcohol (1 part glacial acetic acid and 3 parts 95% alcohol), which results in a little less shrinkage than alcohol alone. The heated solution causes worms to straighten instantly and die in that position, thus avoiding the curled and distorted specimens obtained when using cold fixatives.

Small nematodes and larval stages may be studied advantageously while still alive by placing them in a drop of water under a cover glass for examination under the microscope. Water mounts sealed with Vaseline will last for days. When permanent mounts are desired, the following procedure for fixation produces lifelike specimens. The nematodes are killed by dropping them into 1 to 2 ml of hot (100 C) 0.5% acetic acid in a cavity slide and then fixed by transferring to a mixture of 4% formalin and 1 or 0.5% acetic acid for 24 hours. Specimens may be stored in the fixative and studied subsequently as water mounts or processed for permanent mounts in glycerine or semipermanent ones in lactophenol.

It is important that leeches be properly prepared or they are practically worthless for study. They should be moderately extended, reasonably straight, undistorted, and neither macerated nor overhardened. This is accomplished by stupefying them before fixation by the careful use of a good narcotizing agent, such as nembutal, tricaine methanesulphonate, or 1 of the other recommended reagents. The living leeches are washed, placed in a small amount of water and a little of the anesthetic added. If the solution is too strong the leeches will die quickly, in a contracted stage. When they no longer respond to stimulation, they are straightened on a slide wrapped with tissue or filter paper resting in a petri dish with just enough water to keep the paper moist so it does not adhere to the specimens. Another piece of paper is laid on them and a small amount of the fixative slowly added. After allowing a few minutes for the leeches to partially harden, fluid is added to completely cover them, care being taken that they do not float. Hot 95% alcohol plus a few drops of glacial acetic acid or 10% formalin are satisfactory fixatives. After stiffening, they should be placed straight and without crowding in tubes or vials of suitable length and finally preserved in generous quantities of 85% alcohol or 5% formalin.

For killing arthropods, Boardman's solution is advised. This has the advantage of relaxing the animals so they die with appendages outstretched, a most important feature to the taxonomist. The solution was originally devised for ticks, but works equally well for mites, fleas, and lice. Boardman's is a relaxant only, so the material must be transferred to a preservative. Fly larvae and mosquito larvae are best killed by dropping them for a few seconds, depending upon size, into water heated almost to boiling. Transfer to insect preservative or 70% alcohol plus glycerine for storage.

2. Preservation and Storage

Smaller worms are usually properly fixed in approximately an hour, but exposure to the fixing reagent for several hours or even days will not impair the value of the specimens and insures complete fixation of the larger specimens.

Specimens fixed in Bouin's solution should be washed in several changes of 70% alcohol until the yellow color is lost. This process may be hastened by the addition of a few drops of ammonium hydroxide to the alcohol. Material fixed in 10% formalin should be removed to 5% formalin. For specimens fixed in 70% alcohol + glacial acetic acid, the solution should be replaced with 70% alcohol. The A-F-A solution should likewise be replaced by 70% alcohol. In each case, worms should be left in the killing solution for some hours (3 to 24) before the solution is replaced by 70% alcohol or 5% formalin (only in case the material has been killed in 10% formalin is it preserved in 5% formalin).

If the material is to be stored for some time (months or longer) before being used, it is recommended that 70% alcohol to which 5% glycerine has been added as a protection against loss from evaporation be used as a preservative instead of straight 70% alcohol. Enough glycerine to make a 5% solution may likewise be added to the formalin preservative.

Important: Each vial of material should contain a note bearing the following information: (1) scientific name of host, (2) locality where host was collected, (3) location of parasite within the host, (4) fixing reagent, (5) date (of autopsy), and (6) your name. Specimens without complete accompanying data are worthless.

3. Staining

The following instructions are based upon methods which have proved satisfactory for laboratory use by experienced workers and students commencing research in parasitology. Experience has shown that students need something more in the way of instructions to guide them through the various methods described in the available texts dealing with microtechnique.

When treating specimens with alcohol or other reagents, always use at least 4 times their bulk of the reagent. Transfer of the specimens from 1 liquid to another is made by pouring off the first liquid, after which the second is added immediately. If the specimens are very small or delicate, rather than pouring off the liquid, carefully withdraw it with bulbed pipette under a dissecting microscope. Afterwards the second liquid is promptly added. In either case the liquid, whether poured or pipetted off, should be placed temporarily in a preparation or a petri dish. If specimens have inadvertently escaped with the liquid, recovery of them can be made from the dish, which is impossible if the solution with the specimen has been discarded in the sink. While large specimens may be transferred with a spatula from 1 container to another, it is difficult or impossible with most materials and always involves the risk of damage to the specimen.

Preparation dishes with ground covers are the most satisfactory containers to use in the treatment of specimens. Although less desirable, screw-cap vials of appropriate size may be used in handling small specimens. Containers should always be tightly covered except when changing solutions or when observing specimens. In order that the source of each lot of material may be known, a small pencil-numbered slip of paper should accompany the specimens while in preparation and the same number should be placed in the vial with any remaining material.

Since most structures are relatively transparent in their natural state, their visibility is usually increased by staining specimens with dyes or other color-bearing chemicals. Inasmuch as the various structures absorb certain dyes to different degrees, it is possible to stain some of them conspicuously while others remain less colored, thus giving them various shades of one color; or even to stain certain organs in contrasting colors by combining different colored dyes which have an affinity for different structures.

In the choice of a stain, consideration should be given to its ability to differentiate clearly the anatomy of the specimen. It should (1) be very dependable and easy to use, (2) follow the common fixing agents well, (3) retain optimum staining properties for many years, and (4) not form unsightly precipitates in the material. While a large number of carmine and hematoxylin stains, two of the most widely used classes, have been described, each with its advocates and adversaries, the following few meet practically all needs for whole mounts of helminths. Carmine stains should be used for alcohol-fixed specimens, and hematoxylin for formalin-fixed material. Nematodes and arthropods do not require staining. Important structures show well, in nematodes mounted in lactophenol or glycerine, and even in water mounts of small living forms. However, specimens well stained with carmine are especially useful. Overstaining specimens and then partially decolorizing them is known as regressive staining, in contradistinction to progressive staining, in which the stain once taken by the specimens is not removed. In progressive staining, differentiation is accomplished through the selective affinity of the dye for different organs.

All trace of the fixing reagent should be removed from specimens prior to staining. Wash out formalin in distilled water; A-F-A and Bouin's in 70% alcohol. As the alcohol becomes discolored by the picric acid in Bouin's, it should be replaced with fresh solution. The decoloring process can be hastened by using alkaline 70% alcohol rather than neutral 70% alcohol.

Semichon's Acetic-Carmine: Transfer specimens from 70% alcohol to stain (1 part stock stain diluted with approximately 2 parts of 70% alcohol). Stain for 15 to 30 minutes, depending upon the size of the specimens. Destain in 70% acid alcohol, until the cortical layer is nearly free of stain and reproductive organs are a very light pink.

Grenacher's Alcoholic Borax-Carmine: Place specimens from 70% alcohol into stain and leave for 24 hours or longer, depending upon the size and permeability of the specimens. After being thoroughly stained they are transferred to the destaining solution and treated as described above.

Lynch's Precipitated Method, Using Grenacher's Alcoholic Borax-Carmine: This method gives a much more selective and brilliant stain than that obtained by Grenacher's original method, when the material is treated as follows:

1. Transfer material from 70% alcohol into undiluted Grenacher's borax-carmine and stain for approximately 12 hours.

2. Cautiously add concentrated HCl drop by drop to the dish containing the material and the stain, meanwhile gently agitating the dish, until all the carmine is precipitated out as a brick red flocculent mass. Generally a drop of HCl is required for approximately each 5 ml of stain. (The volume of a preparation dish with a ground glass cover is about 9 ml, a Syracuse watch glass approximately 20 ml.) Allow the worms to remain in the acidulated solution 6 to 8 hours, or preferably overnight.

3. Next place the material in 70% acid alcohol and destain until a light pink color has been ob-

tained. If so much of the precipitated carmine remains in suspension that the material cannot be observed satisfactorily under the binocular dissecting microscope, draw off the liquid with a bulbed pipette and replace with fresh acid alcohol. Repeat the process until practically all of the precipitated carmine has been removed. The destaining process usually requires several hours, depending, of course, upon the size of the specimen.

4. When the material is properly destained, draw off the acid alcohol and replace with neutral 85% alcohol. Change the alcohol 2 or 3 times, but the total time in 85% should not be less than an hour.

Counterstaining with Fast Green: Fast Green, used progressively as a counterstain for the preparation of whole mounts of specimens previously stained with Semichon's acetic-carmine or Grenacher's alcoholic borax-carmine each used regressively, is especially recommended to bring out the ventral glands of monostomes and spines and/or hooks of appropriate groups.

After the material has been stained, destained, and dehydrated through 95% alcohol, add a few drops (1 or 2) of Fast Green stock solution to the dish filled with 95% alcohol. Observe under the dissecting microscope, removing the material immediately to absolute alcohol when properly stained. This step requires constant observation and in all probability will not require more than a minute in the Fast Green solution. Extreme caution should be exercised since Fast Green is not only color fast but fast acting!

Should the specimen become overstained it may be remedied by destaining with a weak 95% alcohol-alkaline solution, which, however, tends to detract from the sharp, bright green contrast otherwise obtained. After counterstaining, dehydrate further in absolute alcohol, clear and mount in the usual manner.

Harris' Hematoxylin: Although the general procedure is the same as when using the carmines, the hematoxylins differ in one important respect: instead of putting the specimens into the stain from 70% alcohol, as with carmines, they are put into the hematoxylins from water.

Hydrate from 70% alcohol by passing through 50% and 35% alcohols, and leave in distilled water (30 minutes or longer in each), stain in Harris' (stock diluted approximately 1:15 with distilled water) 12 to 24 hours. Rinse in distilled water, to remove the free coloring, and upgrade through 35% and 50% alcohol (30 minutes to an hour in each), preparatory to destaining, as with specimens stained in carmine.

Ehrlich's Hematoxylin: Follow the same procedure outlined for Harris' hematoxylin.

Combination Hematoxylin With a Mordant: The combination hematoxylin is prepared by adding 1 ml each of stock Delafield's hematoxylin and stock Ehrlich's hematoxylin to 40 cc of potassium alum solution. This staining solution should be made as used, following the same procedure outlined for Harris' hematoxylin.

4. Destaining and Neutralization

Unless otherwise indicated above, destaining is accomplished through the use of acidulated 70% alcohol. While the acid alcohol is available for class use, it can be made individually by the addition of a drop or 2 of concentrated HCl in a preparation dish filled with 70% alcohol.

Since the carmine stains are alcoholic, specimens, after being rinsed in 70% can be transferred to the acid alcohol from the stain. It is important that specimens be rinsed only momentarily in 70% alcohol after removal from the stain and before placing them into the destaining solution. Allowing specimens to remain in 70% alcohol for any length of time seems to set the stain, making destaining prolonged and difficult. When hematoxylin has been used, specimens should be rinsed in distilled water and then passed through 35% to 50% alcohol (about 30 minutes in each) before being transferred to the acidulated 70% alcohol. The acid alcohol should be changed occasionally and allowed to act until the worms become a light pink against a white background. The cortical layer should be free of stain, but enough stain should remain to color the internal organs. Since no 2 worms behave exactly alike in this respect, the process should be observed carefully in good light under a dissecting microscope. Until the point of proper destaining is learned by experience, it is well to have the instructor check your work at this stage. Too much destaining results in a dull colored specimen, minus the brilliance otherwise obtained. When the proper color has been obtained, specimens stained in hematoxylin should be passed through 70% alkaline alcohol, to restore the bluish color and stop the destaining action, before proceeding with dehydration. *Caution:* If too much neutralizing reagent is used, the worms will turn brownish and satisfactory restaining is impossible.

Table III. FLOW SHEET FOR PREPARING WHOLE MOUNTS

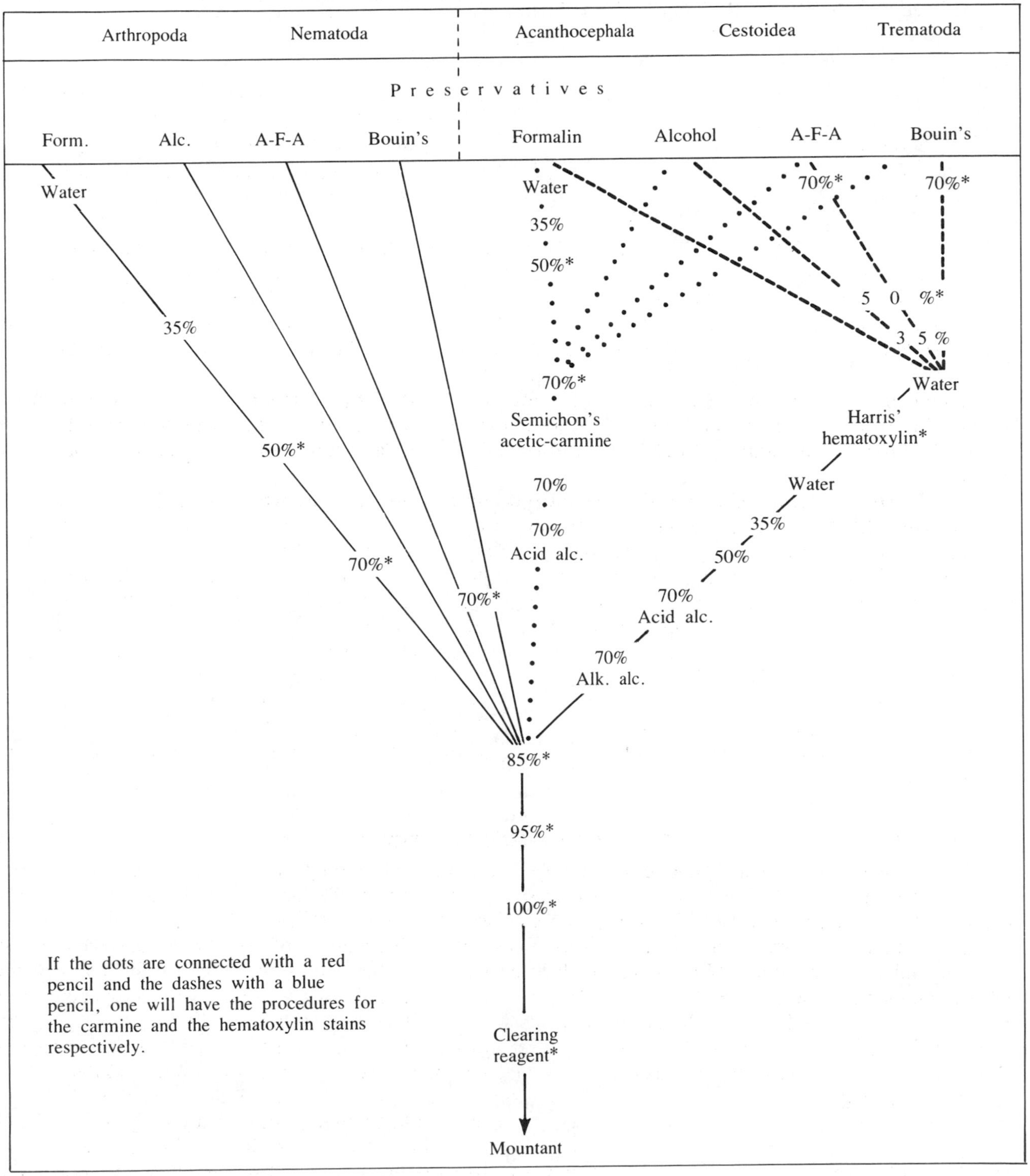

Arthropoda		Nematoda		Acanthocephala	Cestoidea	Trematoda
			P r e s e r v a t i v e s			
Form.	Alc.	A-F-A	Bouin's	Formalin	Alcohol	A-F-A	Bouin's

Water

35%

50%*

70%*

70%*

Water
35%
50%*
70%*
Semichon's
acetic-carmine

70%
70%
Acid alc.

70%
Alk. alc.

85%*

95%*

100%*

Clearing
reagent*

Mountant

70%*   70%*

5 0 %*
3 5 %
Water

Harris'
hematoxylin*

Water
35%
50%
70%
Acid alc.

If the dots are connected with a red pencil and the dashes with a blue pencil, one will have the procedures for the carmine and the hematoxylin stains respectively.

*Solutions in which specimens may be left overnight if necessary.

5. Arranging for Mounting

If specimens are twisted or curved, it is at this point in the process that they must be straightened. This is done by placing them between microscope slides, which are held in position by pressure. A strip of cardboard, of sufficient thickness to prevent too much compression may be inserted at each end of the slide. Keep the specimen on the slide flooded with alcohol, preferably in a petri dish. When the specimen has been oriented and the props are in place, another slide is put flat atop the worm and a clamp(s) or several loops of cotton thread are wound snugly around the slides and the ends tied, while applying some pressure against the faces of the slides with thumb and forefinger. The specimen is then dehydrated through the alcohols in this position in a petri dish or Coplin jar. After an hour or so in 95% alcohol, the worm will be stiffened in the flattened form, after which the slides should be removed and the specimen returned to a preparation dish filled with 95% alcohol. Two changes in absolute alcohol are necessary before clearing. Since only a portion of the specimen, when between the slides, is exposed to the alcohols used in the dehydration, the times normally allowed in each solution should be approximately doubled.

Engorged arthropods, e.g., fleas, sucking lice, ticks, are practically worthless for study unless treated with potassium hydroxide to remove the soft parts so that the chitinized structure, used in making identifications, may be observed more clearly. Such hard parts as the spiracles of fly maggotts, mouth parts, etc. should be removed and left in KOH until the chitin is somewhat bleached and fleshy parts destroyed. Instructions for clearing with the alkali solution are found on p. 283.

6. Dehydration

This consists of replacing the water in the specimen with an anhydrous solution, such as absolute alcohol. Absolute, often called "100%," alcohol, contains about 1.0-2.0% water. The usual steps of 35%, 50%, 70%, 85%, 95%, and absolute alcohol are generally used in dehydration, allowing about an hour in each grade. Using these as a base, steps may be added or deleted as necessary to dehydrate best the material at hand. Replacing the alcohol with water is known as hydration.

The length of time required in the alcohols for hydration and dehydration depends on the size of the specimen and the permeability of its integument. While approximate times are given, no arbitrary rule can be given on this point; each group of specimens should be judged with respect to its special needs. In general, however, Nematoda and Acanthocephala should be handled very slowly, since their integument, no matter how thin, is easily shrunken by rapid dehydration. To avoid damage while dehydrating nematodes and thorny-headed worms, the body wall should be punctured with a small insect needle at several points along its length. It is essential, however, that material be left in absolute alcohol until every trace of water has been removed, otherwise clearing will not be possible.

While it is desirable to have permanent mounts of nematodes for class study, for taxonomic purposes cleared unmounted and/or temporary mounted specimens are often more satisfactory. In the case of the temporary mounts of smaller worms, they may be rolled about under the cover glass, enabling one to view them from all sides. Instructions for making temporary and permanent mounts will be found on p. 281. Leeches and copepods are usually made into permanent mounts for class study, but they should not be so treated for the purpose of taxonomy.

7. Clearing

All traces of water having been removed from the specimen, it is next transferred to a clearing reagent, which renders it transparent and miscible with a resinous mountant. Despite the advantages of terpineol (lilacin), benzene, toluene, wintergreen (methyl salicylate), and beechwood creosote for clearing specimens, xylene is the most widely used for this purpose.

The transfer from absolute alcohol to the clearing reagent should be made gradually in order to avoid the formation of violent diffusion currents which tend to distort specimens. First add enough xylene (or other indicated clearing reagent) to have a mixture of approximately 1/4 xylene and 3/4 alcohol; next a 50-50 mixture; then 3/4 xylene and 1/4 absolute; and finally pure xylene; after which a second change in xylene is made (about an hour in each).

For Acanthocephala, nematodes, and other delicate material which tends to collapse when subjected to sudden change of solutions, additional caution should be taken when going from absolute alcohol to the clearing reagent. This is effected first by puncturing the body wall, as described above, and second by placing the clearing medium under the alcohol. To a preparation dish about half-filled with absolute alcohol and containing the specimens, a sufficient quantity of clearing reagent to fill the dish is introduced at the bottom of the alcohol with a bulbed pipette. Or, first pour in enough clearing reagent to half fill the dish, after which enough absolute alcohol to fill the dish is carefully added with a bulbed pipette. The specimens to be cleared, being now carefully put into the alcohol, float at the interface of the two fluids; the exchange of the fluids occurs gradually, and the objects slowly sink down into the lower layer. When they have sunk to the bottom, the alcohol is pipetted off and the clearing reagent replaced with fresh clearing reagent prior to mounting. Further premounting suggestions to avoid collapsing and subsequent "opacity" will be found on p. 282; they are particularly appropriate for Acanthocephala and nematodes.

8. Mounting

For permanent mounts, "HSR," kleermount, gum damar, or other preferred resinous medium soluble in benzene, toluene, or xylene is satisfactory as a mountant. Small nematodes are commonly mounted in pure glycerine, after which the cover glass is sealed with Thorne's Zut.

Care must be exercised to place the specimen properly on the slide; to select a cover glass of proper size; to add the proper amount of mountant; to prevent the cover glass from tilting to one side; and to prevent the inclusion of air bubbles in the mount. While it is better to use too little rather than too much mountant (in the latter case the specimen will become displaced when the clamp is applied to the cover glass), only by trial and error can one determine the right amount to use.

If the specimen is relatively small, it is best to place it in the center of the slide. Make a guide for this purpose by placing a slide on a piece of white paper, marking around it with a pencil, and then ruling diagonal lines connecting the corners. For mounting, lay the slide on this guide and place the specimen over the intersection of the diagonals.

Arthropods (except fleas) and medium-sized worms (those large enough to orient conveniently and not large enough to require a rectangular cover glass) should be oriented perpendicularly to the long axis of the slide. Specimens flattened dorsoventrally should be mounted so that the ventral surface is up or in contact with the cover glass. Fleas should be mounted on the right side, with the legs pointed away from the technician, so that they will appear in the normal position and facing left when examined under the compound microscope.

A large mount, such as several lengths of a tapeworm (scolex + neck, mature and gravid proglottids), or a *Fasciola hepatica* should be placed lengthwise, to 1 end of the slide, to leave room for the label on the other end. The mount should be placed somewhat away from the end so that approximately 5 mm of the slide will be free of the cover glass.

Differences in the size of various specimens make it impossible to use the same size of cover glass always; the technician's aim should be to use the smallest cover glass that will adequately cover the specimen. Not only does this result in a more attractive mount, but it also means a saving in mountant and cover glasses, both being sold by weight.

The use of 1 or more spring clamps made from discarded watch mainsprings is especially useful for applying pressure and for preventing the cover glass from tilting. These can be manipulated to apply uniform pressure, and they are not subject to accidental displacement. The slide with 1 or more of these clamps attached may be moved freely and even placed on the stage of a compound microscope for examination under low power. These spring clamps replace the vials containing shot, or mercury, or other small weights used by some technicians. Although less satisfactory, a sliver of a glass slide placed under each corner of the cover glass may be used to keep the cover from tilting. Or, pieces of a fine glass rod of appropriate length made from a 3 or 4 mm glass rod heated over a Bunsen burner until it is soft and then quickly pulled apart at the ends may be used for the same purpose.

While spring clamps are in most cases preferred, when dealing with delicate material such as larval helminths pressure should not be applied unless the corners of the cover glass are supported by small bits of a glass slide or, depending upon the thickness of the specimen, slivers of a cover glass.

Use sufficient mountant to spread somewhat beyond the edge of the cover glass. Knowledge of the correct amount of mounting medium for specimens of different volume and cover glasses of different sizes is gained with experience. Air bubbles trapped beneath the cover glass are no cause for serious concern since they will normally disappear after the mount is placed in the drying oven. Any mounting medium naturally shrinks in drying, and if one uses only enough to reach to the edge of the cover, it will eventually shrink back and leave the edge of the cover exposed, or will withdraw unevenly and suck in bubbles of air. If the medium retracts from the edges of the cover while in the drying oven, it should be replaced with new mounting medium so that the entire area between the slide and the cover glass is filled.

As each slide is prepared, the same number that accompanied the specimen in the preparation process is written on it with a wax pencil before placing it in a drying oven. For the label to stick better, the wax-number is removed before adding the label with the required data. Following mounting, place the slide in the drying oven and after the mountant has hardened, which will take a week or more, remove excess medium from the surface of the slide and cover glass by scraping with a safety razor blade. Next, immerse the slide in a Coplin jar filled with 95% alcohol to remove the powdered mountant, then remove it from the alcohol and touch 1 end to a paper towel to drain off the excess alcohol, and finally polish it clean with a soft, lint-free cloth. After the mount is cleaned, it is ready for labeling and study. Whole mounts do not develop their maximum transparency until several months after preparation because of the time required for the mountant to diffuse evenly through the specimen and to harden to an increased refractive index with greater clearing power.

Since the image is inverted under the compound microscope, the anterior end of the animal should be directed toward the technician when adding the label. Place the label on the end of the slide indicated by the instructor and add (printed in India ink) the scientific name of the specimen and host, locality, location, date (of mounting), the corresponding number on the "Mount History" card, and your name or initials. *Caution:* Always be sure to place the label on the same side of the slide as the cover glass!

It is important, as in all scientific work, that utmost cleanliness be observed throughout the process. Exercise caution in using the right reagent, avoid contamination, and be sure that slide and cover glass are clean. Much chagrin and fruitless explanation can be avoided if these instructions are rigorously followed. Your first attempt may not be cause for much pride, but after a few trials you will succeed and have a mount which will pass inspection.

9. Storing Slides

Once slides have been placed in a slide box, it should always be placed on end, so that the slides are in a horizontal position, to prevent the drifting of specimens to the margin of the cover glass, thus becoming relatively useless. Any mountant, even when apparently so hard that it will chip, is really a solid fluid (much like molasses in January). If such slides are stored for long periods on edge, particularly in a warm place, specimens will drift toward the bottom edge of the mount. The practical preventive is to store all such slides so that they remain flat. Observance of this simple rule will eliminate all drifting trouble.

	No........................

MOUNT HISTORY

	Solution	Day	Time
1			
2			
3			
4			
5			
6			
7			
8			
9			
10			
11			
12			
13			

Student.. Mounting Date.............................

Host (CN) .... (SN) ....

Locality..Autopsy Date..............................

Parasite ....

Location ....

Fixed in...Stored in....................................

Autopsied by ....

Stained in...No..........................

Figure 258. Mount history card, showing both sides. Actual size 3″ x 5″.

Special Techniques and Further Notes

Cleaning Slides and Cover Glasses

New slides and cover glasses may usually be satisfactorily cleaned by dropping them, a few at a time, into 95% alcohol in a Coplin jar. After a few minutes they should be removed, 1 at a time, and carefully dried with a soft, lint-free cloth.

Used slides and cover glasses may usually be cleaned by washing in a solution of Bon Ami in a Coplin jar. When dry, wipe clean with a soft, lint-free cloth. If this treatment is insufficient, place for several hours in equal parts of HCl and 95% alcohol in a Coplin jar, keeping the slides well separated, so that the liquid may act on the entire surface of each. Then rinse in water and place them in 95% alcohol, after which they should be dried.

When cleaning a cover glass, grasp it by the edges in 1 hand, cover the thumb and the forefinger of the other hand with the cleaning cloth, and rub both surfaces of the glass at the same time. To avoid breaking the cover, keep the thumb and finger each directly opposite the other.

Always grasp cleaned slides and cover glasses by their edges to avoid soiling their surfaces. It is advisable to clean all the slides and covers to be used at any one time before starting to mount.

Removing Broken Cover Glasses

Place the damaged slide in the freezing unit of a refrigerator. After about 10 minutes the resinous mountant has congealed so that the broken pieces of glass can be easily removed by lifting them off with a dissecting needle or scalpel. After the condensate on the glass disappears, add fresh mountant and another cover glass. This method is much easier and faster than the often used heating method and the even longer procedure of dissolving off the mountant in xylene or other solvent and mounting anew.

1. Protozoa

Blood Smears

It is essential that slides used in making smear preparations be unscratched, noncorroded, and meticulously clean, free from grease, dust, acid, or alkali; that slides be handled by their edges; that the blood be taken as it exudes; that the process be done rapidly so as to prevent coagulation; and that smears be left to dry in a horizontal position away from flies and dust. The finger tip or structure to be pricked is cleaned with 70% alcohol, after which a prick is made with a blood lancet or a sterilized needle. The first drop is wiped off with absorbent cotton or gauze. Mark necessary data with wax pencil on the end of each slide. Blood films should be stained as soon as possible after drying to insure proper staining.

1. *Thin film*: On slide "A" place a drop of blood about one-half inch from the end. Take a second slide "B" and place it on the surface of the first slide at about a 45° angle, as indicated in Figure 259, and move it to the right until contact is made with the drop of blood. The free end of slide "B" may be supported by the third finger. As soon as it touches the blood, the latter will spread. Now push slide "B" toward the left, being careful to keep the edge pressed uniformly against the surface of slide "A."

In this way a thin smear with uninjured host cells and protozoans and/or microfilariae will be obtained. The size of the drop of blood and acuteness of the angle formed between the slides, will determine the thickness of the film, a more acute angle resulting in a thicker film. Allow film to dry thoroughly.

2. *Thick Film:* Four or 5 drops of blood are placed on a slide, spread with a toothpick or the corner of a slide over an area about the size of a dime, and allowed to dry at room temperature. When dry, immerse in distilled water for about 20 minutes to decolorize, then allow it to dry again.

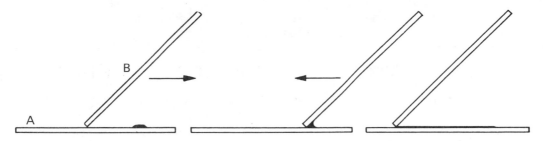

Figure 259. Preparation of thin blood smear.

Staining

Giemsa: Place slide in absolute methyl alcohol for 5 to 10 minutes, then rinse in distilled water. Dilute stain with neutral distilled water (1 drop of stain to each ml water), add slide and stain for about an hour. Wash in distilled water to remove excess stain, then allow slide to dry.

Wright's: Flood film with undiluted stain and leave for about 2 minutes. Add an equal amount of distilled water and leave for about 2 minutes. The stain is then washed in distilled water, after which it is allowed to dry. Determine the time for staining by trial with a series of slides. This is usually about 1 to 4 minutes, but is variable with every batch of stain. The granules in the neutrophils should stain lilac, the eosinophils bright red, and the basophils deep blue.

Smears of Parasitic Protozoa

For protozoans found in the alimentary tract, gallbladder, urinary bladder, or other organ cavities, a small amount of the material should be spread evenly on a slide or cover glass, preferably the latter, with a toothpick; if necessary, add physiological solution. For the forms inhabiting tissues, the latter are cut into small pieces in a little physiological solution, and smeared uniformly over the cover glass. Do not allow the smear to dry. As the edges of the cover glass start to dry, fix in Schaudinn's fluid at room temperature or warmed to 50 C. Float the cover glass, smeared side down in a dish of fixative. After about 1 minute, turn the cover glass over and let it remain on the bottom of the dish for 5 to 10 minutes. Next transfer the cover glass-smear to a Columbia staining dish, containing 50% alcohol (2 changes, 10 minutes each); next to 30% alcohol for 5 minutes, and then to water, which is now placed under gently running tap water for 15 minutes. After rinsing in distilled water, the smear is ready for staining.

While any stains which give good results for cytological and histological work are satisfactory, 1 of the most commonly used stains is Heidenhain's iron hematoxylin. It requires a mordant, iron alum (ferric ammonium sulphate), followed by the dye. The smear from the distilled water is first placed in the mordant for about 3 hours, after which it is washed with running water for 5 minutes and rinsed in distilled water. Place the smear now in a well-ripened 1% solution of hematoxylin for about 3 hours. After rinsing in distilled water, the smear is destained in a 0.25% iron alum in water. The nuclei should show the stain while the cytoplasm should be fairly clear. Proper destaining comes with practice. After destaining, wash in running water about 30 minutes. Upgrade the smear through the alcohol series (3 minutes in each), clear in xylene (10 minutes), and transfer to a slide in a drop of thin mountant. Make sure that the smear side of the cover glass is down.

Isolation of Oocysts of *Toxoplasma gondii* from the Feces of Cats

Unsporulated oocysts are easily recovered from the feces of infected cats. Comminute freshly passed feces in water to form a thick paste and suspend it in about 10 volumes of sugar solution (sp. gr. 1.15). Sieve the suspension through wet cheesecloth or a tea strainer to remove coarse material. Centrifuge at 1,000 to 3,000 rpm for 10 minutes. The oocysts will be at the surface of the solution in the centrifuge tubes from which they may be transferred by means of a bacteriological loop to slides for microscopic examination. For sporulation of the oocysts, pipette or pour off the supernatant solution and wash in water by repeated centrifugation to remove all of the sugar. Put the oocysts in a dish of water kept at room temperature and observe them daily for progress in sporulation. For details consult Dubey et al. (1972).

Sapero and Lawless' Fixative-Stain for
Protozoa and Helminth Eggs

In addition to providing good preservation of both protozoan cysts and trophozoites and helminth eggs, this method is simple enough that collections may be taken in vials by unskilled workers in the field or home, for later examination by skilled personnel, and it provides almost immediate staining. The MIF technique may be used (1) in making direct fecal smears or (2) for collection and preservation of bulk stool specimens.

1. For direct smears, the ingredients, sufficient for some 2 dozen preparations, are combined in a Kahn tube. Since the resulting solution is unstable, it must be made up daily; and the Lugol's used should not be older than 3 weeks. To use the stain solution, add a small amount of fecal sample to one drop of MIF solution and a drop of distilled water, mix thoroughly, and add a cover glass. The completed wet preparation should be thin enough to permit the slide to be tipped on edge without shifting the cover glass.

2. For collecting and preservation of fecal samples when not brought directly to the laboratory, a modification of the above MIF stain is advised. The MIF stain-preservation solution consists of (1) a MF stock solution, stored in a brown bottle, and (2) Lugol's solution not over a week old. To use, add 2.35 ml of stock solution into a Kahn tube and stopper with a cork, and in another such tube add 0.15 ml of the fresh Lugol's and close with rubber stopper. Combine the two solutions immediately before the addition of some 0.25 gm (about twice the volume of a medium-size pea) of fecal material, and comminute thoroughly with wooden applicator. If stoppered well to prevent evaporation, specimens will retain a good stain for some months. For examination remove a drop of the supernatant fluid from the top, place on a slide, and add a cover glass.

The above MIF technique has been modified by Blagg et al., to increase its diagnostic yield for helminth eggs, by the addition of ether to dissolve fats and to float fecal detritus. The MIF-preserved specimen is shaken vigorously for 5 seconds, strained through 2 layers of wet cheesecloth into a 15 ml centrifuge tube to which 4 ml of refrigerated ether is added. Close the tube with rubber stopper and shake vigorously, after which the stopper is removed and the tube left standing for 2 minutes. It is then centrifuged for 1 minute at 1,600 rpm, resulting in the formation of 4 layers: ether at the top, a plug of fecal detritus, a layer of MIF, and a bottom sediment containing helminth eggs. The fecal plug is loosened by an applicator stick, and all but the bottom sediment layer is quickly decanted. Mix the sediment with an applicator stick, and put a drop onto a slide for examination.

Recovery of Helminth Eggs and Protozoan Cysts

Practically all helminth eggs and protozoan cysts are evacuated in the feces, the important exceptions being the kidney worms and *Schistosoma haematobium*, whose eggs are voided in the urine. While various methods of examining feces for helminth ova and their larvae, and/or protozoan cysts, are in vogue, they may be classified as either: (1) the direct fecal smear, or (2) concentration techniques.

1. To make a direct smear, a small portion of the uncontaminated stool is spread on a slide and mixed, using a wooden applicator, with a few drops of physiological solution. A cover glass is added, and the slide is first examined under low power. In the case of doubtful objects and in making the final specific determination, change to high power. It should be remembered that the fecal smear must be thin enough to view all the objects under the cover glass, and that there should not be too much solution, so as to float the cover glass.

The direct fecal film should always be made and examined when protozoan and helminthic infections of the intestinal tract and its appendages are suspected. In heavy infections this technique will reveal the presence of the parasite.

2. While the concentration of cysts and eggs may be accomplished in various ways, all methods involve flotation or sedimentation—with or without centrifugation. The methods depend upon mixing the fecal sample with a liquid, the specific gravity of which is different from that of the products being searched for. Although each author advances particular claims for his method, some fall short in 1 or more of the features desired. Any efficient concentration technique should embody the following: (1) simplicity of operation, (2) be relatively rapid, (3) remove the nonparasitic matter while retaining the parasitic materials, and (4) the parasitic objects obtained should be diagnosable, i.e., not shrunken or so altered in shape as to be unrecognizable.

Zinc Sulfate Flotation

This technique will concentrate protozoan cysts, most helminth eggs, and larvae.

1. Prepare a fecal suspension by mixing 1 part feces (the size of a pea) with about 10 parts of tapwater.

2. Strain suspension through a layer of wet cheesecloth, supported in a small funnel, into a centrifuge tube.

3. Centrifuge for 1 minute at about 2,500 rpm, decant the supernatant, add 2 or 3 ml of water, shake to break up sediment, fill tube with water, centrifuge and decant as before. Repeat until the supernatant fluid is clear.

4. Decant the last supernatant, add 2 or 3 ml of zinc sulfate solution, shake to resuspend the sediment, and add zinc sulfate solution to within about 5 mm from the rim. (The zinc sulfate solution contains 330 gm of chemical in 1 liter of distilled water, adjusted to a specific gravity of 1.18).

5. The tube is centrifuged for 1 minute at about 2,500 rpm and is allowed to stop without interference. Do not remove the tube.

6. Using a wire loop, remove several loopfuls of the surface film and place on a slide. Add a drop of Lugol's iodine solution, mix, apply a cover glass, and examine.

Formalin-ether Centrifugation

The eggs of schistosomes are not easy to find in direct fecal smears, and are usually destroyed by conventional concentration methods. This technique has been developed for the recovery of helminth eggs, including those of schistosomes, and protozoan cysts.

1. Prepare a fecal suspension by mixing 1 part feces (the size of a pea) with about 12 ml of physiological solution.

2. Strain suspension through 2 layers of wet cheesecloth, supported in a small funnel, into a centrifuge tube.

3. Centrifuge for 2 minutes at about 1,800 rpm, decant the supernatant, and add fresh physiological solution. Stopper with the gloved finger and shake well.

4. Wash twice in physiological solution, centrifuging as above, and decanting the supernatant each time. Repeat until the supernatant fluid is clear.

5. Add 10 ml of 10% formalin to sediment, and leave for 5 minutes.

6. Add 3 ml of ether, and resuspend the sediment by vigorous shaking (with gloved finger).

7. Centrifuge as above.

8. Loosen the "plug" at the formalin-ether junction with an applicator, and completely decant the supernatant.

9. Using a Pasteur pipette or a wire loop, transfer sufficient sediment to a slide, add a cover glass, and examine. If not enough fluid drains back from the tube wall to enable one to withdraw sediment, add a few drops of physiological solution and mix with sediment.

At no time, with either technique, is the oil immersion objective to be used with wet mounts unless a No. 1 cover glass is used and the preparation sealed with Vaseline or wax.

Diagnosis of Microfilarial Infections

To 1.5 ml of fresh blood, Knott (1939) recommends adding 10 ml of 2% formalin and centrifuging at 1,500 rpm for 2 to 3 minutes, after which the supernatant fluid is decanted. Transfer a drop of sediment to a slide, add a cover glass, and examine for microfilariae. A modification involves substituting 1% acetic acid for the formalin and, after transferring to slide followed by drying, stain with Wright's or Giemsa's stain.

Harder and Watson (1964) described useful staining techniques for sheathed and unsheathed blood-borne microfilariae. A simple key to human microfilariae, based upon criteria clearly demonstrated by these stains, is included.

Papanicolaou-Hematoxylin and Eosin Procedure
for Sheathed Microfilariae

1. About an inch from 1 end of a slide prepare a thick blood film. Cover a space about the size of a dime with as much blood as will easily spread over this without cracking and peeling when dry.

Spread with a toothpick or the corner of a slide. Ordinary printing can just be read through the wet center of a well-made film. On the opposite end of the film put an identifying mark with a wax pencil or other marking device.

2. Lay the slide flat to dry and have it well protected from insects and dust. Air dry for 8 hours or more.

3. Dehemoglobinize thoroughly in enough distilled water to cover the smears as the slides stand on end in a Coplin jar, thick film downward. The film should be a milky white, which may take several hours.

4. Air-dry slides, standing on end on absorbent paper or a coin pad.

5. Rinse in 95% alcohol, 2 changes.

6. Harris' hematoxylin (stock stain diluted equally with distilled water), 5 minutes.

7. Rinse in alkaline tap water, 2 changes.

8. Acidulated 70% alcohol, 2 quick dips.

9. Running tap water, 1 minute.

10. Rinse in distilled water.

11. Pass through 70% and 80% alcohols, 2 minutes in each.

12. Orange G (1% solution in 90% alcohol), 1.5 minutes.

13. Rinse in 95% alcohol, 2 changes.

14. E.A. 50,* 2 minutes.

15. Rinse in 95% and absolute alcohols, 2 changes in each.

16. Terpineol, 5 minutes.

17. Dry. (The remaining steps can be completed later, but within 2 days gives best results.)

18. Harris' hematoxylin, 20 minutes for *Brugia malayi* and *Wuchereria bancrofti;* 45 minutes for *Loa loa,* longer for older slides. (Proper staining time is gained with experience.)

19. Acidulated 70% alcohol, 1 dip.

20. Running tap water, 1 minute.

21. Rinse in ammonia water (strong ammonia, 0.5 ml in 1,000 ml distilled water).

22. Stain in eosin (saturated solution of Eosin Y in 80% alcohol), 5 to 10 seconds (according to desired intensity).

23. Rinse in 95% and absolute alcohols, 2 changes in each.

24. Clear in absolute alcohol-xylene (equal parts) carbol-xylene (melted phenol crystals, 100 ml in 900 ml xylene) and xylene, and mount.

Modified Gomori Trichrome Procedure for Unsheathed Microfilariae

1. Prepare thick blood film and dry, as above.

2. Dehemoglobinize in magnesium sulphate solution (1 gm crystals in 1,000 ml distilled water) for 1 hour (or longer) until all color disappears from the film.

3. Rinse in distilled water, 2 changes.

4. Harris' hematoxylin, 5 minutes for *Dipetalonema perstans;* 30 minutes for *Mansonella ozzardi.* (Proper staining time is learned with experience.)

5. Distilled water, 10 to 15 dips.

6. Acidulated 70% alcohol, 5 seconds.

7. Tap water, rinse.

8. Ammonia water, rinse.

9. Distilled water, 2 changes.

10. Gomori trichrome stain, 6 minutes.

11. Acetic acid (5.0 ml in 1,000 ml distilled water), 2 minutes.

12. 95% and absolute alcohols, 2 changes in each.

13. Clear in xylene and mount.

---

*This polychrome eosin-azure formulation by Papanicolaou, may be purchased from pharmaceutical dealers or from Ortho Diagnostics, Raritan, N.J. 08868.

Improved Method for Pinworm Diagnosis

Since the eggs of *Enterobius vermicularis* are not ordinarily recovered in the feces, diagnosis of infection is usually based on the recovery of eggs deposited around the anus by migrating gravid females. For this, a length of transparent adhesive cellulose tape held adhesive-side-out on the end of a tongue depressor or similar object by the thumb and index finger, is pressed against the right and left perianal folds and then spread flat on a microscope slide for examination. The buttocks should be spread enough to allow pressing the tape firmly against the line of junction between the moist outer part of the anal canal and the somewhat dry, waxy perianal folds. This is where the egg yield is the highest. A strip of paper bearing identification should be placed between the tape and the slide at 1 end. For microscopic examination, the tape should be grasped at the loose (identification) end and turned back far enough to expose all the surface that had been in contact with the perianal skin. Add 1 drop of toluene and with a straight wooden applicator smooth the tape back onto the slide and level it. Since the toluene is at once a good solvent for the adhesive material on the tape and a good clearing reagent, the air bubbles are allowed to escape as the tape is smoothed out and almost everything except the pinworm eggs becomes cleared, leaving the eggs, if present, the only conspicuous objects in the preparation. Xylene, benzene, ether, chloroform, and N/10 sodium hydroxide are all unsatisfactory for clearing. Until skill is acquired in handling the toluene and tape, a clean razor blade or scalpel may be used to mark off the portion of the tape that had been in contact with the perianal skin and the remainder of the tape removed. With a shorter strip only about 20 mm long to handle, it is easier to get clear smooth preparations.

Double Cover Glass Mounting Technique

For making permanent mounts of helminth eggs and other objects of suitable size, which are difficult or impossible to prepare satisfactorily in a resinous medium, an aqueous solution with the addition of a second cover glass, sealed in a resinous mountant is advised. Number 1 circular cover glasses of 2 sizes, differing 3 mm or more in diameter, are recommended. The chloral-gum solution used is a modification of Berlese's original formula, but it is believed that other similar aqueous media would prove equally satisfactory.

1. Prepare a series of dilutions of chloral-gum in 10% formalin, starting with a 10% solution and increasing concentrations by 2% for each subsequent step, or prepare each solution as needed.

2. Concentrate fecal suspensions that have been thoroughly fixed in neutral formalin until each drop contains an adequate number of eggs. Each ml of this will yield about 10 slides.

3. Pipette 5 ml of the concentrated fecal suspension into a test tube or a vaccine bottle of 15 ml capacity. Next carefully add an equal quantity of 10% chloral-gum-formalin, tilting the receptacle to allow this fluid to run in along the lower side. Each succeeding dilution of the series is heavier than the preceding, and consequently will occupy the lower half of the column of fluid.

4. Cap the containers and place them in a drying oven at about 35 C until the fecal material has completely settled on the bottom. Then remove the clear fluid above the feces with a finely-drawn bulbed pipette, and carefully add an equal volume of the next higher concentration of chloral-gum solution. Continue this process until the eggs are suspended in full-strength chloral-gum medium. When the last sedimentation is complete and the supernatant removed, mix the remaining contents of each container thoroughly, but wait until any bubbles disappear before starting to prepare mounts.

5. Preparations are made by 2 persons working as a team, 1 transferring drops of the mixture to the slides, the other adding the smaller cover glasses. Circular cover glasses are advised since they permit an even flow of material to the outer edges. Slides should be laid out in advance on a warming table, if available, since this results in a better spread of the mountant. With some practice, one will come to judge the size of the drop required to spread completely under the cover without excess. Cover glasses should be added quickly, before the medium hardens and spreads poorly. For transferring the material to slides, a finely-drawn bulbed pipette or a heavy platinum wire loop, about 4 mm in diameter, is recommended.

6. Place mounts in a horizontal position in a drying oven at about 35 C. Following drying, remove any excess mounting medium with a razor blade and finish cleaning with a clean cloth saturated with water. The margin must be free of mountant if a strong seal is to be obtained.

7. Next add a drop or 2 of thin resinous mountant atop the previous cover and apply the larger cover glass, being careful to place the second cover so as to have the same overlapping margin on all sides. The application of a spring clamp is suggested. Return to drying oven and when dry, clean, if necessary, in the manner suggested for whole mounts. Such preparations will last indefinitely if the cover glass is not cracked because there is a firm moisture-tight seal at the edge of the larger cover glass.

A modification of the dual mount involves placing the aqueous medium with the material being mounted between the 2 cover glasses, rather than between the slide and the smaller cover glass as above.

1. Prepare a raised working surface no larger in diameter than the smaller cover glass. A cork of suitable size attached to a base provides desirable stability. Place the smaller cover glass on the cork.

2. Transfer a small drop of the warmed concentrated chloral-gum medium containing the eggs to the center of the cover glass. Keep the stock of chloral-gum warm in a water bath. The mounting drop should be of such a size that it just, and only just, covers the cover glass. Judgment, which comes with practice, is needed to gauge the size of the drop, which, again depends on the warmth of the mixture.

3. Grasp the larger cover glass with cover glass forceps and quickly bring it into contact with the preparation on the cork, keeping the cover glasses as nearly parallel and concentric as possible.

4. When the larger cover glass contacts the chloral-gum, which will spread between the 2 cover glasses, carefully invert so that the smaller cover glass is uppermost. If necessary, center and flatten out the mount with a dissecting needle. The preparation is placed in a warming oven for a few hours to allow the mountant to set.

5. Place a drop or 2 of thin resinous mountant on the smaller cover glass. Carefully lower a slide until it touches the mountant on the cover glass, and quickly invert so that the larger cover glass is uppermost. Center the cover glass, apply a spring clamp and return to the drying oven.

Either method may be used for making permanent whole mounts of small hard-to-clear specimens, such as helminth eggs, mites, and small nematodes. The sealing effect of the resinous mountant prevents the crushing of the specimen which often results from the pulling down of the cover when only a single cover glass is used with chloral-gum and similar water-miscible mountants.

Recovery of Monogenetic Trematodes

For recovering the small Monogenea, Mizelle (1938) recommends freezing fish gills for 6 to 24 hours in an appropriately sized jar. They are then thawed in tap water, the jar capped and shaken vigorously. The liquid, with the freed parasites, is poured into a Syracuse watch glass and alternately diluted and decanted until clear enough for examination under a dissecting microscope. With the use of a bulbed capillary pipette, the parasites are transferred to clear water,

According to Hoffman (1967), *Gyrodactylus* may be removed from fish or gills by placing them in a 1:4,000 solution of commercial formalin for 15 to 45 minutes, which results in nicely extended detached worms.

Due to their small size and the tendency of the worms to become entangled in debris and with other specimens, it is practically impossible to handle the small monogenetic trematodes in containers during the ordinary technical processes requisite to mounting. For such forms, the following procedure in making total mounts is advised. With a bulbed pipette transfer specimens from a mixture of water and glycerine to a slide; then remove most of the water-glycerine with pieces of paper toweling or filter paper. Adherence is secured by immersing the slide in a fixative (A-F-A or Bouin's) contained in a Coplin jar. Following fixation, the material is dehydrated (using Coplin jars) through the alcohols, cleared and mounted in the usual way. Due to the inconsistencies in staining, together with the poor differentiation and the obscuring of structures, staining is not advised.

Examination of Snails for Cercariae

Large numbers of snails can be dredged up from mud and vegetation with various types of equipment used by limnologists for this kind of collecting. Place the material in pails or large trays with sufficient water to cover the contents. Use pond water for snails from fresh water and seawater for marine species. Cover the surface of the water with paper towels, leaving open spaces. Many of the snails will crawl out on the wet towels and up the sides of the containers where they can be collected easily. Cover the containers with a cloth or paper to prevent the snails from escaping. It is necessary to search the sandy and muddy bottoms of habitats carefully for some species. Tea strainers and collecting equipment constructed with screen bottoms and designed to strain out sand and mud are necessary to recover the small species of snails.

Wash and separate snails according to species, and transfer several to wide-mouthed glass bottles. Shell vials may be used for the smaller species. Cover the containers. Change water as necessary and feed on fresh lettuce, if kept beyond 6 days. Examine each container several times daily, in a good light against a dark background. The cercariae will appear as whitish tumbling bodies. When cercariae appear, isolate snails individually, to determine which one(s) is (are) infected. It is necessary to check the containers mornings, afternoons, evenings, and nights because cercariae of many species are not liberated continuously, but are shed only at certain times of the day or night. If snails, especially prosobranch species, are first dried on blotting paper for a few hours and then isolated in water, sometimes cercarial emergence will follow almost immediately.

Cercariae should be studied alive as well as fixed and stained. With a capillary pipette equipped with a rubber bulb transfer specimens to a slide and add a No. 1 cover glass. The addition of a weak solution of a vital stain reveals certain internal structures. Draw sufficient water from under the cover glass by means of absorbent paper to press the cercariae to the point where they are held still but not crushed. In these preparations, flame cells and excretory system show best. The edges of the cover glass should be sealed with Vaseline by running a hot needle around the edge to spread the sealant evenly. Fresh egg albumen, because of its viscosity and physiological properties, is a good medium in which to mount living cercariae for study.

After having studied the cercariae, gently crush the shells of the snails. For crushing a large number of snails, a small bench vise is recommended. Remove the broken shell and tease the viscera apart in physiological solution with dissecting needles, and examine under a dissecting microscope to determine whether cercariae are produced by rediae or daughter sporocysts. The snail's digestive gland and gonad in the apex of the shell are the best sources of trematode larvae. Transfer larval forms in a drop of physiological solution to a slide, apply a cover glass and examine under a compound microscope.

The freed cercariae and rediae or daughter sporocysts containing the cercariae should be fixed, stained, and made into permanent mounts. Label each slide, stating the kind of cercaria, whether produced in redia or daughter sporocyst, and the species of snail.

Vital Staining, or Staining *in Vivo*

Vital (often called supravital) staining, the absorption of coloring by living cells or organisms, with neutral red, Nile blue sulphate, or other suitable stains, is advantageous for distinguishing the number of penetration glands and their secretions, the slender undeveloped ceca, the excretory system, and the presence of spines on those cercariae possessing them. Such stains are also useful for studying pseudophyllidean coracidia. With a finger tip, spread a drop of 0.5% aqueous neutral red on a slide and let dry. Add a drop of water containing cercariae, apply a No. 1 cover glass, and examine. Larvae may be placed directly in the stain if a very dilute solution (0.01% or weaker) is used. Mounts of living larvae last much longer if the edges of the cover glass are sealed with Vaseline.

Artificial Evagination of Tapeworm Cysts

The viable cysts are carefully removed from the host and placed in physiological solution. Each cysticercus should be liberated from its cystic wall, if necessary, and placed in a bile salts solution. Invaginated specimens, upon being placed in the warmed (37 to 38 C) solution of bile salts, generally show definite signs of movement within 20 seconds. There is a gradual pushing out of the neck region as peristalticlike waves move away from the bladder until there is a complete eversion of the scolex.

Collecting Nematodes from Soil

Larval stages of Strongyloididae, suborders Strongylina, Trichostrongylina, and Metastrongylina nematodes occur in the soil and feces, having hatched from eggs passed in the feces of infected hosts. Species of all of these groups occur in domestic and wild animals and some of them in humans. They may be recovered by means of simple apparatus constructed from material readily available in any laboratory.

The Baermann funnel is the apparatus commonly used (Thorne, 1961: 48). Its utility lies in the fact that the larvae are very active and constantly moving. To prepare a Baermann funnel, a short piece of soft rubber tubing is pushed onto the stem of a 6 to 10 inch funnel. The free end of the tubing is closed tightly with a pinch clamp. A tall wide-mouthed jar or a ring stand provides a good support for the funnel, which is filled with water. A piece of loosely woven cloth is spread over the top of the funnel so as to lie loosely on the water and is attached to the rim by means of clothespins, a rubber band, or string. The free margins of the cloth should be arranged over the mouth of the funnel so as not to draw off the water by capillary action. A thin layer of soil from the habitat of infected animals is spread over the slightly submerged cloth. A metal pan similar to a pie tin with a screen bottom may be used to support the cloth in the funnel.

Larvae and free-living worms wriggling through the soil and cloth gravitate into the rubber tubing. They may be recovered in numbers after a few hours by opening the pinch clamp and drawing a few ml of water into a Syracuse watch glass or petri dish. Fecal material mixed with sand, animal charcoal, or sphagnum moss will permit good aeration and facilitate hatching of eggs. Larvae are recovered after a few days by means of the Baermann funnel.

When baermannizing soil from habitats of parasitized hosts, larvae of the parasitic species and both adults and larvae of nonparasitic soil nematodes will be collected, often with the latter in great numbers. While these can be separated by microscopic examination, the task is tedious and demanding. A quick and easy way to separate them is to add a few drops of concentrated HC1 to the dish. The nonparasitic species are killed quickly while the parasitic ones, which must survive passage through the acidic condition of the stomach, remain alive and very active.

All baermannizing apparatus and glassware used should be scrupulously cleansed with boiling water between each operation. Equipment contaminated with larvae can lead to erroneous conclusions.

Inasmuch as larvae of suborders Strongylina and Trichostrongylina nematodes of horses, cattle, and sheep have a tendency to migrate, they may be collected by another simple method described by Whitlock (1956). Moist feces mixed with sterilized sphagnum moss or other diluting material are placed in a small beaker, or similar container. The moss should be loose to enable larvae to move about freely and it should fill the beaker. The beaker is placed in the middle of a large mouthed bottle which is filled with sufficient water to come to the top of the beaker but not run into it. The large bottle is covered. Larvae migrate up the moist sides of the beaker, over the rim, and into the water where they sink to the bottom in great numbers. A small petri dish placed inside a larger one and covered will work equally well. Once the principal is understood, many applications of it will be apparent.

Artificial Digestion of *Trichinella* Larvae from Muscle Tissue

Rats or other hosts, after being skinned and eviscerated, are ground in a food chopper. Digestion is accomplished by placing the ground meat in a modified Baermann apparatus, to which the digestive fluid is added. A sling of 4 layers of commercial cheesecloth is suspended in a large glass funnel by means of clip clothespins to retain the ground meat. A bottle is attached by means of a short section of soft rubber hose to the stem of the funnel, which is supported by a ring stand. The mixture should be left for some 24 hours at a temperature of 37 C, and about 1 liter of solution should be used with 30 gms of meat.

If viable larvae are present, the action of the digestive fluid, together with their own activity, will free them from muscle tissue. The larvae will excyst and by their sinuous movements will work through the cheesecloth sling and fall to the bottom of the bottle. At the end of the digestive period the fluid in the bottle should be poured into a petri dish and examined under a dissecting microscope.

Recovery of Larval Helminths, Including Plerocercoids

A preliminary search for metacercariae and plerocercoids in the viscera and body wall of fresh fish, after both the external and internal examinations have been completed, may be done in the following manner. Fillet the fish and then make incisions to divide the musculature into slices about 1 cm thick, which are carefully inspected. Using small insect needles inserted in dowel handles, carefully dissect out larvae.

Following this cursory examination, the host should be placed in artificial digest fluid (p. 287) at the rate of about 1 gm of tissue to 10 ml of solution, which frees the larvae alive. Not only are small nematode and acanthocephalan larvae, metacercariae, and plerocercoids obtained in this way, but a qualitative-quantitative study with reasonable accuracy can be made if the different organs are removed and digested separately. For the recovery of larvae other than plerocercoids, grinding of tissue is advised, but grinding is contraindicated when plerocercoids are involved lest they not survive the process intact. The tissue and the digest are placed in a graduated cylinder or appropriately sized glass jar and left for a few hours at 37 C, with occasional shaking. In the absence of an oven, the necessary temperature can be maintained by placing the glass container in a water bath.

Following digestion, rinse and remove the pieces of host tissue and decant the supernatant. The sedimentation-decantation is repeated until the material is clear. Pour the sediment into a petri dish and examine in good light under a dissecting microscope. In the case of delicate metacercariae, it is better to use physiological solution in the digest as well as for washing.

Artificial Infection of Laboratory Animals with *Trichinella*

Food should be withheld for 48 to 72 hours from the individually caged animals, after which they are offered the infected meat or given the larvae in solution freed by peptic digestion. If they fail to eat the meat within an hour or so, replace it in the refrigerator, and withhold food for an additional 12 to 24 hours. Animals should be given water throughout fasting. Rabbits and guinea pigs should be infected with digested larvae by means of a stomach tube, a method which may also be used with rats and mice.

Each host should be given a dosage in accordance with weight and the degree of infection of the meat. Twenty larvae, either as free larvae by stomach tube or as in infected meat, per gram of host weight is recommended.

Mounts of Nematodes

Nematodes may be prepared in a number of ways for study. The manner of preparation may be determined by the need or the permanency of the mounts.

Water mounts—Larvae and small adults, both parasitic and nonparasitic forms, should be studied alive in temporary water mounts or dead in fixatives where formalin is used. Such mounts are very valuable and will last for a considerable length of time if the edges of the cover glass are sealed by spreading Vaseline over them with a hot teasing needle.

Temporary mounts—These may be prepared from fixed specimens cleared in phenol-alcohol solution, lactophenol, or glycerine. Temporary mounts have the advantage of permitting the specimens to be rolled under the cover glass for viewing them from various positions.

Phenol-alcohol solution is useful for observing cuticularized structures. Fixed specimens may be transferred to the solution directly from any fixative. Clearing is rapid and complete; however, the degree of transparency can be controlled by drawing 95% alcohol under the cover glass. The phenol must be removed from specimens to be stored by washing several times in alcohol.

Lactophenol acts rapidly but does not over clear. Put a small amount of 0.01 to 0.0025% aqueous solution of cotton blue (China blue) in lactophenol on a slide and warm to 65 to 70 C on a hot plate. By means of a small needle, transfer worms directly into the lactophenol and leave for 2 to 3 minutes. They will be cleared and tinted blue. Mount them in lactophenol tinted with 0.0025% cotton blue. Specimens left in lactophenol for a week or longer tend to macerate and become distorted. If worms are to be preserved for later study, the lactophenol should be washed out through several changes of alcohol.

Glycerine is an excellent mounting medium and is used extensively by nematologists. Nematodes are transferred directly from the fixative into a small amount of the following mixture: 21 ml of 95% alcohol, 1 ml of glycerine, and 79 ml of distilled water.

The dish containing the nematodes is placed on a support in a closed vessel containing 95% alcohol for at least 12 hours at 35 to 40 C, or longer for large nematodes. After this period, the dish is filled with a solution of 5 parts glycerine in 95 parts of 95% alcohol and placed in a partially closed petri dish kept at 40 C until all of the alcohol has evaporated, leaving the worms in pure glycerine.

Temporary or permanent mounts may be made from specimens cleared in this manner. For permanent mounts, glycerine tinted with 0.0025% cotton blue kept in a desiccator should be used as the mountant.

To make permanent mounts, a drop of the dehydrated glycerine is put on a slide and the worms transferred to it. Small bits of glass or a ring of Zut are used as supports for the cover glass to prevent damage by pressure to the specimens. The amount of glycerine should be sufficient to almost reach the edge of the cover glass when it is in place. The completed mount should be sealed with Zut. When properly prepared, these slides will last indefinitely. Such slides should be stored flat to prevent drifting of specimens.

Resinous mounts—While time-consuming, they are not difficult to prepare and are especially suitable for nematodes and Acanthocephala stained with carmine dyes and counter-stained with Fast Green. Preparatory to staining and dehydration, the body wall is punctured in a number of places for better movement of the reagents into and out of the worms. For infiltration, the completely dehydrated worms are transferred from absolute alcohol into a mixture of 1 part xylene and 3 parts absolute alcohol; equal parts xylene and alcohol; 3 parts xylene and 1 part alcohol; several changes of pure xylene; and finally into a very dilute mixture of xylene and the resinous mountant of choice in a Syracuse watch glass containing the mountant. The open dish is covered with a piece of paper to keep out dust and dirt and the xylene allowed to evaporate at room temperature or in a warm oven to a point where the resin is the right consistency for mounting the worms on a slide. It is important to use a sufficiently large volume of the dilute xylene-resin mixture to provide adequate mountant in the dish to cover the worms when the solvent has evaporated. Permanent mounts prepared in this way are valuable for study. Fixed worms may be cut in half so there is free access of the stain, alcohols, clearing agent, and mountant and processed in the usual manner without difficulty. The halves may be mounted on the same or on different slides.

Preparation of *en face* View of Nematodes

For the study of cephalic structures, which are important in the taxonomic study of nematodes, one has to cut off the head end and mount it under a cover glass, orienting it so that the mouth opening is upwards. Such *en face* mounts may be temporary or permanent, and various methods of preparation have been suggested.

The beheading is done under a suitable power of a dissecting microscope with a micro-scalpel on a transparent celluloid slide. Instructions for making these scalpels are given on p. 284. Cut squarely no farther back than 1 body width, or closer in the case of large specimens. Anderson (1958) suggests using glycerine for clearing the worms before decapitating and glycerine jelly for mounting. A small drop of warmed glycerine jelly is placed on a 15 mm cover glass, and under a dissecting microscope the head oriented in it so that the lips are next to the cover glass. Small glass rods or a ring of Zut are put on a slide for support and the cover glass inverted on it. When properly prepared, the glycerine jelly is in contact with both the cover glass above and the slide below. The supporting structure prevents pressure on the head. If necessary, slight shifting of the cover glass while observing the mount under the low or high power of compound microscope will orient the head properly. When the cover glass is ringed with Zut, the mount becomes permanent.

For large specimens, another relatively simple method is suggested. Decapitate the uncleared nematode as described above. Transfer the head to a preparation dish and dehydrate in the usual manner through 100% alcohol. Then transfer the head to a slide and, under a dissecting microscope, orient it so that the lips face the observer. Next place a drop of 4% celloidin (in equal parts of 100% alcohol and ether) directly on the material, promptly reorienting if necessary, allow to harden in the air for about 1 minute, and then drop the preparation into pure chloroform and leave for about 5 minutes. Since the chloroform hardens the celloidin, any excess may be trimmed away with a sharp knife. The preparation is now dropped into xylene to which has been added 10% of terpineol by volume. When completely cleared it is mounted in a resinous medium, a cover glass added, and allowed to dry.

Revealing Nematode Cuticular Structures

If nematodes are left for several hours in aqueous or formalin solutions of cotton blue, particles of stain collect in the depressions, pits and cavities, revealing the phasmids, cervical papillae, amphids, and excretory pores in relief.

Trisodium Phosphate for Softening Helminths and Insects

Van Cleave and Ross (1947) recommend that hardened acanthocephalans and nematodes be placed from water into a solution of 0.25% trisodium phosphate. Freshly preserved specimens become soft, pliable, and translucent almost immediately, while specimens that have become hard, brittle, and unyielding after long preservation may require several hours or even days before they attain the proper degree of softness and pliability. If placed in a warming oven, the process is hastened. It is well to keep treated material under observation, since the use of too strong solutions of trisodium phosphate or weaker solutions over too long a time may render specimens too soft and jellylike for easy handling. When the desired degree of softness and translucency has been reached, specimens should be removed to distilled water to check the action.

Specimens thus treated, washed in distilled water and stained, become much more brilliantly stained than untreated specimens. Furthermore, the treated specimens may be dehydrated, cleared, and mounted in resinous medium without developing opacity, except in rare instances.

Another distinct advantage of the softening effect is the ease with which specimens may be straightened prior to dehydration. When softened, specimens may be folded back and forth on a microscope slide along strips of cardboard or toothpicks and may be slightly compressed and held in place by covering with another slide. Several wrappings of thread or spring clamps provide the pressure and hold the slides together so that the stained specimen may be carried in a petri dish or Coplin jar through the grades of alcohol. In 100% alcohol (95% if counterstaining with Fast Green), the specimen becomes firm enough so that it may be removed from between the slides and yet retain its series of folds, ready for clearing and mounting.

Clearing Arthropods with Potassium Hydroxide

Pierce the abdominal region with a small insect needle while the specimen is still in the preservative, and then transfer to water. Replace the water with 10% KOH and leave for 12 to 24 hours, or until the specimen becomes lighter in color. After rinsing in water, replace with acidulated 70% alcohol, dehydrate, clear, and mount in the usual manner. While not advised for class use, the time in KOH can be shortened considerably by heating the solution containing the specimen. About 10 minutes in the boiling solution should be adequate. This process should be watched since the chitin is also destroyed with overexposure to KOH.

Benton (1955) suggests using Cellosolve (ethylene glycol monoethyl ether) for both dehydration and clearing. Following KOH, specimens are passed through water before Cellosolve is added. The advantage of this method is that it reduces the number of times specimens are handled, which minimizes the chance of damaging them.

Collecting and Mounting Parasitic Mites

Parasitic mites such as *Notoedres, Otodectes, Sarcoptes* et al. are usually deeply imbedded in the host tissue and are therefore often lost or damaged when the attempt is made to separate them. For collecting them, a scalpel or knife blade may be sterilized by passing it through a flame and then cooled by dipping it in water. Mites are collected more easily if the edge of the scraper is first dipped in acetic-glycerine (1% glacial acetic acid in glycerine). A fold of skin showing lesions is held between the thumb and forefinger and the crest of the fold scraped until lymph begins to ooze. Care must be taken to avoid drawing blood. Transfer scrapings from scalpel, using a rotary motion, into a drop of acetic-glycerine on a slide. Add a cover glass and additional acetic-glycerine, if necessary, and observe under the low power of a compound microscope.

The material may be stored in the acetic-glycerine indefinitely and mites may be separated at leisure. Add a small crystal of thymol to discourage molds. Separate parasites with a very fine dissecting needle or finely drawn pipette equipped with a rubber bulb. This is done under a dissecting microscope in good light.

After they have been separated from the debris, the mites should be placed on a slide in almost any of the several well-established chloral-gum preparations. The Double Cover Glass Mounting Technique, described on p. 277, is advised for permanent mounts.

The double cover glass mounts containing dorsoventrally flattened arthropods, as well as trematodes and cestodes, may be mounted between 2 pieces of thin cardboard cut the same size as a microscope slide. Identical holes are cut in each piece of cardboard, the cover glass mount placed over the hole in 1 and the 2 pieces of cardboard glued together by such adhesives as Elmer's Glue. With these simple, easily prepared mounts, the specimens between the cover glasses may be viewed from either the dorsal or ventral side. The hole in the cardboard should be large enough to allow either the high dry or oil immersion objectives to swing into position without striking the paper mount.

Micro-Scalpels

Excellent scalpels for minute dissections can be made from safety razor blades. Grip the edge of a blade obliquely (over the area desired), between the jaws of needle-nose pliers, Cresson insect pinning forceps, or similar tool. With another pair of pliers bend the blade laterally until it snaps off. Since scalpels with shanks are desirable, first break away a corner of the blade, before proceeding down the blade edge. The good pieces, once crudely broken, can be ground to the desired shape on a fine emery wheel or stone. Do not touch the cutting edge. Insert the shank of the finished scalpel in the cleft of a suitable wooden staff, apply Duco or similar cement, clamp in a vise and leave overnight before using.

Spring Clamps

The fabrication of spring clips is accomplished in five distinct operations: (1) preparation of material, (2) annealing, (3) shaping, (4) hardening, and (5) washing.

Preparation of material: The material, consisting of broken watch mainsprings, should first be uncoiled by holding each end, while the spring is pulled back and forth a few times, with the convex surface against a piece of rounded wood. The springs are then cut to the desired length of about 2 5/8" with a pair of sheet metal shears or cutting pliers.

Annealing: The sections are then annealed by placing them in a piece of nongalvanized iron pipe about 1 1/2" inner diameter closed at 1 end, which is placed horizontally on a ring stand, and heated (2 Fisher high temperature burners work nicely) until the mass of springs is a bright cherry red. The heat is then turned off and the mass allowed to cool slowly. This slow cooling allows the molecules to align themselves in one plane, thereby making the metal workable.

Shaping: A small circle (approximately 1/8" diameter) is formed at 1 end with a pair of round-nose pliers (Figure 260, A) and the length of spring then shaped over a triangular piece of wood, planed to the proper shape and size (about 1/2" x 5/8" x 3/4"). The second and third bends (Figure 260, C and B respectively) will have to be finally adjusted by hand to make sure that the arm resulting from the last bend goes beyond that of the first in order to give the spring clip some tension.

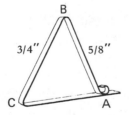

Figure 260. Finished spring clip.

Hardening: The hardening process is effectively accomplished by placing the shaped springs inside the segment of pipe (closed at 1 end as previously), heating the entire mass to a cherry red, and quickly dropping the springs en masse into a container of ordinary No. 10 or 20 motor oil. The purpose of the oil is to cool the springs at the proper rate so that they will be hard but not too brittle. A large pair of pliers should be used in transferring the springs into the oil, and if more than one "batch" of springs is to be hardened the oil should be cooled again by placing the container in cold water.

Washing: The oil used for cooling should be poured off and the finished spring clips washed in kerosene. Any suitable solvent can be used for washing, but kerosene leaves a thin coat of oil on the metal and thereby prevents rusting.

Appendix

Reagents and Solutions

The following list includes reagents and solutions used in the present outline. For convenience, they are listed alphabetically rather than according to use.

Alcohols

Commercial grain alcohol, which is approximately 95% ethyl by volume, is used in making the lower strengths of alcohol. When preparing the different solutions, the procedure given below should be followed. To obtain a given percentage of alcohol through dilution of a higher percentage with distilled water, subtract the percentage required from the percentage of the alcohol to be diluted; the difference is the proportion of distilled water that must be added. Thus, if 70 is the percentage desired, and 95 the percentage to be diluted, then 95 minus 70 equals 25; hence 25 parts of distilled water and 70 parts of 95% alcohol are the proportions used.

This means that in practice one needs only to fill the graduated measuring cylinder to the same number as the percentage required (e.g., 70) with the alcohol being diluted (e.g., 95%) and then fill to the percentage of the latter with distilled water. Thus, one would obtain 95 ml of 70% alcohol, if the measuring cylinder is graduated in milliliters.

Acid Alcohol, Two Percent

Alcohol, 70% . . . . 98 ml

Hydrochloric acid, commercial. . . . . 2 ml

A-F-A (Alcohol-formol-acetic) Fixative

Alcohol, 85% . . . . 85 ml

Formalin, commercial . . . . 10 ml

Acetic acid, glacial . . . . 5 ml

Possibly no other fixing fluid has as many names applied to it. Many modifications in the amount of each constituent have been recommended, and the resulting mixture often bears the name of its sponsor. Evidence indicates, however, that the proportions are not critical, within limits. Water enters into the formula only incidentally through that contained in the formalin and in the diluted alcohol.

Keep a stock solution containing the above proportions of alcohol and formalin. Add the acetic acid to measured quantities of this as needed for use. Thus if one needs 100 ml of the final solution, take 95 ml of stock and add 5 ml of acid; if only 50 ml of this is required, each of the parts should be halved.

Alcohol-glycerine Mixture

Alcohol, 70% . . . . 95 ml

Glycerine . . . . 5 ml

Alkaline Alcohol

Alcohol, 70% . . . . 500 ml

Ammonium hydroxide, concentrated . . . . 0.5 ml

Berlese's Chloral-gum Solution, Modified

This aqueous medium is recommended for small, hard-to-clear specimens. Material can be mounted wet or dry, alive or from alcohol or formalin and the cover glass added. Since clearing takes only a few minutes, it is rapid and efficient. It avoids the necessity of using a fixative; it makes unnecessary the dehydration process of conventional technique and it avoids the loss of specimens due to their small size and the need for concentrating them after each alcohol change. The mounts harden slowly, but with care they may be used almost immediately.

Water, distilled...50 ml

Gum arabic (acacia), clear lumps or powdered ....40 gm

Glycerine ....20 ml

Chloral hydrate...50 gm

Acetic acid, glacial ....3 ml

Dissolve the gum arabic in the cold water, then dissolve the chloral hydrate with gentle heat in a water bath, add the glycerine and glacial acetic acid and filter while warm through several layers of cheesecloth.

Bile Salts Solution

Chocolate covered bile salts (available from druggist).............1 grain

Physiological solution ....30 ml

Grind 1 tablet finely in a mortar and place in the physiological solution. Filter after a few minutes to remove excess chocolate, thus permitting ready observation of the evagination process, by cysticerci placed in the solution.

Bles' Fixative

Alcohol, 70% ....90 ml

Formalin, commercial ....1 ml

Acetic acid, glacial ....3 ml

Boardman's Solution

Alcohol, 20% ....97 ml

Ether ....3 ml

Bouin's (picro-formol-acetic) Fixative

Picric acid, saturated aqueous solution ....75 ml

Formalin, commercial ....25 ml

Acetic acid, glacial ....5 ml

This solution keeps indefinitely, and is probably the best general purpose fixative. Specimens are usually left in it about 24 hours before being transferred directly to 70% alcohol.

Delafield's Hematoxylin

Hematoxylin crystals ....1 gm

Alcohol, absolute ....10 ml

Saturated aqueous ammonia alum
(aluminum ammonium sulphate) ....100 ml

Glycerine ....25 ml

Alcohol, methyl ....25 ml

Dissolve the hematoxylin crystals in the absolute alcohol and add to this solution, a few drops at a time, the ammonia alum. Leave it exposed to the light and air in an unstoppered bottle for several weeks (a month is not too long) to "ripen." Filter, add the glycerine and the methyl alcohol.

Dextrose-salt Solution

Sodium chloride	2.25 gm
Calcium chloride	0.06 gm
Potassium chloride	0.1 gm
Sodium carbonate	0.04 gm
Dextrose	0.62 gm
Water, distilled	1000.0 ml

Digestive Fluid, Artificial

Powdered pepsin, fresh	5 gm
Physiological saline (warmed)	1 liter
Hydrochloric acid, commercial	7 ml

For fish, increase the pepsin to 7 gm and reduce the HCl to 4 ml/liter of saline solution.

Ehrlich's Hematoxylin

Acetic acid, glacial	10 ml
Alcohol, absolute	25 ml
Hematoxylin crystals	2 gm
Glycerine	100 ml
Water, distilled	100 ml
Potassium alum (aluminum potassium sulphate)	10 gm

Mix the glacial acetic acid with the absolute alcohol and add the hematoxylin crystals. When dissolved, add an additional 75 ml of absolute alcohol and the glycerine. Heat the distilled water and add the potassium alum. When dissolved and still warm, add while stirring to the hematoxylin solution. Let the mixture ripen in the light (with an occasional admission of air) until it acquires a dark red color.

Fast Green, Stock

Fast Green, powdered	0.2 gm
Alcohol, 95%	100 ml

Formalin

Commercial formalin is approximately a 40% solution of formaldehyde gas in water. This solution is known as 100% formalin and is diluted accordingly when concentrations of formalin are desired. Thus 5% formalin, which is 2% formaldehyde, contains

Water, distilled	95 ml
Formalin, commercial	5 ml

Giemsa's Stain

The dry powder for preparing the stock solution is available from firms dealing in biological stains. A formula for making a stock solution is:

Giemsa's powder	1 gm
Glycerine	66 ml
Methyl alcohol, absolute	66 ml

To the alcohol-glycerine, add glass beads and the dry powder. Allow the alcohol-glycerine to penetrate the dye for a few minutes and then rotate the flask for about 2 minutes. This mixture is agitated about every 30 minutes until the procedure has been repeated 6 times. When possible, the stain should be made up early in the day so that the final shaking will be completed before the end of the working day. When prepared in this way, the stain is ready for immediate use. Store in a tightly stoppered bottle. The stock solution is diluted just before use with distilled water or dilute buffer solution. A suggested formula is the following:

Giemsa stock solution . . . . .1 ml

Water, distilled . . . . .20 ml

or

Phosphate buffer (approx. 0.1 M, pH 6.5) . . . . 20 ml

Glycerine Jelly

Soak 7 gm of granulated gelatine in 40 ml of distilled water for 30 minutes. Then melt in a warm water bath and filter through several layers of cheesecloth previously moistened with hot water. Finally, dissolve 1 gm phenol in 50 ml of glycerine and add to the gelatine. Stir until the mixture is homogeneous.

Gomori Trichrome

Chromotrope 2R . . . . .0.6 gm

Light Green SF . . . . .0.3 gm

Acetic acid, glacial . . . . 1.0 ml

Phosphotungstic acid . . . . .0.8 gm

Water, distilled . . . . 100.0 ml

Grenacher's Alcoholic Borax-Carmine

Carmine . . . . 3 gm

Borax, c.p. . . . . 4 gm

Water, distilled . . . . 100 ml

Alcohol, 70% (preferably methyl) . . . . 100 ml

Add the carmine and the borax to the water and boil until carmine is dissolved (30 minutes or more), or, better, allow the mixture to stand for 2 or 3 days, with an occasional stirring, until this occurs; then add the alcohol. Allow the solution to stand for a few days and filter.

Harris' Hematoxylin

Hematoxylin crystals . . . . 2 gm

Alcohol, absolute . . . . 20 ml

Water, distilled . . . . .400 ml

Ammonia alum (aluminum ammonium sulphate) . . . . 40 gm

Mercuric oxide . . . . 1 gm

Dissolve the hematoxylin crystals in the absolute alcohol. Heat the distilled water and add ammonia alum. When dissolved and still warm, add, while stirring, the hematoxylin solution. Boil quickly and after a minute of boiling add the mercuric oxide to ripen the solution, which should now be purple. After a minute more of boiling, the flask is plunged into cold water.

Heidenhain's Iron Hematoxylin

Two solutions are used. They are not to be mixed.

Solution I

Ferric ammonium sulphate (iron alum), use only clear violet
crystals.. 2 gm

Water, distilled..100 ml

This acts as mordant and, when diluted, as differentiator.

Solution II

Hematoxylin crystals .... 1 gm

Alcohol, absolute .... 20 ml

Water, distilled...200 ml

First dissolve the hematoxylin in the alcohol, then add the water. Hematoxylin solution must be thoroughly "ripe." This may be accomplished by diluting the stock solution with distilled water until the hematoxylin percentage is approximately 1%, which is ready for immediate use.

Helix Physiological Solution

Sodium chloride ....5.87 gm

Potassium chloride ....0.73 gm

Calcium chloride ....1.99 gm

Sodium bicarbonate ....1.82 gm

Magnesium chloride, hexahydrate ....5.62 gm

Potassium bicarbonate ....0.22 gm

Water, distilled...1000 ml

Insect Preservative

Alcohol, 95% ....53 ml

Ethyl acetate ....15 ml

Benzene .... 5 ml

Water, distilled..27 ml

Kronecker's Solution

Sodium chloride .... 3 gm

Sodium hydroxide ....0.03 gm

Water, distilled...500 ml

Lacto-phenol

Water, distilled..20 ml

Glycerine...40 ml

Lactic acid ....20 ml

Phenol, melted crystals ....20 ml

This solution should be kept in a brown bottle or in a dark place, because exposure to light causes it to turn yellow.

Lugol's Solution

Lugol's consists of potassium iodine, iodine, and water but the amount of each ingredient varies widely with different workers, often resulting in a solution bearing the name of its sponsor. A representative formula consists of 1 gm KI mixed with 0.5 gm iodine, dissolved in 5 ml of distilled water (accompanied by shaking). This is diluted to 50 ml with distilled water.

Merthiolate-Iodine-Formalin (MIF) Stain-preservation Solutions

I. For direct Smear

Tincture of Merthiolate, Lilly No. 99 (1:1,000) . . . . 0.775 ml

Lugol's solution, fresh . . . . 0.10 ml

Formalin (USP) . . . . 0.125 ml

II. For Stain-preservation

1. Stock MF Solution (Stored in brown bottle)

Water, distilled . . . . 250 ml

Tincture of Merthiolate, Lilly No. 99 (1:1,000) . . . . 200 ml

Formalin (USP) . . . . 25 ml

Glycerine . . . . 5 ml

2. Lugol's Solution, fresh

For approximately each 0.25 gm of stool to be processed first introduce 0.15 ml of Lugol's solution followed by 2.35 ml MF stock solution into a Kahn tube or other suitable glass container, then add stool specimen and comminute thoroughly.

Phenol-alcohol Solution

Phenol . . . . 80 ml

Alcohol, absolute . . . . 20 ml

Store in a brown bottle to prevent solution from changing to a dark color.

This solution is useful for clearing specimens quickly, equally well from water or alcohol, or mixtures as occur in various fixatives or preservatives. Its principal use is rapid clearing of nematodes for observing cuticularized structures, such as spicules. Although it overclears quickly, the degree of clearing can be controlled by drawing 95% alcohol under the cover glass, using absorbent paper. Nematodes placed in this solution become round and turgid. If specimens are to be kept, wash several times in 70% alcohol to remove phenol and store in 70% alcohol containing glycerine.

Physiological Solution

Sodium chloride . . . . 0.75 gm

Water, distilled . . . . 100 ml

It is unlikely that any other physiological solution is used more universally for parasitic helminths. Various modifications in the amount of NaCl and the addition of other ingredients, the resulting solution often bearing the name of its sponsor, have been recommended for parasites from the different host taxa. Evidence indicates, however, that the particular solution employed, within limits, is not critical.

Potassium Alum Solution

Potassium alum (aluminum potassium sulphate) . . . . 6 gm

Water, distilled . . . . 100 ml

Ringer's Solution

Sodium chloride . . . . 8 gm

Sodium bicarbonate . . . . 0.02 gm

Potassium chloride . . . . 0.02 gm

Calcium chloride (anhydrous) . . . . 0.02 gm

Dextrose (may be omitted) . . . . 1 gm

Water, distilled . . . . 1000 ml

Schaudinn's Fluid

Mercuric chloride, saturated aqueous solution . . . . 66 ml

Alcohol, 95% . . . . 33 ml

Acetic acid, glacial . . . . 3 ml

The first two can be kept mixed without deterioration, but the acid must be added just before fixation.

Semichon's Acetic-Carmine

Acetic acid, glacial . . . . 100 ml

Water, distilled . . . . 100 ml

Carmine "in excess" . . . . about 1.5 gm

Mix distilled water and acetic acid in an Erlenmeyer flask, and add carmine. The objective is to prepare a saturated solution of carmine, but since any excess is wasted, there is no need to add more than will go into solution. Heat in boiling water bath for 15 minutes, then cool the flask in cold water and filter the contents. This stock stain should be diluted with approximately 2 parts of 70% alcohol before use.

Trisodium Phosphate Solution

First, a saturated stock solution is made as follows:

Trisodium phosphate . . . . 28.3 gm

Water, distilled . . . . 100 ml

Second, for final use prepare as follows:

Trisodium phosphate, stock . . . . 2 ml

Water, distilled . . . . 98 ml

At room temperature this gives approximately a 0.25% solution.

Wright's Stain

The preparation of this stain is rather complicated. It is recommended that the solution be purchased ready-made and used as it comes from the bottle.

REFERENCES

Anderson, R.C. 1958. Methode pour l'examen des nematodes en vue apicale. Ann. Parasitol. Hum. Comp. 33: 171-172.

Benton, A.H. 1955. A modified technique for preparing whole mounts of Siphonaptera. J. Parasitol. 41: 322-323.

Blagg, W., E.L. Schloegel, N.S. Mansour, and G.I. Khalaf. 1955. A new concentration technic for the demonstration of protozoa and helminth eggs in feces. Amer. J. Trop. Med. Hyg. 4: 23-28.

Dubey, J.P., G.V. Swan, and J.K. Frenkel. 1972. A simplified method for isolation of *Toxoplasma gondii* from the feces of cats. J. Parasitol. 58: 1005-1006.

Goodey, T. 1963. Soil and Freshwater Nematodes, rev. ed. John Wiley & Sons, New York, 544 p.

Harder, H.I., and D. Watson. 1964. Human filariasis. Identification of species on the basis of staining and other morphologic characteristics of microfilariae. Amer. J. Clin. Path. 42: 333-339.

Hoffman, G.L. 1967. Parasites of North American Freshwater Fishes. Univ. of California Press, Berkeley, 486 p.

Knott, J. 1939. A method for making microfilarial surveys on day blood. Trans. Royal Soc. Trop. Med. Hyg. 33: 191-196.

Mizelle, J.D. 1938. Comparative studies on trematodes (Gyrodactyloidea) from the gills of North American fresh-water fishes. Illinois Biol. Monog. 17: 1-81.

Sapero, J.J., and D.K. Lawless. 1953. The "MIF" stain-preservation technic for the identification of intestinal protozoa. Amer. J. Trop. Med. Hyg. 2: 613-619.

Thorne, G. 1935. Notes on free-living and plant-parasitic nematodes. Proc. Helminthol. Soc. Washington 2: 96-98.

_____. 1961. Principles of Nematology. McGraw-Hill Book Company, Inc., New York, 553 p.

Van Cleave, H.J., and J.A. Ross. 1947. Use of trisodium phosphate in microscopical technic. Science 106: 194.

Whitlock, H.V. 1956. An improved method for the culture of nematode larvae in sheep feces. Australian Vet. J. 32: 141-143.

Index

NOTES

NOTES

NOTES